Material Research in Atomic Scale by Mössbauer Spectroscopy

T0185504

NATO Science Series

A Series presenting the results of scientific meetings supported under the NATO Science Programme.

The Series is published by IOS Press, Amsterdam, and Kluwer Academic Publishers in conjunction with the NATO Scientific Affairs Division

Sub-Series

I. Life and Behavioural Sciences	IOS Press
II, Mathematics, Physics and Chemistry	Kluwer Academic Publishers
III. Computer and Systems Science	IOS Press
IV. Earth and Environmental Sciences	Kluwer Academic Publishers
V. Science and Technology Policy	IOS Press

The NATO Science Series continues the series of books published formerly as the NATO ASI Series.

The NATO Science Programme offers support for collaboration in civil science between scientists of countries of the Euro-Atlantic Partnership Council. The types of scientific meeting generally supported are "Advanced Study Institutes" and "Advanced Research Workshops", although other types of meeting are supported from time to time. The NATO Science Series collects together the results of these meetings. The meetings are co-organized bij scientists from NATO countries and scientists from NATO's Partner countries – countries of the CIS and Central and Eastern Europe.

Advanced Study Institutes are high-level tutorial courses offering in-depth study of latest advances in a field.
Advanced Research Workshops are expert meetings aimed at critical assessment of a field, and identification of directions for future action.

As a consequence of the restructuring of the NATO Science Programme in 1999, the NATO Science Series has been re-organised and there are currently Five Sub-series as noted above. Please consult the following web sites for information on previous volumes published in the Series, as well as details of earlier Sub-series.

http://www.nato.int/science
http://www.wkap.nl
http://www.iospress.nl
http://www.wtv-books.de/nato-pco.htm

Series II: Mathematics, Physics and Chemistry – Vol. 94

Material Research in Atomic Scale by Mössbauer Spectroscopy

edited by

Miroslav Mashlan
Department of Experimental Physics,
Palacký University, Olomouc, Czech Republic

Marcel Miglierini
Department of Nuclear Physics and Technology,
Slovak University of Technology, Bratislava, Slovakia

and

Peter Schaaf
Zweites Physikalisches Institut,
Universität Göttingen, Göttingen, Germany

Kluwer Academic Publishers

Dordrecht / Boston / London

Published in cooperation with NATO Scientific Affairs Division

Proceedings of the NATO Advanced Research Workshop on
Material Research in Atomic Scale by Mössbauer Spectroscopy
Smolenice, Slovak Republic
1–6 June 2002

A C.I.P. Catalogue record for this book is available from the Library of Congress.

ISBN 1-4020-1196-2 (HB)
ISBN 1-4020-1197-0 (PB)

Published by Kluwer Academic Publishers,
P.O. Box 17, 3300 AA Dordrecht, The Netherlands.

Sold and distributed in North, Central and South America
by Kluwer Academic Publishers,
101 Philip Drive, Norwell, MA 02061, U.S.A.

In all other countries, sold and distributed
by Kluwer Academic Publishers,
P.O. Box 322, 3300 AH Dordrecht, The Netherlands.

Printed on acid-free paper

Printed in the Netherlands.

CONTENTS

V.   SYNCHROTRON RADIATION

VI.   MINERALS

VII.   NEW DEVELOPMENTS

PREFACE

Advanced materials of today's technology deserve thorough knowledge of their preparation and properties. They must be studied on an atomic and/or subatomic scale. Mössbauer spectroscopy is a well-established and rapidly evolving analytical tool, which provides superior possibilities via its different variations to probe the nearest atomic neighbourhood of the resonant spy atoms both from the structural and also from the hyperfine interaction point of view. The diagnostic potential of Mössbauer effect is underlined by its application in a variety of fields of human activities.

Mössbauer spectroscopy offers a unique possibility for hyperfine interaction studies via probing the short-range order of resonant atoms. Materials, which contain the respective isotope as one of their constituent elements (e.g., iron, tin, ...), but also those which even do not contain them, can be investigated. In the latter case, the probe atoms are incorporated into the material of interest in minor quantities (ca. 0.1 at. %) to act as probes on a nuclear level.

This workshop has covered the topics evolving most recently in the field of Mössbauer spectroscopy applied to materials science. During five days, 49 participants from 19 countries discussed the following areas: *Nanoscale Systems, Conversion Electron Mössbauer Spectroscopy, Magnetism, Nanocrystalline Alloys and Metals, Synchrotron Radiation, Minerals,* and *New Developments*. A total of 42 contributions (34 keynote and 8 contributing speeches) reviewed the current state of the art of the method, its applications for technical purposes, as well as trends and perspectives.

A total of 35 papers are included in the present volume. In the chapter *Nanoscale Systems*, two main ways of preparation of fine particles were discussed, namely via mechanosynthesis and thermal processing. Their characterization was done primarily by Mössbauer effect techniques but in combination with complementary methods.

Conversion Electron Mössbauer Spectroscopy offers a possibility to trace the surface layers down to the depth of about 150 nm, thus extending the diagnostic potential of the method. The application of this technique to the study of modified steel surfaces as well as surfaces of nanocrystalline alloys and organic thin films was reviewed with a special emphasis on the basic principles of the method.

Mössbauer spectroscopy is an analytical tool, which provides both structural and magnetic information about the system under investigation. The latter feature was treated in a special section devoted to *Magnetism*. The advantages of the use of external magnetic fields in these studies were pointed out and combined with the information from classical magnetic measurements.

Recently, nanocrystalline materials, which show outstanding features for practical technological applications, have emerged. Investigations dealing with their magnetic and structural properties are discussed in the section *Nanocrystalline Alloys and Metals*. It is shown that structurally different atomic positions can be identified in nanocrystalline multiphase systems. Non-conventional processing of steels and their characterization by the help of other techniques completes this chapter.

A fundamentally different approach to the use of the nuclear resonance is demonstrated in the chapter entitled *Synchrotron Radiation*. Basic principles and applications of nuclear resonance scattering of synchrotron radiation with high temporal and spatial resolution for the study of diffusion processes and magnetic properties are described there.

For thousands of years, minerals have been of great importance for mankind. Examples of applicability of Mössbauer effect studies are presented in Chapter VI, *Minerals*, for mineral magnetism and oxidation processes.

The potential of Mössbauer diffraction, emission Mössbauer spectroscopy, and Mössbauer polarimetry prove the ongoing expansion of this technique in new directions. Chapter VII, *New Developments*, summarises the contributions related to the topics mentioned before as well as the new concepts of data evaluation of complex Mössbauer spectra. An unusual application of the Mössbauer effect in dosimetry finalises this chapter and the whole proceedings.

This NATO Advance Research Workshop continues the tradition of a series of scientific conferences devoted to material research entitled *Mössbauer Spectroscopy in Materials Science* (**msms**). Starting in 1994, the meetings are held on a biannual basis in Slovakia and the Czech Republic in Kocovce (1994), Lednice (1996), Senec (1998), Velke Losiny (2000), and Smolenice (2002). The aims of a **msms** are always to contribute to the critical assessment of the existing knowledge on new important topics, to identify directions for future research, and to promote close working relations among scientists from different countries and with different professional experience. This year, the funding was provided by NATO to cover organizational expenses, travel and living expenses of keynote speakers and to contribute to travel and living expenses of other participants 40% to 50% of whom are from Partner or Mediterranean Dialogue countries. In our opinion, these aims have really been achieved during this stimulating conference. All the participants agreed and supported the idea that this conference series should certainly be continued. In this respect, we gratefully acknowledge the financial support of the NATO Scientific Committee that made such an event and also these proceedings possible.

We would also like to express our deepest gratitude to all the participants for the energy and efforts they put into making this workshop and proceedings very successful.

October 2002

Miroslav Mashlan
Marcel Miglierini
Peter Schaaf

NANOCRYSTALLINE OXIDES AND SULPHIDES PREPARED BY HYDROTHERMAL PROCESSING AND MECHANICAL MILLING

The Use of Mössbauer Spectroscopy in Characterisation

F.J. BERRY[1], A. BOHORQUEZ[1], Ö. HELGASON[2], J.Z. JIANG[3], J.F. MARCO[4], E.A. MOORE[1] AND S. MØRUP[3]

[1] *Department of Chemistry, The Open University, Walton Hall, Milton Keynes, MK7 6AA, United Kingdom*
[2] *Science Institute, University of Iceland, IS-107 Reykjavik, Iceland*
[3] *Department of Physics, Technical University of Denmark, Lyngby DK2800, Denmark*
[4] *Instituto de Química Física 'Rocasolano', Consejo Superior de Investigaciones Científicas, c/Serrano 119, 28006 Madrid, Spain.*

Abstract

The synthesis of nanocrystalline tin-doped α-FeOOH (goethite) by hydrothermal processing of a precipitate and of nanocrystalline iron sulphides by mechanical milling gives particles which are well suited for examination by Mössbauer spectroscopy. The morphology of the tin-doped α-FeOOH crystals is reflected in the [57]Fe Mössbauer spectra and depends on the pH of the precipitating media. The [119]Sn Mössbauer spectra are consistent with a complex microstructure around the octahedrally coordinated tin ions in the goethite structure and are sensitive to the changes in the magnetic structure of α-FeOOH resulting from tin doping. The formation of *ca* 10nm crystals of iron sulphides of composition FeS and FeS$_2$ by mechanical milling of a mixture of powdered iron and sulphur was monitored by [57]Fe Mössbauer spectroscopy. The [57]Fe Mössbauer spectra obtained in applied magnetic fields showed the FeS particles to be antiferromagnetic.

1. Introduction

The properties of materials can be dramatically changed when one of more dimensions are in the 1-100 nm regime. The resulting changes in chemical and physical behaviour has been of intense and growing interest over the past decade and a new research area, that of nanostructured materials, has emerged. Mössbauer spectroscopy is an important technique for the characterisation of these interesting materials and is illustrated here by studies of tin-doped α-FeOOH formed by hydrothermal processing and of iron sulphides formed by mechanical milling.

1

M. Mashlan et al. (eds.), Material Research in Atomic Scale by Mössbauer Spectroscopy, 1–10.

The iron oxyhydroxide of formula α-FeOOH, which is known as goethite, is widely found in natural environments. The structure is composed of double chains of iron in octahedral oxygen coordination, which are further linked by sharing vertices in a three-dimensional framework structure. Given that α-FeOOH is isostructural with α-AlOOH, α-MnOOH and VOOH it is not surprising that isomorphous replacement of some iron by trivalent ions such as aluminium, manganese and vanadium is possible. In contrast there appears to be a sparsity of data on α-FeOOH substituted by tetravalent ions and we report here on the synthesis of tin-doped α-FeOOH by hydrothermal processing and the important role of ^{57}Fe- and ^{119}Sn-Mössbauer spectroscopy in their characterisation.

The iron sulphide of composition FeS has the nickel arsenide structure where iron has six immediate sulphur neighbours. Iron in FeS$_2$, which adopts either the pyrites or marcasite structure, is surrounded by six sulphur atoms. Although mechanical milling has been extensively used for the synthesis of alloys and oxides, there are relatively few reports of the use of the technique for the preparation of sulphides. We report here on the synthesis of iron sulphides by mechanical milling and the use of ^{57}Fe Mössbauer spectroscopy in monitoring the phase evolution and magnetic properties of the materials.

2. Experimental

Tin-doped α-FeOOH was prepared by precipitating aqueous mixtures of iron(III) nitrate (15.35 g, 100 ml) and tin(II) chloride (0.45 g, 10 ml) with aqueous ammonia at either pH 9.5 or 10.3 and hydrothermally processing the suspensions (ca. 260 ml) in a Teflon-lined autoclave at 200 °C and 15 atm pressure for 5 hours. The products were removed by filtration and washed with 95% ethanol until no chloride ions were detected by silver nitrate solution. The products were dried under an infrared lamp. The metal contents were determined by ICP analysis.

Iron sulphides were prepared by milling a 1:1 molar ratio of powdered metallic iron and elemental sulphur in a Fritsch Pulverisette 5 ball mill with tungsten carbide vials and balls in an argon atmosphere. The milling intensity was 200 rotations per minute and a ball to powder weight ratio of 20:1 was used.

X-ray powder diffraction patterns were recorded at 298K with either a Siemens D5000 or a Philips PW3710 diffractometer both using Cu-Kα radiation.

The tin K-edge extended X-ray absorption fine structure (EXAFS) measurements were performed at the Synchrotron Radiation Source at Daresbury Laboratory, UK, with an average current of 200 mA at 2 GeV. The data were collected in transmission geometry on Station 9.2 at 77 K. The raw data were background subtracted using the Daresbury program EXBACK and fitted using the non-linear least-squares minimisation program EXCURV92 which calculates the theoretical EXAFS function using fast curved wave theory.

X-ray photoelectron spectra were recorded with a triple channeltron CLAM 2 analyser using Mg-Kα radiation and a constant analyser transmission energy of 100 eV for the wide scan spectra and 20 eV for the narrow scan spectra. Base pressure in the analyser chamber during the experiments was ca. 2×10^{-8} Torr. All values of binding energy were charge converted to the C 1s signal (284.6 eV) and are accurate to ±0.2 eV.

Relative atomic concentrations were calculated using tabulated atomic sensitivity factors [1].

The Mössbauer spectra were recorded between 17 and 298 K with a conventional constant acceleration spectrometer in transmission geometry. The ^{57}Fe Mössbauer spectra were recorded using a 400 MBq ^{57}Co/Rh source and the ^{119}Sn Mössbauer data were obtained with a 200 MBq Ba^{119}SnO$_3$ source. The drive velocity was calibrated with the ^{57}Co/Rh source and a natural iron foil. All the isomer shift data are reported relative to that of α-Fe at room temperature.

Electron micrographs were recorded from specimens suspended on copper grids using a JEOL 2000 FX electron microscope with an accelerating voltage of 200 KeV.

3. Results and Discussion

3.1. TIN-DOPED α-FeOOH

The X-ray powder diffraction patterns recorded from the light brown powders showed the materials to be related to goethite, α-FeOOH [2]. ICP analysis of the metal contents was consistent with a formulation α-Fe$_{0.86}$Sn$_{0.14}$OOH in the material prepared at pH 9.5 and α-Fe$_{0.88}$Sn$_{0.12}$OOH in that prepared at pH 10.3. The X-ray powder diffraction data were not amenable to refinement to a structural model. The tin K-edge EXAFS (Table 1) were best fitted to a model in which tin was substituted for iron in the goethite structure [3]. Interatomic potential calculations for the substitution of Sn^{4+} ions on a Fe^{3+} site and for the insertion of Sn^{4+} ions into an interstitial site showed that there was a clear energy advantage for substitution over insertion into an interstitial site. The most favoured charge balancing defects of those considered were vacancies on the Fe^{3+} sites.

The X-ray photoelectron spectra recorded from tin-doped goethite showed no evidence for the formation of Fe^{2+} and demonstrated that charge balance is not achieved by the partial reduction of Fe^{3+} to Fe^{2+}. Given that ^{57}Fe Mössbauer spectroscopy also failed to show evidence for the reduction of Fe^{3+} to Fe^{2+} (*vide infra*) and the results of the interatomic potential calculations (*vide supra*) we envisage that charge balance in tin-doped goethite is achieved by the formation of cation vacancies.

Scanning electron microscopy showed the presence of needle-shaped particles up to *ca.* 400 nm in length in the sample precipitated from iron(III) nitrate and tin(II) chloride with aqueous ammonia at pH 10.3. The material prepared at pH 9.5 was composed of rounded and smaller particles (*ca.* 25-50 nm).

The ^{57}Fe Mössbauer spectra (Fig. 1) recorded at 298 K from tin- doped goethite prepared at pH 9.5 and 10.3 were similar to those previously recorded from natural goethite formed by the weathering of oxides, sulfides and silicates [4] and from pure synthetic goethite [5-7].

TABLE 1. Best-fit parameters for tin K-edge EXAFS recorded from tin-doped goethite at 77 K.

Atom type	Coordination number	Distance/Å±1%	$2\sigma^2/\text{Å}^2$
O	6	2.039	0.008
Fe	2	3.120	0.010
O	1	3.297	0.010
Fe	2	3.298	0.011
Fe	4	3.525	0.012
O	4	3.558	0.010
O	2	3.637	0.010
O	2	3.881	0.011
O	2	4.014	0.008
R=15.80			

The origin of the line broadening in the Mössbauer spectra recorded from these types of materials has been the subject of much controversy [5-8]. The spectra were fitted to a model-independent distribution of the magnetic hyperfine fields [9]. All the linewidths of the sextets were kept fixed (0.23 mm s^{-1}) and the isomer shift and quadrupole splitting values were kept constant in the final version of the fitting procedure (Table 2). The magnetic hyperfine field distributions (Fig. 1b) which characterise the samples prepared at pH 9.5 and 10.3 are different, and both the mean magnetic hyperfine field, $B_{<ave>}$, and the hyperfine field at maximum probability, $B_{<max>}$, are larger in the sample prepared at pH = 10.3. In both cases, the chemical isomer shift and quadrupole splitting data are similar and resemble those recorded from pure- and aluminium-doped goethite. [5,6,10] The spectrum recorded from the sample prepared at pH 9.5 showed it to contain a small amount of α-Fe$_2$O$_3$ (less that 3% of total area); these data were excluded in all calculations relating to the results shown in Table 2. The differences between the spectra recorded from the samples prepared at pH 9.5 and 10.3 diminished as the temperature of measurement was decreased. Indeed, the spectra recorded from both samples at 80 K (Fig. 2) are very similar and can be fitted with nearly identical field distributions and hyperfine parameters (Table 2). The spectra recorded from both samples at 17 K were identical and show a smaller distribution in magnetic hyperfine fields, with a maximum value of 50.5 T. A spectrum at 17 K is shown in Fig. 2.

A comparison of the data recorded at 80 K from tin-doped α-FeOOH with those reported for aluminium-doped goethite [10,11] suggests that both tin and aluminium reduce the magnitude of the magnetic hyperfine field. In previous work [10,11] the decrease in the magnetic hyperfine field was attributed to two different effects, the amount of dopant and the crystallinity, and empirical quantitative relationships were proposed. However, the earlier work on aluminium-doped goethite has been recently re-examined [12] and doubt has been cast on the proposed simple relationships between the amount of impurities, the crystallinity, and $B_{<max>}$. In these circumstances it does not seem appropriate to make a quantitative comparison of the influence of tin and aluminium on the magnetic hyperfine field in goethite.

TABLE 2. ^{57}Fe Mössbauer parameters recorded from tin-doped goethite.

T/K	pH of reaction mixture	δ (±0.01)/mm s^{-1}	Δ (±0.01)/mm s^{-1}	$B_{<max>}$/T	$B_{<ave>}$/T
298	9.5	0.37	−0.27	32.9	29.4
298	10.3	0.37	−0.28	36.2	31.6
80	9.5	0.47	−0.26	49.3	48.7
80	10.3	0.47	−0.26	49.6	49.0
17	10.3	0.48	−0.27	50.5	50.2

The ^{119}Sn Mössbauer spectra recorded at 298 K from tin-doped α-FeOOH formed at pH 9.5 and 10.3 are shown in Fig. 3a. The data were also fitted to a model independent distribution of magnetic hyperfine fields (Fig. 3b). Initial inspection suggests an absence of detailed structure and both spectra can be satisfactorily fitted to a broadened doublet with a wider linewidth than that characteristic of tin dioxide. However, the fitting of the data to a narrow doublet and a magnetic hyperfine field distribution with a maximum magnetic hyperfine field of 3-4 Tesla, reflecting the magnetic structure deduced by ^{57}Fe Mössbauer spectroscopy, is also appropriate (Fig. 3b). The room temperature data alone preclude unequivocal determination of the most superior fit. Support for the fit involving a magnetic interaction on the tin nucleus is obtained from the spectrum recorded at 17 K (Fig. 3a). In this spectrum a component with a larger magnetic hyperfine field can be resolved and the field distribution determined from the spectrum is shown in Fig 3b. The ^{119}Sn Mössbauer hyperfine parameters are collated in Table 3. It appears that *ca.* 20% of the total tin content can be assigned to the spectral component with a maximum at 6.5 T and it might be reasonable to associate this with those tin ions which have only iron nearest cation neighbours. The other part of the magnetic hyperfine field distribution may reflect a more complicated environment about tin characterised by other factors such as vacancies or other tin ions. Taken together, the results suggest that dopant tin can be a sensitive means of probing the magnetic structure of goethite. The ^{57}Fe Mössbauer spectroscopy results at 298 K (Fig.1, Table 2) indicate values of $B_{<max>}$ of 33 and 36 T and $B_{<ave>}$ of 29 and 32 T. At 17 K all these values are 50 T (Fig. 2). Assuming that the Sn^{4+} ions are distributed on the Fe^{3+} sites and that the magnetic hyperfine field sensed by Sn^{4+} is due to the antiferromagnetic structure of goethite, then the observation of similarly large increases in the magnitude of the magnetic hyperfine field in the ^{119}Sn Mössbauer spectra as the temperature decreases from 298 to 17 K (Fig. 3) is indicative of the sensitivity of tin to changes in the magnetic structure of goethite. The low field peak at about 1 T in the distribution of the spectrum recorded at 17 K exceeds the magnitude expected from a tin dioxide doublet (a value of Δ of 0.06 mm s^{-1} for the doublet equates to *ca* 0.4-0.5 T), although the calculation procedure cannot distinguish between a low field sextet and a paramagnetic doublet. This, together with the results from X-ray powder diffraction which showed no evidence for the presence of tin dioxide, supports the conclusion that tin has been incorporated within the goethite structure.

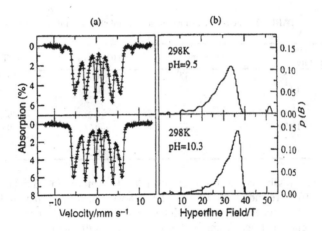

Figure 1. (a) ^{57}Fe Mössbauer spectra recorded at 298 K from tin-doped goethite prepared at pH 9.5 and 10.3; (b) data fitted to a model independent distribution of magnetic hyperfine fields.

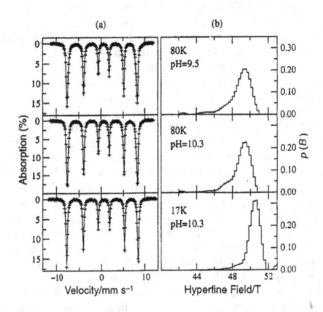

Figure 2. (a) ^{57}Fe Mössbauer spectra recorded at 80 and 17 K from tin-doped goethite prepared at pH 9.5 and 10.3; (b) data fitted to a model independent distribution of magnetic hyperfine fields.

3.2. IRON SULPHIDES

The investigations [13] of the milled reaction mixture of iron and sulphur showed the evolution of iron sulphide phases as a function of milling time. X-ray powder diffraction showed the appearance of a mixture of FeS and FeS_2 after only 11 hours of milling. After 43h peaks corresponding to elemental iron and sulphur had disappeared and those due to FeS predominated. Milling for longer periods of time resulted in only the appearance of reflections due to FeS in the X-ray powder diffraction pattern. A particle size of *ca.* 7 nm was determined by the Scherrer method from the X-ray powder diffraction peak width after 67h of milling.

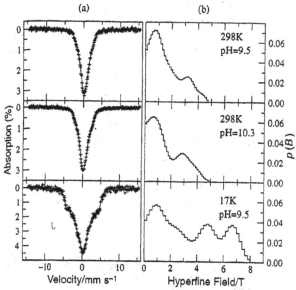

Figure 3. (a) ^{119}Sn Mössbauer spectra recorded at 298 and 17 K from tin-doped goethite prepared at pH 9.5 and 10.3; (b) data fitted to a model independent distribution of magnetic hyperfine fields.

TABLE 3. ^{119}Sn Mössbauer parameters recorded from tin-doped goethite.

T/K	pH of reaction mixture	$\delta(\pm0.01)$/mm s^{-1}	$\Delta(\pm0.01)$ /mm s^{-1}	$B_{<max>}$/T	$B_{<ave>}$/T
298	9.5	0.15	0.04	0.7 (3.1)^a	1.5
298	10.3	0.16	0.04	0.7 (2.9)^a	1.6
17	9.5	0.24	−0.03	0.8 (4.8 and 6.6)^a	3.3

^{a}Value for high field components in the distribution

The ^{57}Fe Mössbauer spectra recorded at 298 K from samples collected after different milling times are shown in Fig. 4. The results endorsed those recorded by X-ray powder diffraction. After only 1 hour a sextet due to α-iron was visible but, after 19 hours of milling a quadrupole split absorption (*ca.* 12%) δ = 0.32 mm s^{-1}; Δ = 0.62 mm s^{-1}, characteristic of FeS_2 in the pyrites modification together with a sextet (*ca.* 14%), δ = 0.7 mm s^{-1}; H = 27 T had developed. After 43 hours of milling the sextet compound was dominant with only traces of α-iron and the FeS_2 being detectable and, after longer periods of milling, the sextet was the only component in the spectra. The broad spectral lines were indicative of a distribution of magnetic hyperfine fields and four sextets were used to fit the spectra. Three of the sextets had magnetic fields in the range 30-25 T, similar to that of large particle FeS with the nickel arsenide structure (H = 31.5 T), whilst the fourth with very broad lines had a magnetic field of *ca.* 20 T. The Rietveld structure refinement of the X-ray powder diffraction data was consistent with the presence of cationic vacancies. Such vacancies could be responsible for the observed slightly smaller magnetic hyperfine field. It should also be noted that the Mössbauer data indicates that the process in the mill involves the reaction of iron and sulphur to form both FeS and FeS_2 and also the reaction of iron with FeS_2 to form FeS.

The ^{57}Fe Mössbauer spectra recorded at 80K and 14K from the sample made by milling the reactants for 113 hours (Fig. 5) showed the disappearance at low temperature of the broad component with an average hyperfine magnetic field of 20 T which was observed in the spectrum recorded at 298 K. It is therefore possible that this sextet is indicative of small particle superparamagnetic FeS. The application of a magnetic field of 0.7 T perpendicular to the gamma-ray direction at 80 K and 300 K induced negligible effects on the spectra (Fig. 5). For ferro- or ferri-magnetic materials the relative intensities of lines 2 and 5 would increase in the presence of an applied magnetic field. Hence the results are consistent with the nanocrystalline iron sulphide particles being antiferromagnetically ordered.

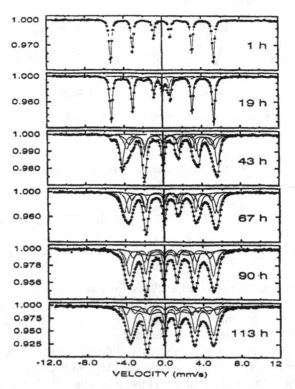

Figure 4. ^{57}Fe Mössbauer spectra of samples collected from the mixture of iron and sulphur after the indicated milling times and recorded at 298 K.

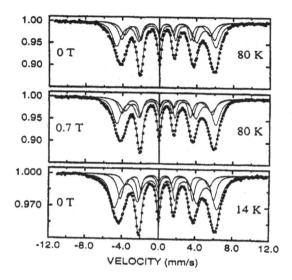

Figure 5. ^{57}Fe Mössbauer spectra of the sample milled for 113 h obtained at 80 K with and without an applied magnetic field of 0.7 T, and also recorded at 14 K.

Acknowledgements

We thank the EPSRC for beamtime at the SRS at Daresbury Laboratory. We thank Colciencias and the Universidad del Valle (A.B.) and The Danish Council for Technical Research for financial support.

References

1. Wagner, C.D., Davis, L.E., Zeller, M.V., Taylor, J.A., Richmond, R.M., and Gale, L.H. (1981) Empirical Atomic Sensitivity Factors for Quantitative-Analysis by Electron-Spectroscopy for Chemical-Analysis, *Surf. Interface Anal.* **3**, 211-225.
2. Berry, F.J., Helgason, Ö., Bohorquez, A., Marco, J.F., McManus, J., Moore, E.A., Mørup, S., and Wynn, P.G. (2000) Preparation and characterisation of tin-doped alpha-FeOOH (goethite), *J. Mater. Chem.* **10**, 1643-1648.
3. Manceau, A., and Combes, J.M. (1988) Structure of Mn and Fe Oxides and Oxyhydroxides - A Topological Approach by Exafs, *Phys. Chem. Miner.* **15**, 283-295.
4. Govaert, A., Dauwe, C., Plinke, P., De Grave, E., and De Sitter, J. (1976) *J. Phys.* **37**, C6-825.
5. Mørup, S., Madsen, M.B., Franck, J., Villadsen, J., and Koch, C.J.W. (1983) A New Interpretation of Mossbauer-Spectra of Microcrystalline Goethite - Super-Ferromagnetism or Super-Spin-Glass Behavior, *J. Magn. Magn. Mater.* **40**, 163-174.
6. Koch, C.J.W., Madsen, M.B., and Mørup, S. (1985) Evidence for Microcrystallinity in Large Particles of Goethite, *Surf. Sci.* **156**, 249-255.
7. Bocquet, S., Pollard, R.J., and Cashion, J.D. (1992) Dynamic Magnetic Phenomena in Fine-Particle Goethite, *Phys. Rev. B* **46**, 11657-11664.
8. Coey, J.M.D., Barry, A., Brotto, J.-M., Rakoto, H., Mussel, W.N., Collomb, A., and Frucherd, D. (1995) Spin-Flop in Goethite, *J. Phys. Cond. Mater* **7**, 759-768.

9. Wivel, C., and Mørup, S. (1981) Improved Computational-Procedure for Evaluation of Overlapping Hyperfine Parameter Distributions in Mossbauer-Spectra, *J. Phys. E.*, **14**, 605-610.

10. Murad, E., and Johnston, J. H. (1987) in G.J. Long (ed.) *Mössbauer Spectroscopy Applied to Inorganic Chemistry Vol 2*, Plenum New York, p53.

11. Barrero, C.A., Vandenberghe, R.E., De Grave, E., and da Costa, G.M. (1996) in I Ortalli (ed.) *Proceedings of the International Conference on the Applications of the Mössbauer Effect*, SIF, Bologna, p717.

12. Vandenberghe, R.E., Barrero-Manese, C.A., de Costa, G.M., Van Sax, E., and De Grave, E. (2000) Mossbauer characterization of iron oxides and (oxy)hydroxides: the present state of the art, *Hyperfine Interactions*, **126**, 247-259..

13. Lin, R., Jiang, J.Z., Larson, R.H., Mørup, S., and Berry, F.J. (1998) *Hyperfine Interactions (C)* **3**, 45.

MECHANOSYNTHESIS OF NANOSTRUCTURED MATERIALS

G. LE CAËR[1], S. BEGIN-COLIN[2] AND P. DELCROIX[2]

[1] *Groupe Matière Condensée et Matériaux, U.M.R. CNRS 6626, Université de Rennes-I, Campus de Beaulieu, Bâtiment 11 A, F-35042 Rennes Cedex, France*
[2] *Laboratoire de Science et Génie des Matériaux et de Métallurgie, U.M.R. CNRS 7584, Ecole des Mines, F-54042 Nancy Cedex, France*

1. Introduction

In Lilliput, Gulliver noted that "there are some laws and customs in this Empire very peculiar" [1], an observation which might be appropriate for nanophased materials too. Nanophased materials behave indeed differently from their macroscopic counterparts because their characteristic sizes are smaller than the characteristic length scales of physical phenomena occurring in bulk materials. Hereafter, we shall focus on consolidated nanomaterials or on powdered materials whose particle sizes are large as compared to the mean crystallite size which is required to be lower than 100 nm, a conventional limit used for structural materials.

Nanostructured materials are bulk solids with a nanometer-scale microstructure in which the chemical composition, the atomic arrangement and/or the size of the building blocks (e.g. crystallites) forming the solid vary on a length scale of a few nanometers throughout the bulk [2]. Nanostructured materials have then a significant fraction of atoms residing in defect environments (grain boundaries, interfaces, interphases). The volume fraction associated with grain boundaries is for instance of ~ 20 % for a grain boundary thickness of ~ 0.7 nm [3] and a grain size of 10 nm. They are produced by a large variety of methods, among which high-energy ball-milling [4] has attracted much attention. This interest stems from a fortunate combination of technical simplicity and of complexity both of phenomena occurring during grinding and of mechanosynthesized materials. Mechanical alloying (MA), which is a dry and high-energy milling process, has been used to synthesize all kinds of materials, often with non-equilibrium structures: amorphous, quasicrystalline and nanocrystalline phases, extended solid solutions, alloys of immiscible elements, all sorts of compounds and composites. Mechanical alloying of mixtures of powders of pure elements or of powders of already partially combined elements differs from grinding of materials whose chemical composition remains the same during milling but whose structure or 'microstructure' is expected to evolve. High-energy ball-milling is indeed a way of inducing phase transformations in solids: amorphization or polymorphous transformations of compounds, disordering of ordered alloys [4-6] are examples among others. High-energy ball-milling is moreover a means of modifying the reactivity of as-milled solids or of inducing chemical reactions between ground reactants at temperatures and at rates at which they would normally not

11

M. Mashlan et al. (eds.), Material Research in Atomic Scale by Mössbauer Spectroscopy, 11–20.
© 2003 *Kluwer Academic Publishers. Printed in the Netherlands.*

occur ([4,7] and references therein). High-energy ball-milling offers thus supplementary degrees of freedom in the choice of possible routes for synthesizing new materials. After having evoked some basic features of high-energy ball-milling, we shall present a few selected examples to illustrate its flexible use in the synthesis of nanophased materials and to stress the influence of the competition between damaging and restoring processes [5-7] on the final state of ground powders.

2. High-energy Ball-milling

Powders to be mixed are introduced in the required proportions in a vial together with balls usually made of hardened steel. The vial is generally sealed in a neutral atmosphere. The vial, filled in total to about 50%, is then vigorously shaken by a "high-energy" mill [4] to produce collisions between balls or between vial and balls with impact speeds of the order of some m/s and shock frequencies of about some hundreds Hz. Planetary mills, vibratory mills and ball mills [4] are the most widely used in Laboratories. In a planetary ball mill, a rotating disc bears vials which rotate in an opposite direction. Both rotation speeds are of the order of some hundreds rpm.

2.1. THE 'BAKER' TRANSFORMATION IN HIGH-ENERGY MILLS

The repeated fracturing and welding processes of particles trapped during collisions causes matter to be exchanged permanently between them and mix elements in a way reminiscent of chaotic transformations of the 'baker' type [8]. For instance, particles of ductile powder mixtures are subjected to a severe plastic deformation followed by their fracture and welding which result in the formation of composite particles with a layered microstructure. The latter becomes finer and finer and convoluted when milling proceeds. During collisions, powder particles are subjected to high stresses, which are of the order of some GPa in a planetary mill, for times of the order of milliseconds. Average strain rates are $\sim 10^3$-10^4 s^{-1}. The waiting period between such efficient trapping events is typically of the order of tens to hundreds of seconds. Temperature rise, shear with dislocations and non-conservative vacancies generated by it and the presence of a high density of grain boundaries, contribute all to produce an efficient interdiffusion of elements which mixes them down to the atomic scale in the layered nanometer-sized microstructure. The temperature rise ΔT of powders during MA is believed to be less than 300 K in metallic systems [4]. In MA of extended Cr-Sn solid solutions formed by mechanical alloying in a vibratory mill, ΔT is at least of 230 K as microstructures consist of alternate layers of chromium and of tin which melts during shocks [9]. In the most energetic working conditions of an AGO-2 planetary ball mill, with a ball speed of about 8 m/s and a specific injected power P_S of the order of 3 W/g (P_S <~1W/g for a Fritsch planetary P7 or P5 mill), the ball temperature can reach 900 K when the vials contain only balls and no powder [10]. In the case of brittle materials, the temperature rise of the powder surface plays a major role and thermal activation is required for MA to occur [4]. In the latter case, a granular-type microstructure is observed.

In that way, powders evolve progressively, in most cases to some final stationary state. The milling duration ranges typically between some hours and some tens of hours for most high-energy ball-mills. Final alloy powders consist generally of submicronic or micronic particles. When crystalline, every particle is however made of a large number of nanograins with a typical mean size of ~ 10 nm, i.e ~ 10^6 grains/$(\mu m)^3$, and may thus be considered as an equiaxed polycrystal with a high density of grain boundaries. The typical average powers injected in ground materials, ~ 10^{-4} eVat^{-1}s^{-1} for usual high-energy mills, are comparable with those involved in various processes applied to solid materials, for instance ion implantation [5-7].

2.2. TWO NOTABLE TIMES IN MECHANICAL ALLOYING

Given milling conditions, t_m is defined as the current milling time and the synthesis time t_f as the *minimum* time needed to enter a stationary state in which all kinds of powder characteristics remain unchanged when milling for $t_m \geq t_f$ (hardness, magnetic properties, average grain sizes, crystallographic structures,..).

Besides the minimum synthesis time, t_f, we introduced a second characteristic milling time t_{cm} we termed "chemical mixing time" [9], $t_{cm} \leq t_f$. To define it, let us consider for instance a binary mixture of elemental powders with a mean composition $A_{1-x}B_x$. Powder particles are initially either pure A or pure B. The time t_{cm} is then the *minimum* milling time at which *all powder particles* have essentially the final chemical composition $A_{1-x}B_x$. The particle scale is relevant because t_{cm} can be measured, for instance using a microprobe [9], and because it is the shortest possible synthesis time, $t_f \geq t_{cm}$. Even if they have the expected composition, powder particles may be made up of various phases whose structures and compositions still evolve for $t_{cm} \leq t_m \leq t_f$. For $t_m \leq t_{cm}$, mixing occurs simultaneously at different scales with a broad distribution of characteristics from particle to particle which must be born in mind when interpreting all kinds of experimental results. Both t_f and t_{cm} were measured for $Fe_{1-x}T_x$ elemental powder mixtures ground in a planetary ball-mill, with x=0.50, 0.70, 0.72 for T=V, Cr and Mn respectively [9]. The synthesis time depends strongly on the considered Fe-T alloy while chemical mixing times are estimated to be similar whatever T, $t_{cm} \approx$ 3-4 h, in the milling conditions described in [9]. The times, $\approx$ 4 h, needed to observe a stationary hardness and stationary shapes of hyperfine magnetic field distributions for T=V, Cr [9] are both of the order of t_{cm}. The chemical mixing time has been shown recently to play also a role in the mechanical alloying of Fe-Ge alloys [11]. The time t_f needed to reach a final stationary state is obtained to be of the order of $\approx$ 8-10 h for V and Cr, a time much larger than t_{cm}, while it is of the order of t_{cm}, being $\approx$ 5-6 h, for Mn. At moderate temperatures, thermally activated atomic jumps tend to unmix Fe-Cr alloys ([12,13] and references therein) or to order Fe-V alloys ([14] and references therein) while mechanically forced atomic jumps induced by shear tend to mix these elements. The long synthesis times reflect the fundamental role of these competing mechanisms in the establishment of dynamical stationary states [5-6]. By contrast both types of jumps contribute to mix Fe and Mn, as shown by the large equilibrium solubility of Fe in Mn, and shorten the synthesis time.

2.3. SOME CHARACTERISTICS OF MECHANOSYNTHESIZED MATERIALS

All kinds of solids, almost without restriction, from metallic to ionic have been or may be synthesized or transformed by high-energy ball-milling. The characterization of defects of nanostructured materials is a key to the understanding of their properties, notably their mechanical properties. The nature of grain boundaries in nanostructured materials, as revealed for instance by molecular dynamics, is still in dispute between the defenders of models assuming amorphous structure and those who conclude that they are similar to grain boundaries of coarse-grained materials [15-17]. From transmission (TEM) electron microscopy observations, Li et al. [17] conclude in favor of the latter model but they emphasize that grain boundaries are curved because of grain size and that grain peripheries are slightly strained. The first observation of disclination dipoles in milled Fe has been reported very recently [18]. These defects may play a major role in the formation of nanosized grains, increase the stored elastic energy and contribute to mechanical strengthening. Besides nanograined materials prepared for instance by a direct combination of ground elemental powders or simply by grinding an already formed compound, attractive materials can also be obtained from chemical reactions which take place between ground reactants. Some reactions, for instance the mechanochemical reduction of $SmCl_3$, $SmCl_3+3Na->Sm+3NaCl$ [19], which would never occur at elevated temperatures due to unfavorable thermodynamical conditions, are even rendered possible. Mechanochemical reactions may be put at work to prepare for instance dispersions of nanometric grains within matrices or nanostructured intermetallic compounds [4,7]. Redox or displacement reactions like:

oxide or chloride of A + B -> oxyde or chloride of B + A

(ex: A= Ti, V, Cr, Fe, Ni, Cu, Zn, Zr, Ta, Gd, Er, Sm, B= Na, Mg, Al, Si, Ca, Ti, Ni) yield ultrafine magnetic metal powders, cermets, etc..[4].

3. Synthesis and Coating of Monocrystalline Materials

3.1. MECHANICALLY ACTIVATED SYNTHESIS OF BORIDE NANORODS

Metal diborides, more particularly those made from transition metal, are hard materials with high melting points and high conductivities which are of interest in many fields (cutting tools, reinforcement of ceramics and metals, electron emitters..). The recent discovery of bulk superconductivity in magnesium diboride has been a great boost to the development of synthesis methods of diborides, more particularly those with the simple AlB_2 type structure. The latter structure is hexagonal and consists in a repeated stacking of a graphite-like B layer and of a close-packed Al layer which results in space-filling by triangular prisms of Al atoms whose centers are occupied by boron atoms. However the high melting points of these materials makes it difficult to produce single-phased borides and single crystals of good quality. Mechanical alloying has been widely used to synthesize borides, starting from elemental powder mixtures or from mixtures of oxide and boron powders. Milling is eventually followed by annealing treatments [4].

Barraud et al. [20] have synthesized diborides MB_2 (M=Zr, Hf), either in the form of monocrystalline rods with submicronic to micronic lengths and a diameter of about 100 nm (figure 1) or in the form of facetted submicronic crystals. For that purpose, MCl_4+B+Mg and MCl_4+B powder mixtures are first mechanically activated during some hours in a planetary ball-mill and are then annealed at 1100°C under argon atmosphere for times varying between 1h and 5h. Facetted grains of HfB_2 are obtained from the powder mixtures of MCl_4+B+Mg while monocrystalline rods of HfB_2 are synthesized from the powder mixtures of MCl_4+B. The monocrystalline rods grow parallel to the c-axis of the MB_2 structure from iron-rich grains of the activated powder (figure 1) and are defect-free. Contamination by steel milling tools may thus be beneficial, more particularly when Fe-rich nanoparticles catalyze reactions during annealing of mechanically activated powders.

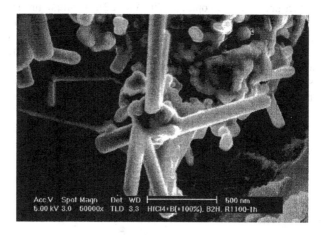

Figure 1. Scanning electron micrograph of HfB_2 nanorods formed from a mixture of $HfCl_4$ and B mechanically activated for 2h and annealed for 1 h at 1100°C [20].

3.2. FRACTURELESS TRANSFORMATION OF GROUND ANATASE TiO_2

Titanium dioxide occurs in nature in three crystalline phases: anatase, brookite and rutile. Other polymorphs, which are formed in equilibrium conditions at high-pressure and at high-temperature, include the orthorhombic α-PbO_2 type-structure, often called TiO_2-II. We chose to describe the polymorphic transformation of anatase induced by grinding because it yields unusual ground materials, it stresses the role of mechanical properties of milled materials and it evidences the importance of a milling parameter. The starting commercial powder consists of single-crystal anatase particles with sizes ranging from 100 nm to 300 nm. Continuous grinding of anatase TiO_2 was performed in a planetary ball mill (Fritsch Pulverisette 7) under air, using different nature of milling tools (alumina, zirconia, steel) for various disc rotation speeds and powder to ball weight ratios. The typical room-temperature compressive strength of coarse-grained TiO_2 is 900 MPa. Anatase transforms by high-energy ball-milling into TiO_2-II ([21-22] and references therein). The formation of metastable TiO_2-II occurs since the first minutes of milling *without fracture* of anatase particles as deduced from particle size measurements, from scanning and from TEM observations of ground powders. TEM evidences the formation of TiO_2-II crystallites, with an average diameter of ~10 nm, at the surface of strain-hardened monocrystalline particles of anatase. Anatase particles are thus coated with a TiO_2-II superficial layer which thickens regularly with milling time. The volume fraction of TiO_2-II increases accordingly to a maximum of ~65% which is largely independent on milling conditions. Rutile is observed to form for longer milling times. For all investigated experimental conditions, the rate constant k of the

16

transformation was obtained by different methods, for instance from the fraction of anatase A remaining after short milling times t_m: $A \propto (1-(1-k\ t_m)^3)$. Local models based on the Hertz impact theory were used to evaluate characteristics such as the powder volume Vp trapped between two colliding balls or between a ball and a vial wall, the maximal contact pressure, the contact radii...etc [22]. A unique linear dependence of k on the average power injected per unit volume of powder trapped during a collision was found (figure 2) while three lines with different slopes were obtained when the average power injected per unit mass of powder was used instead [22]. The former milling parameter appears as physically meaningful because the dynamically confined aggregate of particles, whose volume is here of the order of $5\ 10^{-10}$ m^3, constitutes the very 'reactor' which is at the heart of the complexity of mechanical alloying.

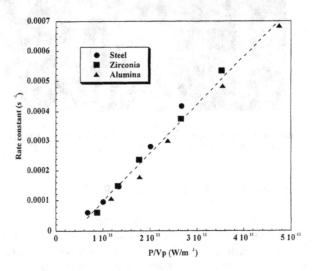

Figure 2. Rate constant as a function of power transferred by unit volume of trapped powder.

4. Competing Mechanisms at Work

Milled materials whose stationary structure results from a dynamical equilibrium between conflicting processes may exhibit original structures at a nanometer scale or may be used to speed up low-temperature transformations of alloys. We describe first the unmixed microstructure reached in ball-milled and in mechanosynthesized $Fe_{0.30}Cr_{0.70}$ alloys and then the ordering of nearly equiatomic Fe-V alloys during low-temperature annealings of disordered as-milled alloys.

4.1. STEADY STATE STRUCTURE OF BALL-MILLED $Fe_{0.30}Cr_{0.70}$

Detailed magnetic characterizations of MA-$Fe_{0.34}Cr_{0.66}$ synthesized by high-energy ball-milling of mixtures of chromium and iron powders are reported in [12]. They evidence the existence of a dominant superparamagnetic contribution from Fe-rich clusters at 400

K above the Curie temperature of the classical homogeneous alloy with the same composition, $T_c \ll \approx 380$ K [23]. X-ray diffraction patterns, TEM and Mössbauer spectra at 7 K show no evidence of the formation of amorphous phases which is known to be non-magnetic at least down to 5 K. This indicates that oxygen contained in the starting powders does not play a significant role on the steady state achieved in these experimental conditions, as further confirmed by the results of annealing experiments. The final ground MA-$Fe_{0.34}Cr_{0.66}$ alloys were concluded to have complex structures with intragranular composition fluctuations of ≈ 0.1 in amplitude at a scale of few nm which reflect the strong trend of Fe and Cr towards segregation. To show that the latter microstructure is a true dynamical equilibrium state, we ball-milled an as-cast $Fe_{0.30}Cr_{0.70}$ alloy in the same milling conditions as those used above for mechanically alloyed powders of elemental Fe and Cr. Advantage was taken from a brittle-ductile transition located between room-temperature and liquid-nitrogen temperature to obtain coarse powder particles from a bulk as-cast alloy. The incorporation of steel from milling tools into ground powders was found to be negligible in such experimental conditions. The magnetic moment at 5K is for instance 1.8 μ_B/Fe atom for the as-cast alloy [24], in perfect agreement with the moment measured by Fischer et al. [23]. The reduced contamination explains why the Fe content is smaller in milled as-cast alloys than in alloys which were mechanosynthesized from elemental $Fe_{0.30}Cr_{0.70}$ mixtures. Magnetic ac susceptibilities (figure 3) were measured with a commercial susceptometer between 5 K and 350 K with an ac field of 2 Oe at four frequencies ranging from 10 Hz to 10 kHz. The experimental results were similar for all investigated frequencies.

When T decreases, the ac susceptibility $\chi(T)$ of the as-cast alloy (fig. 3) exhibits a broad transition around ≈ 290 K, then reaches a maximum at ≈ 200 K and decreases when T decreases down to 5 K. These characteristics reflect the influence of short-range order, namely a strong tendency to clustering, on magnetic properties as clearly shown by the effect of annealing on Fe-Cr alloys with compositions similar to those investigated here [23]. The ac susceptibility of the MA alloy (fig. 3) exhibits a clear ferromagnetic transition at ≈ 170 K, the Curie temperature of an alloy with a Cr content $x \approx 0.75$, then it reaches a maximum at ≈ 150 K and decreases when T decreases down to 50 K. It decreases more markedly below 50 K, as expected from the reentrant transition from a ferromagnetic phase to a cluster-glass-like state for $x \approx 0.75$ [12].

The susceptibility of the as-cast alloy milled for 4h30 (fig. 3) is very close to that of the MA alloy with a small temperature shift likely due to the composition difference. Finally, the susceptibility of the alloy ground for 10h shows a transition at the same temperature as the alloy ground for 4h30, but it is more abrupt for the former alloy. This suggests that the distribution of Fe contents in Fe-rich clusters become narrower with milling time. As discussed in [12], the overall behaviour of $\chi(T)$ of as-milled samples is dominated by the contribution of Cr-rich zones. It complements the information drawn from high-temperature magnetization which is dominated by superparamagnetic Fe-rich clusters.

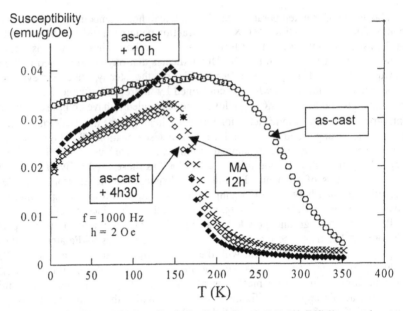

Figure 3. ac susceptibility of $Fe_{0.30}Cr_{0.70}$ 1) mechanically alloyed (crosses, MA 12h [12]) 2) as-cast (empty circles) 3) as-cast and milled for various times (empty diamonds : 4h30, solid diamonds : 10h) [24].

In summary, both mechanically ground and mechanically alloyed $Fe_{0.30}Cr_{0.70}$ alloys exhibit similar stationary microstructures with intragranular composition fluctuations of ≈ 0.1 in amplitude at a scale of few nm. They differ from those found for a macroscopic phase separation and are reminiscent of the steady state patterns described by Enrique and Bellon in systems driven by competing dynamics [25]. The latter show a labyrinthine morphology when phase separation takes place at a finite-length scale.

4.2. LOW-TEMPERATURE ORDERING OF MILLED Fe-V ALLOYS

Owing to the slow kinetics of the $\alpha \rightarrow \sigma$ transformation in concentrated Fe-V alloys, the σ-phase formation can be bypassed kinetically, and a disordered bcc phase can be obtained at low temperatures. A chemical ordering transformation into an ordered B2 phase (CsCl structure) can then occur on the bcc lattice during annealing at low temperatures. The high density of defects in mechanically alloyed disordered bcc Fe-V alloys, especially vacancies and grain boundaries, enhance the kinetics of ordering, enabling studies at low temperatures for practical annealing times [14]. The disorder in the alloy is dominated by antisite defects, namely atoms located on the wrong sublattices, as evidenced by Mössbauer spectroscopy [14].

Figure 4 shows the evolution of the size of ordered domains as a function of the long-range order (LRO) parameter S obtained from neutron diffraction patterns of as-milled $Fe_{0.53}V_{0.47}$ ($0 \leq S \leq 0.94$) annealed at 723 K for various times up to 1176h. The size of ordered domains increases with S up to $S \approx 0.7$, a value close to the steady-state value,

S=0.72. Consistent with the latter observation, fig. 4 shows that the fraction of antisite Fe atoms, f_s, determined from Mössbauer spectra decreases with LRO. This observation is consistent with the Monte Carlo results [26] that the presence of antiphase boundaries may be responsible for a transient population of antisite atoms. A limited decrease in f_s occurs at late times when the nanosized crystals are single domains without antiphase boundaries. The steady-state LRO parameter, S=0.72, is smaller than the equilibrium value, S=0.90, because of residual antisite defects that do not anneal out on the time scale of the experiments [14]. At the higher annealing temperature of 873 K, however, once the domain size grows to the crystal size, S then increases from 0.64 to 0.75, which is close to the value, S=0.80, expected in thermodynamic metastable equilibrium [14]. The elimination of antisite defects occurs at a later stage of the ordering process once the crystals are single ordered domains.

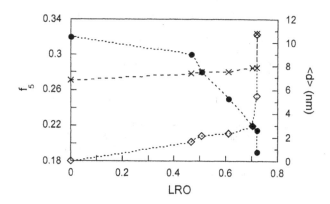

Figure 4. Size of ordered domains and of crystallites (diamonds, crosses, right scale) and f_s (full circles, left scale) as a function of LRO for $Fe_{0.53}V_{0.47}$ annealed at 723 K (LRO=0: as-milled state) [14].

In summary, reordering results first from the elimination of antiphase boundaries and second from the removal of the last antisite defects thanks to long-range diffusion which is however very slow in ordered domains at low temperatures. Grain boundaries probably prevent order from propagating from one grain to another. A notable feature of ordered nanocrystals, in spite of differences in their methods of synthesis, their microstructures (isolated clusters or not) and their ordered structures, is that the steady-state value of the LRO parameter is generally lower than that of bulk samples,. stabilizing typically at 80% of the bulk value. This suppressed order parameter is a kinetic effect associated with the low temperatures of annealing.

Acknowledgements

We thank all colleagues from various laboratories who contributed to our investigations of mechanically alloyed materials and more particularly Prof. B. Fultz (Cal.Tech.), C. Bellouard and our PhD students R. de Araujo Pontès, E. Barraud, T. Girot and T. Ziller.

20

References

1. Swift, J. (1726) *Gulliver's travels*, http://www.on-line.literature.com/swift/gulliver/, chapter 6.
2. Gleiter, H. (2000) Nanostructured materials : basic concepts and microstructure, *Acta mater.* **48**, 1-29.
3. Caro, A. and Van Swygenhoven, H. (2001) Grain boundary and triple junction enthalpies in nanocrystalline metals, *Phys. Rev. B* **63**, 134101 (5).
4. Suryanarayana, C. (2001) Mechanical alloying and milling, *Progr. Mater. Sci.* **46**, 1-184.
5. Martin, G. and Bellon, P. (1996) Driven Alloys, *Solid State Physics* **50**, 189-331.
6. Chaffron, L., Le Bouar, Y. and Martin, G. (2001) Driven phase transformations: a useful concept for wear studies ?, *C.R. Acad. Sci. Paris Série IV* **2**, 749-759.
7. Gaffet, E. et al. (1999) Some recent developments in mechanical activation and mechanosynthesis, *J. Mater. Chem.* **9**, 305-314.
8. Shingu, P.H., Ishihara, K.N. and Otsuki, A. (1995) Mechanical alloying and chaos, *Mater. Sci. Forum* **179-181**, 5-10.
9. Le Caër, G., Ziller, T., Delcroix, P. and Bellouard, C. (2000) Mixing of iron with various metals by high-energy ball-milling of elemental powder mixtures, *Hyp. Int.* **130**, 45-70.
10. Kwon, Y.S., Gerasimov, K.B. and Yoon, S.K. (2002) Ball temperatures during mechanical annoying in planetary mills, *J. Alloys Comps*, in press.
11. Cabrera, A.F. and Sanchez, F.H. (2002) Mössbauer study of ball-milled Fe-Ge alloys, *Phys. Rev. B* **65**, 094202-(9).
12. Delcroix, P., Ziller, T., Bellouard, C. and Le Caër, G. (2001) Mechanical alloying of an $Fe_{0.30}Cr_{0.70}$ alloy from elemental powders, *Mater. Sci. Forum* **360-362**, 329-336.
13. Dubiel, S.M and Inden, G. (1987) On the miscibility gap in the Fe-Cr system : a Mössbauer study on long term annealed alloys, *Z. Metallkde.* **78**, 544-549.
14. Ziller, T., Le Caër, G., Isnard, O., Cénédèse, P. and Fultz, B. (2002) Metastable and transient states of chemical ordering in Fe-V nanocrystalline alloys, *Phys. Rev. B* **65**, 024204-(14).
15. Keblinski, P., Wolf, D., Phillpot, S.R. and Gleiter, H. (1999) Structure of grain boundaries in nanocrystalline palladium by molecular dynamics simulation, *Scripta mater.* **41**, 631-636.
16. Van Swygenhoven, H. (2002) Grain boundaries and dislocations, *Science* **296**, 66-67.
17. Li, D.X., Ping, D.H., Huang, J.Y., Hu, Y.D. and Ye, H.Q. (2000) Microstructure and composition analysis of nanostructured materials using HREM and FEG-TEM, *Micron* **31**, 581-586.
18. Murayama, M., Howe, J.M., Hidaka, H. and Takaki, S. (2002) Atomic-level observation of disclination dipoles in mechanically milled, nanocrystalline Fe, *Science* **295**, 2433-2435.
19. Alonso, T., Yinong, L., Dallimore M.P. and McCormick, P.G. (1993) Low temperature reduction of $SmCl_3$ during mechanical milling, *Scripta metall. mater.* **29**, 55-58.
20. Barraud, E., Bégin-Colin, S. and Le Caër, G. (2002) Nanorods of HfB_2 and of ZrB_2 from mechanically-activated powder mixtures, in preparation.
21. Girot, T., Devaux, X., Bégin-Colin, S., Le Caër, G. and Mocellin, A. (2001) Initial stages of the transformation of single-crystal anatase particles during high-energy ball-milling, *Phil. Mag. A* **81**, 489-499.
22. Girot, T. (2001) Ph.D. thesis, Institut National Polytechnique de Lorraine, Nancy.
23. Fischer, S.F., Kaul, S.N. and Kronmüller, H. (2002) Critical magnetic properties of disordered polycrystalline $Cr_{75}Fe_{25}$ and $Cr_{70}Fe_{30}$ alloys, *Phys. Rev. B* **65**, 064443-(12).
24. Delcroix, P., Bellouard, C. and Le Caër, G. (2002) Stationary unmixed structure of mechanically alloyed and of ball-milled $Fe_{0.30}Cr_{0.70}$, in preparation.
25. Enrique, R.A. and Bellon, P. (2001) Compositional patterning in immiscible alloys driven by irradiation, *Phys. Rev. B* **63**, 134111(12).
26. Fultz, B. and Anthony, L. (1989) Vacancy trapping on lattice with different coordination numbers, *Philos. Mag. Lett.* **59**, 237-241.

IRON(III) OXIDES FORMED DURING THERMAL CONVERSION OF RHOMBOHEDRAL IRON(III) SULFATE

R. ZBORIL[1], M. MASHLAN[1], L. MACHALA[1] AND P. BEZDICKA[2]
[1]Departments of Inorganic and Physical Chemistry, and Experimental Physics, Palacky University, Svobody 26, 771 46 Olomouc, Czech Republic.
[2]Institute of Inorganic Chemistry, Academy of Sciences, CZ-25068 Řež, Czech Republic.

Abstract

The mechanism of thermal decomposition of rhombohedral iron(III) sulfate in air depends significantly on the conditions for diffusion of SO_3 (temperature, thickness of the powdered layer, particle size). The influence of particle size on the reaction mechanism was studied at 600 °C using ^{57}Fe Mössbauer spectroscopy and XRD. Corundum-type α-Fe_2O_3, bixbyite-type β-Fe_2O_3, and orthorhombic ε-Fe_2O_3 were identified as solid conversion products. Time dependence of the relative contents of individual polymorphs (x-Fe_2O_3/ΣFe_2O_3) is a suitable means for monitoring the mechanism of their formation and thermal transformation during the reaction process. The quantitative Mössbauer data obtained from the corresponding spectral areas demonstrate that different transformations occur in the surface layer and in the bulk of sulfate particles. Particles of β-Fe_2O_3 formed after loosening of SO_3 from the surface layer of sulfate particles are relatively stable at 600 °C as documented by the very slow structural change to hematite. The formation of complicated ε-Fe_2O_3 structure is probably related with the slow diffusion of SO_3 from the bulk of sulfate particles. The isochemical transformation of ε-Fe_2O_3 to hematite occurs more quickly due to its lower thermal stability in comparison with β-Fe_2O_3.

1. Introduction

Fe_2O_3 shows various structural and magnetic properties and thus it seems to be a convenient material for a general study of polymorphism and phase transformations. Four basic Fe_2O_3 polymorphs are known: hexagonal corundum structure "alpha", cubic bixbyite structure "beta", cubic spinel structure "gamma" and orthorhombic structure "epsilon".

α-Fe_2O_3 is the most researched and most frequent polymorph existing in nature as a mineral hematite. Hematite has a rhombohedrally centered hexagonal structure of the corundum type (the space group $R\bar{3}c$) with a close-packed oxygen lattice in which two

21

M. Mashlan et al. (eds.), Material Research in Atomic Scale by Mössbauer Spectroscopy, 21–30.
© 2003 Kluwer Academic Publishers. Printed in the Netherlands.

thirds of the octahedral sites are occupied by Fe^{III} ions [1-4]. Hematite shows very interesting magnetic characteristics [1-7]. At low temperatures ($T<260$ K) hematite is antiferromagnetic with the spins oriented along the electric field gradient axis. At the Morin temperature, around 260 K, a reorientation of spins by about 90° takes place whereby the spins become slightly canted to each other (by approximately 5°), resulting in destabilization of their perfect antiparallel arrangement and the existence of weak (parasitic) ferromagnetism between the Morin and Neel temperatures. The Neel temperature of magnetic transition at which hematite loses its magnetic ordering and above which it shows paramagnetic features is frequently understood to be 950 K. In addition, ultrafine nanoparticles of α-Fe_2O_3 (<10 nm) show superparamagnetism, as a result of an increase in the relaxation time due to the decreasing size of particles [8-12].

The beta polymorph of iron(III) oxide has a body-centered cubic "bixbyite" structure with $Ia\bar{3}$ space group, and two non-equivalent octahedral sites of Fe^{III} ions in the crystal lattice. The cubic unit cell contains 32 Fe^{III} ions, 24 of which have a C_2 symmetry (d-position) and 8 have a C_{3i} symmetry (b-position) [13-20]. β-Fe_2O_3 is magnetically disordered at room temperature and it behaves as paramagnetic, a feature that notably distinguishes it from alpha, gamma and epsilon polymorphs. The Neel temperature below which β-Fe_2O_3 shows antiferromagnetic features has been observed by different authors to be between 100-119 K [13,17,20].

γ-Fe_2O_3 is an inverse spinel with a cubic unit cell (space group $P4_132$). It contains, as for Fe_3O_4, cations in tetrahedral A and octahedral B positions, but there are vacancies ($\square$), usually in octahedral positions to compensate for the increased positive charge. Stoichiometry can be formally described by the formula $Fe^A(Fe_{5/3}\square_{1/3})^BO_4$ [1,2,4,6]. Owing to the spinel structure with two sublattices, maghemite is a typical representative of the ferrimagnetic materials. The ultrafine particles of maghemite show superparamagnetism [21-25]. γ-Fe_2O_3 was one of the first and experimentally most researched materials for the conception of the theory of superparamagnetic relaxation.

ε-Fe_2O_3 can be marked as the "youngest" of iron(III) oxides, due to the fact that its structure was completely described only in 1998 by Tronc *et al.* [26]. ε-Fe_2O_3 is orthorhombic with space group $Pna2_1$. The structure derives from a close packing of four oxygen layers. It is isomorphous with $AlFeO_3$, $GaFeO_3$, κ-Al_2O_3 and presumably ε-Ga_2O_3. Three non-equivalent anion (A, B, C) and four cation (Fe_1, Fe_2, Fe_3, Fe_4) positions are in the orthorhombic structure. While the position Fe_4 is tetrahedrally coordinated, the other three positions are octahedrally coordinated. From the magnetic features viewpoint, ε-Fe_2O_3 is a non-collinear ferrimagnet with the Curie temperature of around 480 K [26-29].

In our previous studies [30,31], we proved the formation of different Fe_2O_3 polymorphs during thermal conversion of rhombohedral $Fe_2(SO_4)_3$ and that the conversion mechanism was found to be strongly dependent on the temperature. From the viewpoint of the study of Fe_2O_3 polymorphism, decomposition process is the most interesting at 600 °C because three polymorphs (alpha, beta, epsion) appear in conversion products. In the present study, the influence of particle size of rhombohedral iron(III) sulfate on its conversion mechanism was monitored just at 600 °C using the ^{57}Fe Mössbauer spectroscopy in combination with XRD.

2. Experimental Details

Rhombohedral $Fe_2(SO_4)_3$ was prepared by $Fe_2(SO_4)_3 \cdot 5H_2O$ (Sigma-Aldrich) dehydration at 300 °C in a nitrogen atmosphere. The fraction of particles with the size of 1-5 μm was prepared by grinding iron(III) sulfate in the vibrating ball mill for 1 hour (R1). The fraction of larger particles with diameters between 160-180 μm was obtained from the original material using sieves (sample R2). Powdered samples of $Fe_2(SO_4)_3$ (R1, R2) were heated in air at 600 °C for different periods of time (10-300 minutes).

The transmission Mössbauer spectra were collected using a Mössbauer spectrometer in constant acceleration mode with a $^{57}Co(Rh)$ source. A cryostat with closed He-cycle (Janis Research Company, USA) was used as a basis for a refrigeration system that allowed work in the temperature range of 12-300 K. The phase composition of samples was monitored by XRD using θ-2θ conventional equipment (SEIFERT-FPM, Germany) with Cu-Kα wavelength.

3. Results and Discussion

To analyze the influence of particle size of rhombohedral $Fe_2(SO_4)_3$ on its thermal behavior, both samples with non-equivalent particle size (R1 and R2) were heated at 600 °C for different periods of time and analyzed using XRD and Mössbauer spectroscopy. XRD data revealed the presence of α-Fe_2O_3, β-Fe_2O_3 and ϵ-Fe_2O_3 in all heated samples. With increasing calcination time, the gradual increase of intensities of lines corresponding to hematite was observed in both samples. The significant broadening of the β-Fe_2O_3 and ϵ-Fe_2O_3 lines indicates that both metastable polymorphs are formed as ultrafine particles. XRD pattern of sample R1 heated at 600 °C for 75 minutes (Figure 1) demonstrates the phase composition of the conversion products.

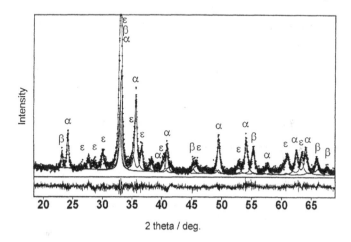

Figure 1. XRD pattern of sample R1 heated at 600 °C for 75 minutes.

All Mössbauer spectra (except for sample R1 heated at 600 °C for 300 minutes) exhibit a superposition of four magnetically split components (sextets) along with a central paramagnetic doublet. Additionally, a singlet line with an isomer shift parameter of about 0.50 mm/s appears in Mössbauer spectra of samples heated for shorter periods of time, evidently a result of the presence of the initial ferric sulfate [30-32]. Hyperfine parameters of the narrow central doublet (IS_{Fe}=0.36-0.37 mm/s; QS=0.72-0.74 mm/s) correspond closely to those reported for β-Fe_2O_3 [13,15,30,32]. The sextet with the highest hyperfine field (~50-51 T) corresponds to iron ions in the hematite structure. The other three magnetically split subspectra may be attributed to different sites in the orthorhombic structure of ϵ-Fe_2O_3. The low values of hyperfine field and isomer shift parameters indicate one tetrahedral Fe^{III} site (B=25-27 T; IS_{Fe}=0.16-0.18 mm/s). Octahedral sites were fitted with two sextets having similar isomer shift values (0.34-0.37 mm/s) and considerably different magnetic hyperfine fields (44-45 T versus 38-39 T). In fact, three non-equivalent octahedral sites are in the ϵ-Fe_2O_3 structure, however two of them are characterized by close B values (~45 T), affecting their strong overlapping. These Mössbauer data are in very good accordance with hyperfine parameters of ϵ-Fe_2O_3 prepared by the thermal transformation of maghemite nanoparticles [26,33]. Room temperature Mössbauer spectra of thermally treated samples are shown in Fig.2 and the corresponding parameters are listed in Table 1 for sample R1 and in Table 2 for sample R2.

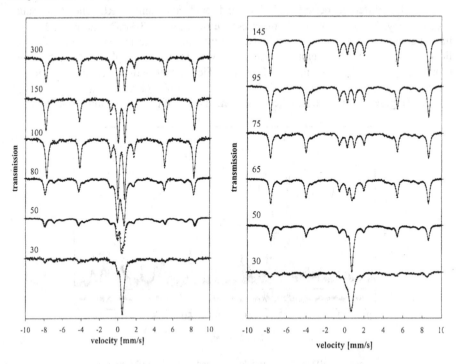

Figure 2. RT Mössbauer spectra of samples R1 (left) and R2 (right) thermally treated at 600 °C for different periods of time.

TABLE 1. Mössbauer parameters of sample R1 thermally treated at 600 °C for different periods of time.

time [min.]	phase	IS_{Fe} [mm/s]	QS [mm/s]	ε_Q [mm/s]	B [T]	A [%]
30	$Fe_2(SO_4)_3$	0.49	-	-	-	45.8
	β-Fe_2O_3	0.36	0.72	-	-	19.1
	ε-Fe_2O_3, Fe_1, Fe_2	0.37	-	-0.21	44.6	10.0
	ε-Fe_2O_3, Fe_3 site	0.36	-	-0.02	39.4	6.2
	ε-Fe_2O_3, Fe_4 site	0.16	-	-0.14	26.5	4.8
	α-Fe_2O_3	0.37	-	-0.22	50.8	14.1
50	$Fe_2(SO_4)_3$	0.48	-	-	-	19.2
	β-Fe_2O_3	0.37	0.73	-	-	27.5
	ε-Fe_2O_3, Fe_1, Fe_2	0.37	-	-0.22	44.5	13.0
	ε-Fe_2O_3, Fe_3 site	0.35	-	-0.06	39.2	8.7
	ε-Fe_2O_3, Fe_4 site	0.18	-	-0.13	26.1	6.0
	α-Fe_2O_3	0.37	-	-0.21	50.7	25.6
80	$Fe_2(SO_4)_3$	0.49	-	-	-	5.0
	β-Fe_2O_3	0.36	0.73	-	-	33.3
	ε-Fe_2O_3, Fe_1, Fe_2	0.37	-	-0.21	44.3	9.8
	ε-Fe_2O_3, Fe_3 site	0.35	-	0.0	38.7	6.8
	ε-Fe_2O_3, Fe_4 site	0.17	-	-0.16	25.9	5.1
	α-Fe_2O_3	0.37	-	-0.21	50.6	40.0
100	$Fe_2(SO_4)_3$	-	-	-	-	-
	β-Fe_2O_3	0.36	0.73	-	-	35.0
	ε-Fe_2O_3, Fe_1, Fe_2	0.35	-	-0.21	44.5	7.0
	ε-Fe_2O_3, Fe_3 site	0.35	-	0.04	38.7	4.1
	ε-Fe_2O_3	0.18	-	-0.09	26.6	3.4
	α-Fe_2O_3	0.37	-	-0.21	50.8	50.5
150	$Fe_2(SO_4)_3$	-	-	-	-	-
	β-Fe_2O_3	0.36	0.74	-	-	35.6
	ε-Fe_2O_3, Fe_1, Fe_2	0.38	-	-0.19	45.2	4.5
	ε-Fe_2O_3, Fe_3 site	0.35	-	0.03	38.6	2.1
	ε-Fe_2O_3, Fe_4 site	0.17	-	-0.10	26.9	1.5
	α-Fe_2O_3	0.37	-	-0.23	51.3	56.3
300	$Fe_2(SO_4)_3$	-	-	-	-	-
	β-Fe_2O_3	0.36	0.74	-	-	34.4
	ε-Fe_2O_3, Fe_1, Fe_2	-	-	-	-	-
	ε-Fe_2O_3, Fe_3 site	-	-	-	-	-
	ε-Fe_2O_3, Fe_4 site	-	-	-	-	-
	α-Fe_2O_3	0.37	-	-0.22	51.3	65.6

IS_{Fe}-isomer shift related to metallic iron, ε_Q-quadrupolle shift, QS-quadrupole splitting, B-hyperfine magnetic field, A-relative spectrum area.

TABLE 2. Mössbauer parameters of sample R2 thermally treated at 600 °C for different periods of time.

time [min.]	phase	IS_{Fe} [mm/s]	QS [mm/s]	ε_Q [mm/s]	B [T]	A [%]
30	$Fe_2(SO_4)_3$	0.50	-	-	-	50.0
	β-Fe_2O_3	0.37	0.72	-	-	6.5
	ε-Fe_2O_3, Fe_1, Fe_2	0.37	-	-0.19	44.1	9.6
	ε-Fe_2O_3, Fe_3 site	0.35	-	0.02	38.4	6.7
	ε-Fe_2O_3, Fe_4 site	0.17	-	-0.04	26.0	6.0
	α-Fe_2O_3	0.36	-	-0.19	50.4	21.2
50	$Fe_2(SO_4)_3$	0.49	-	-	-	29.3
	β-Fe_2O_3	0.36	0.73	-	-	9.3
	ε-Fe_2O_3, Fe_1, Fe_2	0.37	-	-0.20	44.2	12.2
	ε-Fe_2O_3, Fe_3 site	0.35	-	0.02	38.4	7.5
	ε-Fe_2O_3, Fe_4 site	0.18	-	-0.05	25.6	5.7
	α-Fe_2O_3	0.36	-	-0.21	50.5	36.0
65	$Fe_2(SO_4)_3$	0.50	-	-	-	8.0
	β-Fe_2O_3	0.36	0.73	-	-	12.7
	ε-Fe_2O_3, Fe_1, Fe_2	0.36	-	-0.22	44.2	14.6
	ε-Fe_2O_3, Fe_3 site	0.36	-	0.0	38.4	8.6
	ε-Fe_2O_3, Fe_4 site	0.18	-	-0.10	25.7	6.0
	α-Fe_2O_3	0.36	-	-0.21	50.5	50.1
75	$Fe_2(SO_4)_3$	-	-	-	-	-
	β-Fe_2O_3	0.36	0.73	-	-	13.5
	ε-Fe_2O_3, Fe_1, Fe_2	0.36	-	-0.20	44.4	14.0
	ε-Fe_2O_3, Fe_3 site	0.35	-	0.02	38.8	8.3
	ε-Fe_2O_3	0.18	-	-0.15	26.1	6.6
	α-Fe_2O_3	0.37	-	-0.21	50.6	57.6
95	$Fe_2(SO_4)_3$	-	-	-	-	-
	β-Fe_2O_3	0.36	0.73	-	-	13.3
	ε-Fe_2O_3, Fe_1, Fe_2	0.36	-	-0.21	44.4	12.7
	ε-Fe_2O_3, Fe_3 site	0.34	-	0.01	38.8	6.8
	ε-Fe_2O_3, Fe_4 site	0.16	-	-0.17	25.9	5.0
	α-Fe_2O_3	0.36	-	-0.31	50.6	62.2
145	$Fe_2(SO_4)_3$	-	-	-	-	-
	β-Fe_2O_3	0.36	0.73	-	-	12.5
	ε-Fe_2O_3, Fe_1, Fe_2	0.37	-	-0.22	44.7	7.7
	ε-Fe_2O_3, Fe_3 site	0.35	-	0.06	38.9	5.4
	ε-Fe_2O_3, Fe_4 site	0.17	-	-0.14	26.4	4.3
	α-Fe_2O_3	0.36	-	-0.21	50.8	70.1

The determination of the content of individual polymorphs from the corresponding spectral areas (see Table 1 and 2) allows not only an assessment of the influence of particle size on the polymorphous composition of Fe_2O_3, but also provides an opportunity to understand the reaction mechanism. To consider the conversion course also in its initial stage we heated samples R1 and R2 at 600 °C for 10 minutes. RT Mössbauer spectra and XRD patterns of both samples confirmed the low conversion degree (< 10%), thereby making it impossible to precisely determine the relative contents of Fe_2O_3 polymorphs. In order to obtain a precise data, we dissolved ferric sulfate in 0.01M sulphuric acid and Fe_2O_3 phases were isolated by filtration. Figure 3 shows RT Mössbauer spectra of the separated phases and the results of their mathematical deconvolution including the values of the spectral areas corresponding to different Fe_2O_3 polymorphs.

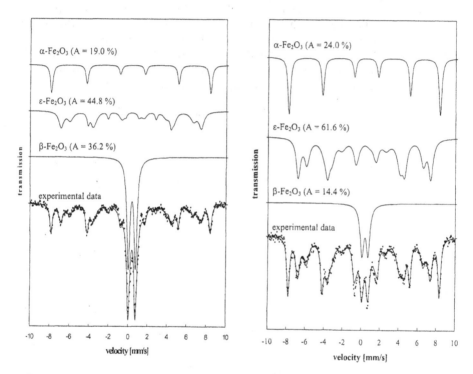

Figure 3. RT Mössbauer spectra of samples R1 (left) and R2 (right) thermally treated at 600 °C for 10 minutes, the residual ferric sulfate was chemically removed.

If we combine the results of the short-time calcinations (Fig.3) with the previous Mössbauer data (Tab.1 and Tab.2) the reaction process could be easily monitored using time dependence of the relative contents of the individual polymorphs (see Fig.4). The low relative content of hematite in the initial conversion stage proves that this thermally stable polymorph is not a primary conversion product. Gradual increase of its relative content is compensated by the corresponding decrease in the percentages of metastable polymorphs, particularly of ε-Fe$_2$O$_3$. The relative content of the bixbyite β-Fe$_2$O$_3$ structure in the mixture of Fe$_2$O$_3$ phases (β-Fe$_2$O$_3$/ΣFe$_2$O$_3$) stays almost constant with increasing calcinations time due to its quick formation and the high thermal stability. RT Mössbauer spectra revealed the considerably higher relative content of the beta phase in samples R1 (β-Fe$_2$O$_3$/ΣFe$_2$O$_3$~35%) in comparison with R2 samples (β-Fe$_2$O$_3$/ΣFe$_2$O$_3$~14%). This fact indicates that β-Fe$_2$O$_3$ is formed from the surface layer of the sulfate particles, the latter providing ideal conditions for the liberation of gaseous SO$_3$. The so formed bixbyite structure is well defined, as documented by the narrow lines in Mössbauer spectra (Γ~0.27-0.33 mm/s). Concerning the orthorhombic ε-Fe$_2$O$_3$, the highest relative content was recorded in the initial stages of the reaction and it gradually decreased with the reaction time as a result of the occurring isochemical structural change of ε-Fe$_2$O$_3$ to α-Fe$_2$O$_3$. The higher relative contents of ε-Fe$_2$O$_3$ in R2 of larger particles (with the higher ratio of bulk atoms/surface atoms) supports the assumption that this polymorph is formed from the bulk of sulfate particles. The different decomposition mechanisms related with the non-equivalent conditions for liberation of SO$_3$ on the surface and in the bulk of sulfate particles result also in the lower degree of the structural ordering of the "bulk phase" as documented by significant broadening of the spectral lines corresponding to ε-Fe$_2$O$_3$ (Γ~0.4-0.7 mm/s).

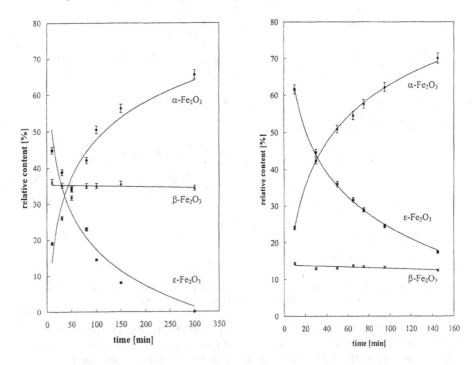

Figure 4. Time dependence of the relative contents of the different F$_2$O$_3$ forms (x-Fe$_2$O$_3$/ΣFe$_2$O$_3$) for samples R1 (left) and R2 (right) thermally treated at 600 °C.

4. Conclusions

The crystal structure of ferric oxide formed by thermal decomposition of rhombohedral ferric sulfate is strongly determined by conditions for diffusion and liberation of sulphur trioxide. The unique mixture of Fe_2O_3 polymorphs (alpha, beta, epsilon) formed during this conversion is a result of the non-equivalent reaction routes occurring on the surface and in the bulk of the particles. Thus the phase composition of samples (content of individual polymorphs) depends significantly on the particle size of $Fe_2(SO_4)_3$ and on the degree of its conversion.

The β-Fe_2O_3 forms from the surface layer of the particles as demonstrated by an increase in its relative content with decreasing particle size of sulfate. Kinetic data obtained at 600 °C for two types of particles with different sizes indicate that ϵ-Fe_2O_3 is the direct product of "bulk decomposition" of rhombohedral ferric sulfate.

Acknowledgements

Financial support from the Grant Agency of The Czech Republic under Projects 202/00/0982 and 202/00/D091 is acknowledged.

References

1. Cornel, R.M., and Schwertmann, U. (1996) *The Iron Oxides. Structure, Properties, Reactions and Uses*, VCH, Weinheim.
2. Mitra, S., (1992) *Applied Mössbauer Spectroscopy, Series: Physics and Chemistry of the Earth*, Vol. 18, Pergamon, Oxford.
3. Catlow, C.R., Corish, J., Hennesy, J., and Mackrodt, W.C. (1988) Atomistic simulation of deffect structures and ion-transport in alpha-Fe_2O_3 and alpha Cr_2O_3, *J. Am. Ceram. Soc.* **71**, 42-49.
4. Bowen, L.H., De Grave, E, and Vandenberghe, R.E. (1993) Mössbauer effect studies of magnetic soils and sediments, in G. J. Long, and F. Grandjean (eds.), *Mössbauer Spectroscopy Applied to Magnetism and Materials Science*, Vol.1, Plenum Press, New York, pp. 132-141.
5. Fysh, S.A., and Clark, P.E. (1982) Aluminous hematite - a Mössbauer study, *Phys. Chem. Miner.* **8**, 257-267.
6. Vandenberghe, R.E., de Grave, E., Landuydt, C., and Bowen, L.H. (1990) Some aspects concerning the characterization of iron oxides and hydroxides in soils and clays, *Hyperfine Interact.* **53**, 175-193.
7. Murad, E., and Schwertmann, U. (1986) Influence of Al substitution and crystal size on the room-temperature Mössbauer spectrum of hematite, *Clay. Clay Miner.* **34**, 1-6.
8. Rancourt, D.G., Julian, S.R., and Daniels, J.M. (1985) Mössbauer characterization of very small superparamagnetic particles – application to intra-zeolitic alpha Fe_2O_3 particles, *J. Magn. Magn. Mater.* **49**, 305-316.
9. Dormann, J.L., Cui, J.R., and Sella, C (1985) Mössbauer studies of Fe_2O_3 antiferromagnetic small particles, *J. Appl. Phys.*, **57**, 4283-4285.
10. Bødker, F., Hansen, M.F., Koch, C.B., Lefmann, K., and Mørup S. (2000) Magnetic properties of hematite nanoparticles, *Phys. Rev. B* **61**, 6826-6838.
11. Rancourt, D.G., and Daniels, J.M. (1984) Influence of unequal magnetization direction probabilities on the Mössbauer spectra of superparamagnetic particles, *Phys. Rev. B* **29**, 2410-2414.
12. Jing, J., Zhao, F., Yang, X., and Gonser, U. (1990) Magnetic relaxation in nanocrystalline iron oxides, *Hyperfine Interact.* **54**, 571-575.
13. Ikeda, Y., Takano, M., and Bando, Y. (1986) Formation of needle-like alpha-iron(III) oxide particles grown along the c-axis and characterization of precursorily formed beta-iron(III) oxide, *Bull. Inst. Chem. Res., Kyoto Univ.* **64**, 249-258.

30

14. Ben-Dor, L., Fischbein, E., Felner, I., and Kalman, Z. (1977) β Fe₂O₃: Preparation of thin films by chemical vapor deposition from organometallic chelates and their characterization. *J. Electrochem. Soc.* **124**, 451-457.

15. Muruyama, T., and Kanagawa, T. (1996) Electrochromic properties of iron oxide thin films prepared by chemical vapor deposition, *J. Electrochem. Soc.* **143**, 1675-1677.

16. Gonzales-Carreno, T., Morales, M.P., and Serna, C.J. (1994) Fine beta-Fe₂O₃ particleswith cubic structure obtained by spray pyrolysis, *J. Mater. Sci. Lett.* **13**, 381-382.

17. Bauminger, E.R., Ben-Dor, L., Felner, I., Fischbein, E., Nowik, I., and Ofer, S. (1977) Mössbauer effect studies of β Fe₂O₃, *Physica B* **86-88**, 910-912.

18. Ben-Dor, L., and Fischbein, E. (1976) Concerning the β phase of iron(III) oxide. *Acta Cryst. B* **32**, 667.

19. Wiarda, D., Wenzel, T., Uhrmacher, M., and Lieb, K. P. (1992) Hyperfine interaction of ^{111}Cd impurities in Mn₂O₃, Mn₃O₄ and β-Fe₂O₃, *J. Phys. Chem. Solids* **53**, 1199-1209.

20. Wiarda, D., and Weyer, G. (1993) Mössbauer investigations of the antiferromagnetic phase in the metastable beta-ferric oxide, *Int. J. Mod. Phys. B* **7**, 353-356.

21. Pascal, C., Pascal, J.L., Favier, F., Elidrissi Moubtassim, M.L., and Payen, C. (1999) Electrochemical synthesis for the control of γ-Fe₂O₃ nanoparticle size. Morphology, microstructure, and magnetic behavior, *Chem. Mater.* **11**, 141-147.

22. Ennas, G., Musinu, A., Piccaluga, G., Zedda, D., Gatteschi, D., Sangregorio, C., Stanger, J.L., Concas, G., and Spano, G. (1998) Characterization of iron oxide nanoparticles in an Fe₂O₃-SiO₂ composite prepared by a sol-gel method, *Chem. Mater.* **10**, 495-502.

23. Cannas, C., Gatteschi, D., Musinu, A., Piccaluga, G., and C. Sangregorio (1998) Structural and magnetic properties of Fe₂O₃ nanoparticles dispersed over a silica matrix, *J. Phys. Chem. B* **102**, 7721-7726.

24. Serna, C.J., Bødker, F., Mørup, S., Morales, M.P., Sandiumenge, F., and Verdaguer S. (2001) Spin frustration in maghemite nanoparticles, *Solid State Commun.* **118**, 437-440.

25. Martinez, B., Roig, A., Obradors, X., Mollins, E., Rouanet, A., and Monty, C. (1996) Magnetic properties of γ-Fe₂O₃ nanoparticles obtained by vaporization condensation in a solar furnace, *J. Appl. Phys.* **79**, 2580-2586.

26. Tronc, E., Chanéac, C., and Jolivet, J.P. (1998) Structural and magnetic characterization of ε-Fe₂O₃, *Solid State Chem.* **139**, 93-104.

27. Schrader, R., and Büttner, G. (1963) Eine neue Eisen(III)-oxidphase: ε-Fe₂O₃, *Z. Anorg. Allg. Chem.* **320**, 220-234.

28. Dézsi, I, and Coey, J.M.D. (1973) Magnetic and thermal properties of ε-Fe₂O₃, *Phys. Status. Solidi A* **15**, 681-685.

29. Dormann, J. L., Viart, N., Rehspringer, J. L., Ezzir, A., and Niznansky, D. (1998): Magnetic properties of Fe₂O₃ particles prepared by sol-gel method, *Hyperfine Interact.* *112*, 89-92.

30. Zboril, R, Mashlan, M., Krausova, D., and Pikal, P. (1999) Cubic β-Fe₂O₃ as the product of thermal decomposition of Fe₂(SO₄)₃, *Hyperfine Interact.* **121-122**, 497-501.

31. Zboril, R., Mashlan, M., and Petridis, D. (2002) Iron(III) Oxides from Thermal Processes-Synthesis, Structural and Magnetic Properties, Mössbauer Spectroscopy Characterization, and Applications, *Chem. Mater.* **14**, 969-982.

32. Zboril, R., Mashlan, M., Krausova, D. (1999) The mechanism of β-Fe₂O₃ formation by the solid state reaction between NaCl and Fe₂(SO₄)₃, in M. Miglierini and D. Petridis (eds.), *Mössbauer Spectroscopy in Materials Science*, Kluwer Academic Publishers, Dordrecht, pp. 49-56.

33. Barcova, K., Mashlan, M, Zboril, R., Martinec, P., and Kula, P. (2001) Thermal decomposition of almandine garnet: Mössbauer study, *Czech. J. Phys.* **51**, 749-754.

PHASE ANALYSIS OF THE Fe-C NANOPOWDER PREPARED BY LASER PYROLYSIS

B. DAVID[1], M. VONDRÁČEK[1], O. SCHNEEWEISS[1], P. BEZDIČKA[2], R. ALEXANDRESCU[3] AND I. MORJAN[3]
[1]Institute of Physics of Materials ASCR, Žižkova 22, CZ-61662 Brno, Czech Republic
[2]Institute of Inorganic Chemistry ASCR, CZ-25068 Řež, Czech Republic
[3]National Institute for Lasers, Plasma and Radiation Physics, Bucharest, Romania

1. Introduction

Iron-carbide-based compounds and composites present a particular interest due to their structural and magnetic properties. Different carbide phases are known, such as the orthorombic cementite (Fe_3C), the monoclinic Hägg carbide (Fe_5C_2), the hexagonal Fe_2C. Another metastable phase Fe_7C_3 was synthesized only when the precursor iron oxides contained small amounts of Ba [1]. A new class of nanopowder iron-carbide materials were synthesized in the past few years by Rice and coworkers [2,3] and by Ecklund and coworkers [4]. Both groups applied the method of laser pyrolysis from gas phase reactants [5], in which a powerful CW CO_2 laser is used to heat a mixture of gases through the absorption of the laser radiation by at least one of the gas components. By using this technique, the production of other carbides and composites (like SiC, Si/C/N, TiC, Ti/Si/C) were reported [6,7]. The catalytic and magnetic properties of iron-based bulk materials could be highly enhanced if particles in the nanometer size were obtained. Using the laser induced pyrolysis of a sensitized system containing iron pentacarbonyl vapors, γ-Fe particles [8] and iron-oxide-based nanopowders [9] were recently reported. In a previous work we presented preliminary results on the synthesis of iron-carbide-based nanopowders by the laser pyrolysis of gas mixtures containing iron pentacarbonyl vapors and ethylene as carrier gas [10].

In this work an analysis of the magnetic properties of FeC nanostructures is also presented. This analysis is of major importance since it is known that magnetic properties are sensitive to the huge surface/volume ratio of nanoparticles. The main points are the orientation disorder of surface spins and the overwhelming role of surface anisotropy [11].

2. Experimental

The experimental apparatus used for laser synthesis of nanosized powders was quite

M. Mashlan et al. (eds.), Material Research in Atomic Scale by Mössbauer Spectroscopy, 31–40.

standard and similar in its main features and most details to the one previously described [7,10]. It is presented in Figure 1. The CO_2 laser radiation (maximum output power 130 W, $\lambda = 944$ cm^{-1}) orthogonally intersects the reactant gas stream which is admitted to the center of the reaction cell through a nozzle (3 mm diameter). The reactant gas was confined to the flow axis by a coaxial Ar stream (outer tube with 10 mm diameter). In order to prevent the NaCl windows from being coated with powder they were continuously flushed with Ar. The interaction of reactant gas with the laser

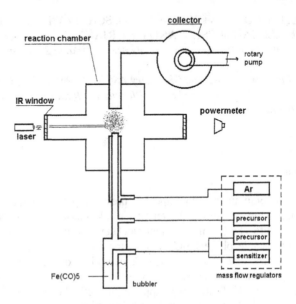

Figure 1. Experimental apparatus.

beam results in a reddish-yellow flame. The nucleated particles formed during reaction are entrained by the gas stream to the cell exit where they are collected in a trap, closed with a microporous filter in the direction of the rotary pump. The flow of the hydrocarbon (C_2H_4) used as carrier gas for Fe(CO)$_5$ vapors (aprox. 25 torr at room temperature), was independently controlled. The total pressure in the reactor was maintained by adjusting the pumping speed through a needle valve. The iron-carbide powders were ultrafine and brown to black in color. In Table 1, the synthesis parameters are presented. Previously reported results on iron carbide fine powders produced by the same laser technique [4] used as carbon donor ethylene. From Table 1, we observe that the run, labeled IC follows this trend. Studies of Fe(CO)$_5$ sequential decarbonylation by laser pyrolysis [12] and by using a sensitized mixture revealed the fast removal of carbonyl ligands (first bond energy of 41.5 Kcal/mol [12]), with the formation of metallic iron and CO:

$$Fe(CO)_5 \rightarrow Fe(CO)_x + (5-x)CO, \quad x = 0 \div 4 \tag{1}$$

Indeed, we should mention that in all experiments, besides iron carbides, different quantities of metallic iron were also formed, which subsequently, if not passivated, spontaneously burned in the atmosphere after withdrawal. To avoid this process, the as-synthesized powders were passivated in-situ by maintaining the particles inside the trap in about 5% O_2 in argon atmosphere during 24 hours.

TABLE 1. Experimental parameters for the iron-carbide-based powder synthesis.

Run	Carrier gas (bubbler)	$\Phi_{carrier}$	$\Phi_{Fe(CO)5}$ **	p
		sccm	sccm	torr
IC	C_2H_4	250	50	150

** $\Phi_{Fe(CO)5}$ was calculated according to the formula $\Phi_{Fe(CO)5} = \Phi_{carrier} \cdot [p_{Fe(CO)5} / (p - p_{Fe(CO)5})]$, where $p_{Fe(CO)5}$ represents the $Fe(CO)_5$ vapour pressure at room temperature.

3. Results and Discussion

3.1. TEM, XRD AND IR EXAMINATION

TEM analysis of IC specimen (Figure 2) revealed that this sample contains only small fractions of the iron carbides Fe_3C/Fe_7C_3 and/or α-Fe, with a mean size 5÷8 nm; the major fraction consisted of big crystallized particles (20÷40 nm). From the electron diffraction patterns (not presented here; published in [13]) we identified the phase α-Fe_2O_3 (due to the rings associated with d = 2.70Å (I=100) and d = 2.47Å (I=80) [14]). The possible presence of Fe_3O_4 [15] could not be excluded.

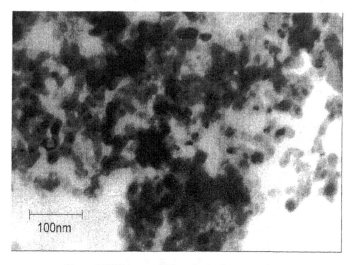

100nm

Figure 2. TEM image of IC synthesized nanopowder.

The XRD pattern of IC powder [13] confirms these assumptions. It is worth mentioning that for sample IC a large amount of pure iron was obtained in the as-synthesized powders. The large quantity of ethylene (see Table 1) which plays not only the role of carbon donor for carbide formation, but also the role of the sensitizer, could help – by the increased absorbed laser energy – to a rapid rise of the reaction temperature, leading to a very fast decarbonylation and formation of bare Fe atoms. It was observed that very finely divided Fe particles are covered with a rather thick layer of iron oxide crystallites [16]. If we refer now to the diffractogram of IC powder sample (not presented), one may observe that the oxidic features of sample IC are clearly evidenced. However, the sample presents a strong and wide peak around $2\theta = 44.5$ deg., which is common as well to both iron carbides (Fe_3C/Fe_7C_3) as to pure α-Fe. Even if difficult to distinguish from this diffractogram, it seems that α-Fe ($2\theta \sim 65.2$ deg.) could prevail.

The IR spectrum of IC sample [13] is consistent with the metallic character of iron carbides [4]. The large band located between 350 and 700 cm^{-1} which peaks around 550 cm^{-1} could be ascribed to either Fe_2O_3 or/and Fe_3O_4.

3.2. MAGNETIC MEASUREMENTS

Thermomagnetic curve of **as-prepared** nanopowder (Figure 3) was measured by a vibrating sample magnetometer at $p = 1$ Pa and $H_{ext} = 50$ Oe. The rate of the temperature change of heating and subsequent cooling was 4 K / min. On the measured

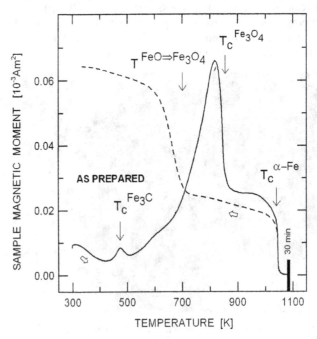

Figure 3. Thermomagnetic curve of IC sample.

curve Curie temperatures $T_C(Fe_3C) = 475$ K, $T_C(Fe_3O_4) = 853$ K and $T_C(\alpha\text{-Fe}) = 1045$ K were identified. Only Curie temperature of Fe_3C meaningfully differs from its tabulated value 488 K. On the curve corresponding to cooling from 1073 K down to 300 K (dashed line) the increase of the sample magnetic moment can be seen, that corresponds to the creation of ferrimagnetic Fe_3O_4 phase from high temperature paramagnetic FeO phase. FeO is known to be thermodynamically stable above 843 K: below this temperature FeO changes into Fe_3O_4 and α-Fe [17]. Fe_3C phase is not present in the sample after this measurement, because Fe_3C decomposes into α-Fe and C at $T \geq 873$ K [18,19]. Hysteresis loops are shown in Figure 4. Hysteresis loop of **annealed** sample (= as-prepared after the measurement of its TM curve) indicates changes in the phase composition and also particle/grain coarsening.

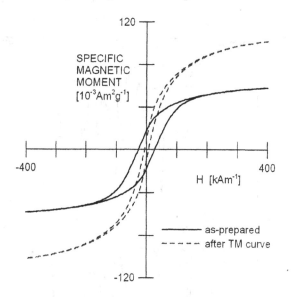

SPECIFIC
MAGNETIC
MOMENT
$[10^{-3} Am^2 g^{-1}]$

H $[kAm^{-1}]$

——— as-prepared
– – – – after TM curve

Figure 4. Hysteresis loops of IC sample.

3.3. MÖSSBAUER ANALYSIS

Mössbauer spectra were measured in standard transmission geometry with ^{57}Co source. Isomer shift was evaluated against α-Fe. Spectra were fitted with Confit software package [20] using transmission integral.

MS spectra of the **as-prepared IC nanopowder** are in Figure 5. Spectrum at 300 K can be well fitted with 11 sextets, 1 distribution and 4 dublets. α-Fe has sextet shown with full line with hyperfine field $B_{hf} = 33,1$ T (16 atomic %). The sextet with $B_{hf} = 20,9$ T (13 at%) we assign to Fe_3C, although nanoparticle Fe_3C could be also fitted with two sextets [4]. Five sextets (B_{hf}: 11,9 T, 14,3 T, 16,6 T, 18,6 T, 19,8 T) could be ascribed to Fe_7C_3 [4] and Fe_5C_2 [21]. But there is a problem: two sextets of Fe_7C_3 (has 3 sextets) overlap two sextets of Fe_5C_2 (has also 3 sextets). Each of these 5 sextets is in the range (2÷5) at%. Two sextets (B_{hf}: 24,9 T, 25,8 T) could be probably associated with Fe_2C [21]. Again – they have 2 at% each. These 7 carbide sextets were together with Fe_3C sextet added into one component shown with dashed line. The last 2 sextets (B_{hf}: 44,0 T, 47,9 T) and one gaussian distribution (B_{hf} = 39,1 T; width = 2,57 mm/s, QS = 0,00 mm/s, IS = 0,36 mm/sec) were added into one component shown with bold line (32 at%). We assign the distribution with oxide α-Fe_2O_3 or/and Fe_3O_4 in the form of very small particles or surface coating (see 20 K spectrum discussion below). Stochiometry of these oxides is questionable. Two sextets (B_{hf}: 44,0 T, 47,9 T; 2 at% each) should correspond to crystalline Fe_3O_4, although for instance in [22] are fields of 45,3 T and 49,1 T stated. Four dublets (IS//QS: 0,33//1,02 mm/sec, 0,22//0,30 mm/sec,

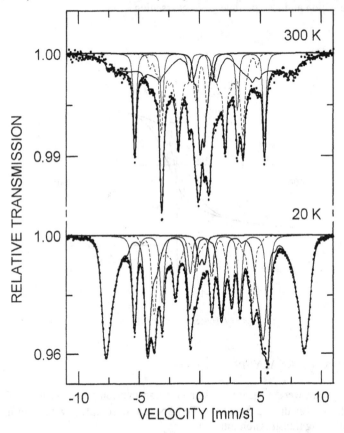

Figure 5. Mössbauer spectra of the as-prepared IC nanopowder.

0,31//1,52 mm/sec, 0,28//0,76 mm/sec) are added again into one component shown with bold line. We ascribe these dublets to superparamagnetic particles with Fe^{3+} [22] - or at least three of them (see 20 K spectrum discussion below). This component has 15 at%. Disappearance of this dublet component in low temperature spectrum (Figure 5) is the evidence of superparamagnetism. But in the spectrum measured at 20 K still occours one dublet (IS = 0,17 mm/s, QS = 0,42 mm/sec, 2 at%) shown with bold line. One can see, that the oxide component has considerably grown. It consist of two sextets (B_{hf}: 46,4 T, 48,8 T; 3 and 4 at%) and two gaussian distributions (B_{hf} = 47,0 T; width = 1,24 mm/s, QS = - 0,24 mm/s, IS = 0,56 mm/sec, 18 at%; B_{hf} = 51,1 T, width = 0,43 mm/s, QS = 0,00 mm/s, IS = 0,45 mm/sec, 31 at%). First distribution – because of its negative value of QS – we ascribe to α-Fe_2O_3 [22] and second to Fe_3O_4 [22]. Carbide component – shown with dashed line – consist of Fe_3C sextet (B_{hf}: 25,3 T, QS = 0,00 mm/s, IS = 0,30 mm/sec, 12 at%) [4] and four other sextets (B_{hf}: 14,6 T, 20,5 T, 23,6 T, 30,3 T; 2, 2, 6 and 1 at%). These four sextets we assign to Fe_7C_3 [4] and Fe_5C_2 [23]. The last component belongs to α-Fe (B_{hf} = 34,1 T; IS = 0,10 mm/sec, 20 at%) and is drawn with full line. Fe_2C is hard to see here but it may modify the value of hyperfine field of α-Fe.

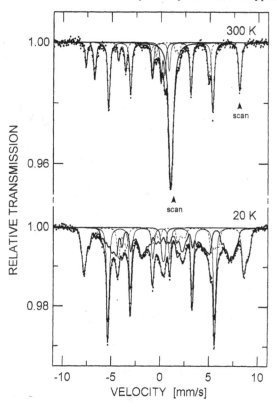

Figure 6. Mössbauer spectra of annealed IC nanopowder.

Mössbauer spectra of **annealed nanopowder** (annealed = as-prepared sample after the measurement of TM curve) are shown in Figure 6. Spectrum at 300 K can be easily fitted with 3 sextets, 1 dublet and 1 singlet. Fe_3O_4 has 2 sextets (B_{hf} = 49,1 T, IS = 0,26 mm/s, 5 at%; B_{hf} = 46,0 T, IS = 0,66 mm/s, 9 at%). They match their tabulated values [24] very well. Sum of them is shown with dashed line. α-Fe has its characteristic sextet (B_{hf} = 33,1 T; IS = 0,01 mm/sec, 20 at%) drawn with ful line. There is also one dublet

(IS = 0,19 mm/s, QS = 0,38 mm/sec, 4 at%) needed to fit successfully the spectrum (full line). Singlet (IS = 1,04 mm/s, 61 at%) corresponds to $Fe_{(1-x)}O$ (bold line). To prove that this singlet really comes from $Fe_{(1-x)}O$, we have measured MS spectra at low temperature: $Fe_{(1-x)}O$ is antiferromagnetic under $T_N \approx 200$ K (depends on X) [25]. In the spectrum measured at 20 K again occours dublet (IS = 0,17 mm/s, QS = 0,30 mm/sec, 4 at%). $Fe_{(1-x)}O$ component shown with bold line is a sum of 10 sextets (B_{hf}: 15,3 T, 16,6 T, 34,0 T, 35,4 T, 35,7 T, 35,7 T, 38,9 T 39,2 T, 39,7 T 43,4 T; altogether 41 at%). These sextets have isomer shift in the range (0,8÷1,6) mm/s. Fe_3O_4 component shown with dashed line includes 5 sextets (B_{hf}: 47,4 T, 50,7 T, 50,8 T, 51,0 T, 53,0 T; altogehter 28 at%) with isomer shift in the range (0,4÷1,0) mm/s. But it is rather hard to fit low temperature spectra of $Fe_{(1-x)}O$ [26] and Fe_3O_4 [22]. The last sextet (B_{hf} = 33,9 T; IS = 0,10 mm/sec, 27 at%) corresponds to α-Fe. Mössbauer temperature scan for increasing temperature from 20 K to 300 K and velocities indicated in Figure 6 is shown in Figure 7. It can be seen there that Neel temperature of magnetic phase transition of $Fe_{(1-x)}O$ from paramagnetic into antiferromagnetic state is close to 200 K.

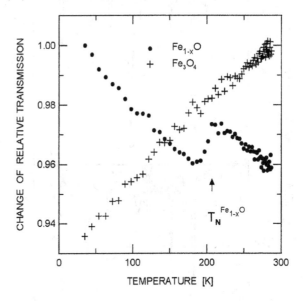

Figure 7. MS temperature scan 20 K → 300 K of annealed IC nanopowder.

To support this Mössbauer analysis, we have also performed XRD analysis of annealed IC nanaopowder. X-ray powder diffraction was performed with Siemens D5005 (Bruker AXS, Germany) using CuKα radiation (40 kV, 45 mA) and diffracted beam monochromator. Qualitative analysis was performed with Diffrac-Plus software package (Bruker AXS, Germany, version 6.0) and JCPDS PDF-2 database [27]. For quantitative analysis of XRD patterns we used PowderCell for Windows, version 2.3 with structural models based on ICSD database [28]. Computer fitted diffractogram – measeured at room temperature – is in Figure 8. There are the same main three phases – α-Fe, Fe_3O_4, $Fe_{(1-x)}O$ – as found with the help of Mössbauer analysis above.

Nevertheless it is correct to state here that paramagnetic dublet (IS = 0,19 mm/s, QS = 0,38 mm/sec, 4 at%) occuring in MS spectra of annealed IC nanopowder remains unassigned. And also the dublet (IS = 0,17 mm/s, QS = 0,42 mm/sec, 2 at%) in the low temperature spectra of as-prepared IC nanopowder.

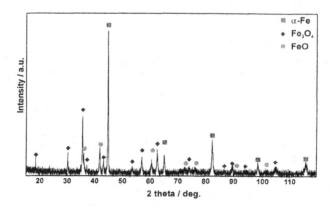

Figure 8. X-ray diffractogram of annealed IC nanopowder.

4. Conclusions

In the as-prepared IC nanopowder was besides Fe_3C also α-Fe identified. Each of carbides Fe_7C_3, Fe_5C_2 or Fe_2C are present at only a few percent concentration. There were also iron oxide phases (α-Fe_2O_3 and Fe_3O_4 nonstochiometric) identified in the nanopowder. Certain amount of oxide particles exhibit superparamagnetic behaviour. This oxidation behaviour may be attributed to the amount of iron nanostrucures obtained during the synthesis process by using large ethylene flows (high temperatures). Performed Mössbauer analysis is in correspondence with XRD, TEM and IR analysis of as-synthesized IC nanopowder and also with our measurement of the thermomagnetic curve.

In the annealed material (annealing took place during measurement of thermomagnetic curve in VSM cell) α-Fe, Fe_3O_4, $Fe_{(1-x)}O$ were clearly detected with the help of Mössbauer analysis and X-ray diffraction.

Acknowledgements

This work was supported by the Czech Ministry of Education, Youth and Sport (COST523-OC523.20) and by the Academy of Sciences of the Czech Republic (K1010104).

References
1. Tajima, S. and Hirano, S.-I. (1990) Synthesis and magnetic-properties of Fe7C3 particles with high saturation magnetization, *Japanese Journal of Applied Physics* **29**, p. 662-668.

40

2. Fiato, R.A., Rice, G.W., Misco, S. and Soled, S.I. (1987) United States Patent, 4637753.
3. Rice, G.W., Fiato, R.A. and Soled, S.I. (1987) United States Patent, 4659681.
4. Bi, X.-X., Granguly, B., Huffman, G.P., Huggins, F.E., Endo, M. and Ecklund, P.C. (1993) Nanocrystalline alpha-Fe, Fe3C, and Fe7C3 produced by CO2-laser pyrolysis, *Journal of Materials Research* **8**, p. 1666-1674.
5. Haggerty, J.S. and Steinfeld, J.I. (ed.) (1981) *Laser-Induced Chemical Processes*, Plenum Press, New York.
6. Borsella, E, Botti, S., Fantoni, R., Alexandrescu, R., Morjan, I., Popescu, I.C., Makris, T.D., Giorgio, R. and Enzo, S. (1992) Composite Si/C/N powder production by laser-induced gas-phase reactions, *Journal of Materials Research* **7**, p. 2257-2268.
7. Alexandrescu, R., Borsella, E., Botti, S., Cesile, M.C., Martelli, S., Giorgi, R., Turtu, S. and Zappa, G. (1997) Synthesis of TiC and SiC/TiC nanocrystalline powders by gas-phase laser-induced reaction, *Journal of Materials Science* **32**, p. 5629-5635.
8. Zhao, X.Q., Zheng, F., Liang, Y., Hu, Z.Q. and Xu, Y.B. (1994) Preparation and characterization of single-phase beta-Fe-nanopowder from cw CO2-laser induced pyrolysis of iron pentacarbonyl, *Materials Letters* **21**, p. 285-288.
9. Alexandrescu, R., Huisken, F., Morjan, I., Ehbrecht, M., Kohn, B., Petcu, S. and Crunteanu, A. (1998) Iron-Oxide-Based Nanoparticles Produced by Pulsed Laser Pyrolysis of Fe(CO)5, *Materials Chemistry and Physics* **55/2**, p. 115-121.
10. Alexandrescu, R., Cojocaru, S., Crunteanu, A., Morjan, I., Voicu, I., Diamandescu, L., Vasiliu, F., Huisken, F., Kohn, B. (1999) Preparation of iron carbide and iron nanoparticles by laser-induced gas phase pyrolysis, *J. Phys. IV France* **9**, Pr8-537-544.
11. Hochepied, J.F. (2001) *Online Nanotechnologies Journal*, vol.2 (1), art 1.
12. Lewis, K.E., Golden, D. M. and Smith, G.P. (1984) Organometallic bond-dissociation energies - laser pyrolysis of Fe(CO)5, Cr(CO)6, Mo(CO)6, AND W(CO)6, *Journal of American Chemical Society* **106**, p. 3905-3912.
13. Alexandrescu, R., Cojocaru, C.S., Crunteanu, A., Morjan, I., Voicu, I., Huisken, F., Kohn, B., Vasiliu, F. and Fatu, D. (1998) Synthesis of iron-carbide-based nanopowders by laser photoinduced reactions from gaseous precursors in T.S. Sudarshan, K.A. Khor and M. Jeandin (eds.), *Surface Modification Technologies XII*, ASM International, Materials Park, Ohio, p. 463-470
14. ASTM 6-0502.
15. ASTM 7-0322.
16. Haneda, K. and Morrish, A.H. (1978) *Surface Science* **7**, p. 584.
17. Fender, B.E.F. and Riley, F.D. (1969) *J. Phys. Chem. Solids* **30**, p. 793.
18. Akamatu, A. and Sato, K. (1949) *Bull. Chem. Soc. Jpn.* **22**, p. 127.
19. Kehrer, V.J. and Leidheiser, H. (1954) *J. Phys. Chem.* **58**, p. 550.
20. Žák, T. (1999) Confit for Windows 95 in M. Miglierini and D. Petridis (eds.), *Mössbauer Spectroscopy in Materials Science*, Kluwer Academic Publishers, Dordrecht, p. 385-390.
21. Cohen, R.L. (1980) *Applications of Mössbauer Spectroscopy, volume II*, Academic Press, New York
22. Greenwood, N.N. and Gibb, T.C. (1971) *Mössbauer Spectroscopy*, Chapmann and Hall Ltd., London
23. Le Caër, G., Dubois, J.M., Pijolat, M., Perrichon, V. and Bussière, P. (1982) *J. Phys. Chem.* **86**, p. 4799.
24. Cohen, R.L. (1976) *Applications of Mössbauer Spectroscopy, volume I*, Academic Press, New York
25. Maddock, A.G. (1997) *Mössbauer Spectroscopy, Principles and Application of the Techniques*, Harwood Publishing, Chichester
26. Greenwood, N.N. and Howe, A.T. (1972) *J. Chem. Soc. Dalton Trans.*, p. 110.
27. JCPDS PDF-2 database, ICDD Newtown Square, PA, U.S.A. release 51, 2001.
28. ICSD database FIZ Karlsruhe, Germany, release 2002/1, 2002.

ON THE APPLICATIONS OF MÖSSBAUER SPECTROSCOPY TO THE STUDY OF SURFACES

J.R. GANCEDO, M. GRACIA AND J. F. MARCO
Instituto de Química Física "Rocasolano", CSIC, Serrano 119, 28006-Madrid, Spain

Abstract

Mössbauer Spectroscopy is a very useful and well-known technique for the study of mainly Fe-containing materials. Most of the measurements with this spectroscopy are performed in the transmission mode, while for the study of surfaces the back-scattering mode is the preferred option. Backscattered radiation contains "resonant" and "non-resonant" radiations, the first originated as a consequence of the relaxation process, which follows resonant absorption (Mössbauer Effect), being the signal, and the second being due to the photoelectric and Compton interactions, the noise. The relaxation process is rather complex and involves both nuclear and atomic relaxations, producing conversion and Auger electrons, and gamma and X-ray photons, all of them of well defined energies, whose probability to leave the surface is related to their nature, energy and track within the solid. The detection of any of these emitted radiations provides a valuable tool to obtain spectra with varying depth resolution.

1. Introduction

Mössbauer Spectroscopy (MBS), a nuclear technique based on the Mössbauer Effect [1], is so named after Rudolf Mössbauer, who discovered this effect during the late 50's. It consists of resonant emission and the absorption of γ-rays by atomic nuclei without energy loss due to nuclear recoil in neither in the absorbing nor in the emitting nuclei. Recoil of either the emitting or absorbing nuclei would lead to an energy loss by excitation of phonon levels of the solid matrix and, consequently, to loss of the resonance.

Minute changes in the nucleus energy levels, due to the so-called hyperfine interactions between the nucleus and the electric and magnetic fields surrounding it, are usually sufficient to prevent the resonant absorption when, as is usual, the natural line width of nucleus levels is smaller than the energy shift caused by the hyperfine interactions. Fortunately the condition of resonance can be re-established by shifting the energy of the γ-quanta by the Doppler Effect, that is, by a relative controlled motion between emitter and absorber. This procedure of increasing and decreasing the energy of either the emitted or absorbed quanta is the basis of Mössbauer Spectroscopy,

M. Mashlan et al. (eds.), Material Research in Atomic Scale by Mössbauer Spectroscopy, 41–52.
© 2003 *Kluwer Academic Publishers. Printed in the Netherlands.*

Mössbauer spectra are plots of absorption of a given γ-ray as a function of the relative velocity between emitter and absorber.

According to theoretical predictions, the Mössbauer Effect is restricted to nuclear transitions of low energies: above 150 keV, even working at a few degrees Kelvin and with very rigid solids the chances of recoilless events are negligible.

Although the Mössbauer Effect has been observed in a total of 109 nuclear transitions of 89 isotopes corresponding to 45 different elements [2], in practical terms, mainly due to the source half-lives, transition energies and isotopic abundances, MBS remains easily feasible in only a few of them: ^{57}Fe, ^{119}Sn, ^{151}Eu, ^{121}Sb, and ^{125}Te. The 14.4 keV transition of ^{57}Fe is the best suited among all the possible ones. Furthermore, since iron is an abundant transition metal with complicated chemistry and great technological interest, it is not surprising that the vast majority of the MBS work has been carried out with ^{57}Fe [3].

A Mössbauer spectrum contains both chemical and structural information that is derived from three parameters, called hyperfine parameters:

- The *isomer shift* IS, due to the monopole electric interaction, reflecting the total charge at the nucleus, provides information on the oxidation state and coordination of the Mössbauer atom. It appears as a shift of the "centre of gravity" of the spectra.

- The *quadrupole splitting*, QS, due to the quadrupole interaction, is a consequence of the electrical field gradient originated by any spatial charge distribution with a symmetry lower than cubic, which could be generated by the last shell electrons and the crystal lattice. It provides information on the coordination and lattice distortion at the Mössbauer atom site.

- The *hyperfine magnetic field*, ΔH, due to the dipole interaction, provides information on the magnetic interactions between nuclei and magnetic fields inherent to and/or imposed to the solid under study.

Typical patterns exhibiting the separate and cooperative joint effects of the three different hyperfine parameters on the ^{57}Fe spectra are depicted in Fig 1.

The variation of the hyperfine parameters with external variables such as temperature, pressure or applied magnetic field can give relevant further information on the chemical and structural properties of the solid containing the Mössbauer absorber. Since each compound has a characteristic set of Mössbauer parameters at a given temperature, this technique can be used as "fingerprint" method for compound and phase identification. This method has proved very useful for iron oxides and hydroxides.

MBS has advantages, which make it a powerful tool in many fields of research. Some of these advantages can be summarized as follows:

- It is a non-destructive technique.
- It is specific for the isotope under study without interference with any other element.
- It is suitable for the study of both crystalline and amorphous materials.
- It allows carrying out "in situ" studies.
- It is possible to use the Mössbauer isotope as a probe by substituting a fraction of another similar element in the material under study.
- It can be used as a "fingerprint" technique.

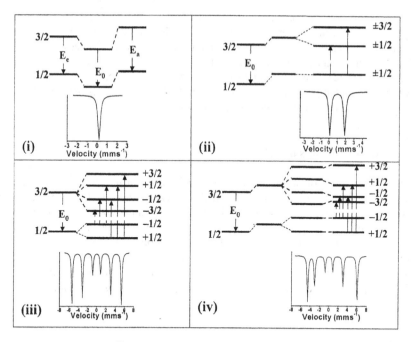

Figure 1. Energy levels of ^{57}Fe atomic nucleus experiencing hyperfine interactions and the corresponding resulting spectra (i) monopole electric interaction, (ii) monopole+quadrupole electric interaction, (iii) dipolar magnetic interaction, (iv) monopole + quadrupole electric + dipolar magnetic interaction with quadrupole electric<<dipolar magnetic.

MBS has, however, severe limitations:
- Feasibility for very few elements.
- Low sensitivity (lower limit: *ca.* 1% by weight).
- Long acquisition times (hours-days).
- Too few parameters for chemical identification purposes.
- Poor accuracy in complex mixtures.
- The use of ionizing radiation sources.

2. Methodology

MBS measurements can be performed in transmission (absorption) and/or backscattering (reemission) modes. For bulk studies the transmission mode represents as the most appropriate choice.

The backscattered radiation contains both resonant and non-resonant components. The term "resonant" is applied to that fraction of outcoming radiation originating from recoilless absorption of radiation impinging on the absorber, and is composed mostly of fluorescent γ-rays, conversion and Auger electrons, and X-rays. Photoelectric and

Compton absorption are the "usual" processes by which any low and medium energy γ-quanta interact with matter, and are the origin of the abundant non-resonant backscattered radiation forming a background that masks resonant signals.

Both resonant and non-resonant radiations are attenuated and/or degraded in their path through matter. Since this attenuation is a function of depth, energy and nature of the re-emitted radiation, the detection of a particular type of resonant backscattered radiation can be used to obtain surface-related information.

3. The Study of Surfaces by MBS

Any technique used for surface studies has to be used in manner such that the bulk contribution is largely minimized. With MBS this can be achieved in two ways, either by using a negligible amount of bulk, which implies a rather complicated experimental set up, or by selecting the most surface sensitive components of the resonant backscattered radiation, the electrons. The first option, transmission, has been very seldom taken into consideration for surface studies for obvious reasons. Electron spectroscopies, such as AES, XPS, etc, are very commonly used for surface science studies mainly because of the large interacting ability of electrons with matter.

4. CEMS

CEMS is the acronym of **C**onversion **E**lectron **M**össbauer **S**pectroscopy. An excited level, E_γ of a nucleus can release its energy either by the emission of a γ-ray or by a non-radiative process called *internal conversion* that consists in the ejection of one electron, called *conversion electron*, of a particular atomic shell. The energy balance between E_γ and the binding energies of the K, L, etc, shells results in the given kinetic energies for the corresponding, K-, L-conversion electrons. The probability of internal conversion is expressed as the *total conversion coefficient* α, $\alpha = \dfrac{N_e}{N_\gamma}$, and decreases with increasing atomic number Z ($\alpha_K \propto Z^3$) and the shell number. For K-shell electrons, $\alpha_K = \dfrac{N_K}{N_\gamma}$, this *partial coefficient* being the highest, i.e. $\cong 8.4$ for the K-conversion electrons for the 14.4 keV ^{57}Fe excited level. In other words, K-conversion electrons are nearly nine times more abundant than the fluorescent 14.4 keV γ-rays when the excited ^{57}Fe relaxes.

The de-excitation scheme of a ^{57}Co source and the corresponding ^{57}Fe absorber is depicted in Fig 2, revealing that the emission of conversion electrons is followed by a cascade of "resonant" events, due to the filling of the shell holes left by the ejection of the conversion electrons. Other photons and electrons are present in the outcoming radiation as a consequence of different γ quanta (136.3, 122 and 14.4 keV) delivered by the source. The probability of this unwanted, "non resonant" radiation is a function of the atomic number Z ($\propto Z^n$, where n=4-5, and $\propto Z$ for photoelectric and Compton effects, respectively) and the concentration of nuclei constituting the absorber, thus light elements are less "noisy" than the heavier ones. In any case, photons, mainly X-rays and electrons, either conversion or Auger, are abundant in the backscattered "resonant" radiation. As stated above, the distance that all these radiations can travel within a solid medium depends on both their nature (photons or electrons) and energy. Resonant photons are very seldom used because of their too long range. The much shorter range electrons are preferred for obtaining surface information.

Since CEMS is an electron spectroscopy, it benefits from both the advantages and disadvantages of electron techniques. Remarkable advantages are:

- Electrons rapidly lose their energy traveling through matter; therefore those which escape from solids come from the near surface region.
- Electrons are easily detectable and their energy is easy to analyze.
- Electrons are easily distinguishable from photons for detection purposes.
- It is possible to evaluate the electron escape depth as a function of their escape angle and energy.

  The main disadvantages are:
- The electron paths through matter are not straight.
- Windowless detectors have to be used.
- Background electrons of different origins can mask the signal.

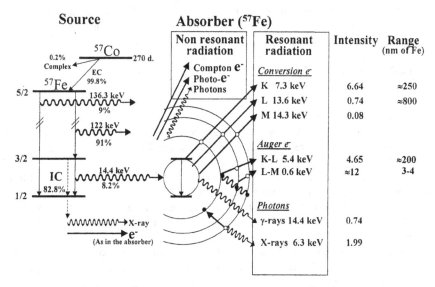

Figure 2. Decay scheme of a ^{57}Co source and de-excitation scheme of the ^{57}Fe absorber. Intensities are based on 100 initial decaying nuclei of ^{57}Co and a complete resonant absorption of the 8.2 γ-ray photons of 14.4 keV.

CEMS methodology has been reviewed by several authors [4-11] emphasizing different aspects. Nomura *et al.* [10] have published a very comprehensive study. Tatarchuck and Dumesic's [7] review is very illustrative on the basics and feasibility of the technique, while Gancedo *et al.*[9,11] have been more concerned with the methodological aspects of the technique.

CEMS can be performed in two different modes: Integral (ICEMS) or Depth Selective (DCEMS).

4.1. ICEMS

ICEMS stands for Integral Conversion Electron Mössbauer Spectroscopy. The basis of this technique is to register all electrons backscattered by the surface whatever the kinetic energy they carry. This approach has important implications:

1. A wide range of electron energies is expected. The low limit, virtually zero, is mainly due to those electrons generated far down the surface plus the contribution of Auger electrons involving the upper shell levels (LLM and MVV). The upper limit is not well defined, since the more energetic electrons have two sources: on the one hand there are Auger electrons of various energies (KLL, KLM) and K-, L-, and M-conversion electrons, all of them resonant, the 7.3 keV K-conversion electrons, being the most abundant by far. On the other hand there are non resonant electrons, generated by photoelectric and Compton interactions of the already mentioned incoming γ-radiation (14.4, 122 and 136.3 keV), which also undergo attenuation and contribute towards masking of the resonant signal.
2. The maximum practical depth analyzed is mainly determined by the range of 7.3 keV conversion electrons, estimated to be around 300 nm.
3. Under favorable conditions (low Z surface constituents), the technique allows to obtain spectra of samples with surface concentrations down to 10^{14} atoms of $^{57}Fe/cm^2$.

4.1.1. *Details of experiment*
Due to low energy of the electrons to be registered, windowless counting is mandatory. Three types of detectors are currently used: proportional counters [4,12], parallel plate avalanche counters (PPAC) [13,14], and electron multiplier devices [16,17].

Gas-flow proportional counters, with two or three wire anodes and using mixtures of He-methane or He-CO, are the most widely used. PPAC's filled with acetone vapor are employed when the gas mixtures are to be avoided. Both types of counters are very easy to build in any average mechanical workshop. Construction details are given under references [9,14,15].

Proportional counters can yield some energy resolution and, consequently, a rough depth analysis [18,19]. Under very careful setting of their high voltage, proportional counters can be operated at very low temperatures (10 K) in the absence of quencher gas [20,22], since it would freeze. The absence of quenching gas necessitates a critically low HV in order to prevent electric arcing, resulting in a low multiplication factor, and requiring a high-gain low-noise preamplifier. Thus, the simplest option at low temperatures is to use electron multipliers [23] (channeltrons and channelplates), properly shielded from the source radiation, a requirement which implies a poor geometrical arrangement, leading to low counting rates even with rather active Mössbauer sources. Maintaining the electron multiplier at a given negative floating potential versus the absorber can yield a certain degree of energy discrimination capable of yielding a modest depth resolution [24].

A problem, of which spectroscopists should be aware of, is the simple fact that a surface bombarded by photons ejects electrons, and therefore could become increasingly positively charged. Unless an adequate flow of electrons from the sample holder to the absorber surface is achieved, a difficult task when working with highly insulating samples, the initial operating conditions of the counter may become altered and the counting severely distorted.

ICEMS has been successfully applied in many fields of research including thin films [25], Langmuir-Blodgett films [28], electrochemistry [29,30], corrosion layers [31,32], adhesion [33], ion implantation [34], surface modification [35], etc.

The "deep penetration" of ICEMS can be an asset of the technique for the study of certain surface processes. Thus, the rate of a corrosion reaction can be monitored by the evolution of the iron phases appearing in the spectra as the reaction is in progress, since the α-iron contribution decreases with corrosion time, as an increase in the contribution of the oxidized phases is observed. Fig. 3 depicts the spectra of a film of α-^{57}Fe at increasing periods of exposure to a corrosive environment [36]. Fig. 4 depicts the spectra of a thin film of α-^{57}Fe, deposited on a silicon wafer and covered by a thin layer of TiN after exposure to humid and polluted SO_2 atmospheres. Parallel XPS measurements, shown in the left part of the figure, indicate that the protecting layer of TiN remains unaltered for a longer period of time following the complete oxidation of the underlying iron film. These experiments permit a good insight into the mechanism of protection and degradation of TiN thin films [37].

4.2. DCEMS

DCEMS stands for **D**epth-selective **C**onversion **E**lectron **M**össbauer **S**pectroscopy. Depth selectivity is achieved by registering the spectrum with electrons coming from solids with a given energy. This technique yields, to a certain extent, depth profiles by recording spectra after very careful tuning of the electron energy.

The depth selectivity can be theoretically expressed by weight functions $T_E\theta_{(x)}$, which give the probability that electrons originated at a depth x leave the surface with energy E and an emission angle θ. Calculation procedures of these functions have been described [38-42]. Unfortunately, irrespective of the quality of the calculations, there are two significant drawbacks inherent to the system. On the one hand, the non-resonant electrons show after attenuation the full range of energies, which ensures their overlapping with any given electron energy of the resonant ones that can be selected for detection, causing a severe masking of the signal. On the other hand, the trajectories of electrons are not straight, leading to a high degree of uncertainty of the depth estimation.

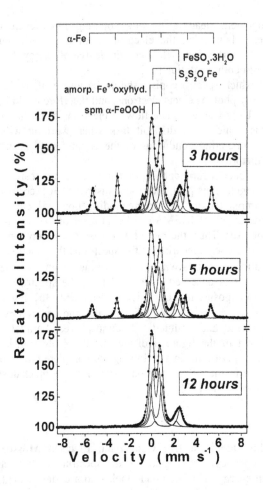

Figure 3. ICEM spectra recorded from an [57]Fe film at different corrosion stages [36].

A simple and a rewarding approach consists of the collection of the free energy-loss 7.3 keV K-conversion electrons. In this way the resulting MBS spectra contain information relative to a layer close to surface, and whose thickness depends on the take-off angle of electrons, thus when collecting those leaving the surface at nearly normal angles, a depth of *ca.* 5-10 nm can be inspected, whereas collecting them around grazing angles the thickness of the inspected layer decreases to about 2.5 nm.

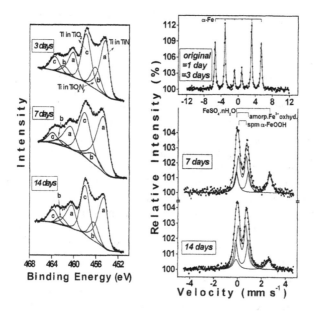

Figure 4. Left: Ti 2p XP spectra recorded from a TiN layer deposited on Fe after different exposure periods to a SO_2-polluted humid atmosphere. Right: Corresponding ICEM spectra recorded from the coated Fe.

Tuning the electron analyzer to energy corresponding to LLM-Auger electrons, $\cong$ 580 eV, whose mean free path is very small, 1-2 nm, a great surface sensitivity can be achieved, this approach being known as AEMS (**Auger Electron Mössbauer Spectroscopy**).

Another approach [43] consists of obtaining the Mössbauer spectrum with the resonant electrons of the lowest energy, below 15 eV, whose mean free path is comparable to that of the LLM Auger electrons. These low energy electrons have to be accelerated in order to be efficiently detected, due to very low counting efficiency of channel plates and channeltrons for electrons of energies below 100 eV. The technique is called LEEMS (**Low Energy Electron Mössbauer Spectroscopy**.

With AEMS or LEEMS, the surface sensitivity is enhanced, but it should be kept in mind that secondary effects and a resonant background from deeper layers are present, making interpretation of the results a delicate task. Thus, the use of any of these two techniques only reports small advantages over the use of the K-conversion electrons that have suffered no energy loss, when the iron-containing layers to be inspected are a few nanometers thick. An excellent example of the application of ICEMS, DCEMS, AEMS and LEEDS, plus a fruitful discussion of the experimental details and results obtained is given in Ref. [44].

4.2.1. Details of experiment

The possibility of performing "semi quantitative" DCEMS, taking advantage of the limited energy resolution of proportional counters, has been explored [19].

Two interrelated parameters are crucial for obtaining good DCEMS measurements: energy resolution and counting rate. The energy resolution, $\Delta E/E$, has to be better than 2-3%, a figure that can be achieved only by magnetic or electrostatic deflection. Deflection of charged particles implies a good definition of the particle trajectories, the origin being the sample itself the end the counting device (channeltron, electron multiplier or channelplate), and the intermediate path being a function of the applied field. Thus, it is easy to understand that in the designing of an analyzer a compromise has to be reached between energy resolution and transmission, this compromise being much more severe than in the case of XPS and β-spectroscopy, where the number of events is not so critically low. This is the reason why there are no commercially available electron analyzers for DCEMS. As for the counting rate, in order to increase the number of events it is mandatory to perform the experiments with ^{57}Fe enriched samples and to use strong ^{57}Co sources ($\geq$100 mCi), both requirements being not only expensive, but troublesome. Successive improvements have been achieved up to the quite impressive performance of the Darmstadt group's *"Orange type magnetic spectrometer"*, with a transmission of $\approx$40% of 2π for an energy resolution of 1%, when collecting the electrons coming from samples 10 mm in diameter, (i.e with a high luminosity), which allows the possibility of obtaining spectra of non-enriched samples in acceptable registering times [45].

It would be evident by now that DCEMS is a complex, laborious technique, requiring not only experimenters well versed in the design and construction of scientific equipment, but a skilled workshop as well. Therefore it is not surprising that the number of ICEMS studies far outweighs that of DCEMS works.

As far as the applications of the technique are concerned, in any problem where the knowledge of the state of iron (oxidation state, coordination, magnetic interactions) in a surface could be relevant, DCEMS is, in principle, the proper choice. The situation is not so simple in the author's experience. Disappointing, and even misleading, results can be obtained after a laborious and lengthy sample preparation and measurement. It is advisable, before embarking on a DCEMS measurement, to characterize the samples to be studied by means of faster surface techniques such as ICEMS, XPS, AES, SEM, AFM, etc, in order to ensure that they deserve a DCEMS study.

References

1. Goldanskii, V.I., and Makarov E.F. (1968) in V.I. Goldanskii and R.H. Herber (eds.), *Chemical Applications of Mossbauer Spectroscopy*, Academic Press, New York-London, p. 1.
2. Stevens, J.G., Bowen, L.H., and Whatley, K.M. (1986) Mossbauer spectroscopy *Anal. Chem.* **58**, 250-264R.
3. Stevens J.G. and Stevens V.E. (1983) *Mossbauer Effect Ref. Data J.* **6**, 51.
4. Tricker M.J., (1981) in *Mossbauer Spectroscopy and its Chemical Applications, Advances in Chemistry*

Series, 194. J.G. Stevens and G.K. Shenoy (eds.). Am. Chem. Soc., Washington, DC, p. 63.

5. Liljequist D., (1983) in *Scanning Electron Microscopy III.* R.M. Albrecht, I.B. Shelburne and J.D. Meakin (eds.), SEM Inc. AMF O'Hare, Chicago, p. 997.

6. Friedt J.M. (1984) Tech. Ing. **P2 608** 1.

7. Tatarchuk B.J. and. Dumesic J.A. (1984) in *Chemistry and Physics of Solid Surfaces,* Vol. 5 (Springer Ser. Chem. Phys., Vol. 35). R. Vanselow and R. Howe (eds.), Springer- Verlag, Berlin-Heidelberg, p. 65

8. Sawicki J.A. (1986) in *Industrial Applications of the Mössbauer Effect.* G.J. Long and J.G. Stevens (eds.), Plenum Press, New York-London, p. 83.

9. Gancedo J.R., Gracia M. and Marco J.F. (1991) CEMS methodology, *Hyperfine Interactions* **66,** 83-93.

10. Nomura K., Ujihira Y. and Vértes A. (1996) Applications of conversion electron Mössbauer spectrometry (CEMS), Journal *of Radioanalytical and Nuclear Chemistry* **202,** 103-199.

11. Gancedo J.R., Gracia M., Marco J.F. and Tabares J.A. (1998) Corrosion studies by GEMS. Facing the experiment, *Hyperfine Interactions* **111,** 83-92.

12. Cook D. C. (1986) A gas-flow proportional counter for the Mössbauer study of surfaces between 100-K and 400-K, *Hyperfine Interactions* **29,** 1463-1466.

13. Weyer, G. (1976) in *Mössbauer Effect Methodology,* Vol 10., I.J. Gruverman and C.W. Seidel. (eds.) Plenum Press, New York, p. 301.

14. Gancedo J.R. and Gracia M. (1986) CEMS in non conducting surfaces with a parallel plate avalanche counter, *Hyperfine Interactions* **29,** 1097-1100.

15. Hanzel D., Griesbach P., Meisel W. and Gütlich P. (1992) Optimization of a conversion electron Mössbauer-spectroscopy gas-flow He/CH4 proportional counter, *Hyperfine Interactions* **71,** 1441-1444.

16. Sawicki J.A. and Sawicka B.D. (1983) Experimental-techniques for conversion electron Mössbauer-spectroscopy, *Hyperfine Interactions* **13,** 199-219.

17. Amulyavichyus A.P. and Yu Davidonis R. (1986) Electron detector for Mössbauer-spectroscopy based on microchannel plates, *Instrum. Exp. Tech. (Engl. transl.)* **29,** 590-593.

18. Nakagawa H., Ujihira Y. and Inaba M. (1982) Possibility of depth selective CEMS with a gas-flow proportional counter, *Nucl. Instr. and Meth.* **196,** 573-574.

19. Kuprin A.P. and Novakova A.A. (1992) Depth-selective Fe-57 CEMS (DCEMS) on samples with rough surfaces using a gas-flow proportional counter, *Nucl. Instrum. Methods Phys, Res., Sect. B* **62,** 493-504.

20. Kishimoto S., Isozumi Y., Katano R. and Takekoshi H. (1987) Operation of helium-filled proportional counter at low-temperatures (4.2-295K), *Nucl. Instr. and Meth. Phys. Res.* **A262,** 413-418.

21. Fukumura F, Katano R, Kobayashi T., Nakanishi A. and Isozumi Y. (1991) Helium-filled proportional counter operated at low-temperatures higher than 13-K, *Nucl. Instr. and Meth. Phys. Res.* **A301,** 482-484.

22. Fukumura K., Nakanishi A. and Kobayashi T. (1994) Hydrogen-filled proportional counter operated at low-temperatures and its application to CEMS, *Nucl. Instr. and Meth. Phys. Res.* **B86,** 387-389.

23. Sawicki J.A., Tyliszczak T. and Gzowski O. (1981) Conversion electron Mössbauer-spectroscopy with the use of channel electron multipliers operating at low-temperatures, *Nucl. Instr. and Meth.* **190,** 433-435.

24. Sato H. and Mitsuhashi M. (1990) A new detection system with a microchannel plate for energy-selected conversion electron Mössbauer-spectroscopy, *Hyperfine Interactions* **58,** 2535-2540.

25. Kajcsos Zs., Meisel W, Kuzmann E., Tosello C., Gratton m.L., Vértes A., P. Gütlich and Nagy D.L. (1990) ICEMS and DCEMS study of Fe layers evaporated onto Al and Si, *Hyperfine Interactions* **57,** 1883-1888.

26. Meisel W. and Gütlich P. (1991) Formation, phase-composition, and transformations of Langmuir-Blodgett multilayers and monolayers containing Fe, *Hyperfine Interactions* **69,** 815-818.

27. Meisel W., Tippman-Krayer P., Möhwald H. and P.Gütlich P. (1991) CEMS XPS study of iron stearate Langmuir-Blodgett layers, *Fresenius J. Anal. Chem.* **341,** 289-291.

28. Meisel W., Faldum T., Sprenger D. and Gütlich P. (1993) Phase-composition of Fe-containing Langmuir-Blodgett layers after thermal-treatment in a reactive atmosphere, *Fresenius J. Anal. Chem.* **346,** 110-113.

29. Kordesch M. E., Eldridge J., Scherson D. and Hoffnann R.W. (1984) A new Mossbauer conversion electron detection technique for insitu studies on iron, *J. Electroanal. Chem.* **164,** 393-597.

30. Meisel W., Stumm U., Thilmann C., Gancedo J.R. and Gütlich P. (1988) A CEMS-study of the passive layer on iron and steel, *Hyperfine Interactons* **41,** 669-672.

31. Kobayashi T., Fukumura K., Isozumi Y.and Katano R. (1990) Magnetic-properties of corrosion products investigated by CEMS at low-temperatures near 4.2K, *Hyperfine Interactions* **57,** 1923-1928.

32. Marco, Dávalos J., Gracia M. and Gancedo J.R. (1994) Corrosion studies of iron and its alloys by means of Fe-57 Mössbauer spectroscopy, *Hyperfine Interactions* **83,** 111-123.

33. Eynatten G. von, Nothhelfer K. and Dransfeld K. (1991) Light-induced chemical-reactions on interfaces between metallic iron and polymers, *Hyperfine Interactions* **69,** 759-762.

34. Sawicki J.A, in *Industrial Applications of the Mössbauer Effect.* G.J. Long and J.G. Stevens (eds.). Plenum

Press, New York-London (1986) p. 83.

35. Carbucicchio M., Casagrande A. and Palombarini G. (1990) Competitive reactivity of iron towards boron and silicon in surface treatments with powder mixtures, *Hyperfine Interactions* **57**, 1769-1774.

36. Gracia M., Marco J.F., Gancedo J.R., Exel W. and Meisel W.(2000) Surface spectroscopic study of the corrosion of ultrathin Fe-57-evaporated and Langmuir-Blodgett films in humid SO2 environments, *Surf. Interface Anal.* **29**, 82-91.

37. Marco J.F., Gracia M, Gancedo J.R, Agudelo A., Exel W., Meisel W. and Hanžel D. (2001) Corrosion of ultrathin iron layers and titanium nitride coated iron, *Hyperfine Interactions* **134**, 37-51.

38. Proykova A. (1979) *Nucl. Inst. and Meth.* **160**, 321.

39. Bonchev T.S.V., Minkova A., Kushev G. and Grozdanov M. (1977) *Nucl. Inst. and Meth.* **147**, 481.

40. Liljequist D. (1978) *Nucl. Inst. and Meth.* **155**, 529.

41. Liljequist D. (1981) The analysis of natural Fe-57 abundance absorbers in conversion electron Mossbauer-spectroscopy, *Nucl. Inst. and Meth.* **179**, 617-620.

42. Salvat F. and Parellada J. (1984) Theory of conversion electron mössbauer-spectroscopy (CEMS), *Nucl. Inst. and Meth. Phys. Res.* **229**, 70-84.

43. Klingelhofer G. and Kankeleit E. (1990) Conversion electron mössbauer-spectroscopy with very low-energy (0 to 15 ev) electrons, *Hypefine Interactions* **57**, 1905-1910.

44. Klingelhofer G. and Meisel W. (1990) Study of very thin oxide layers by conversion and Auger electrons, *Hyperfine Interactions* **57**, 1911-1918.

45. Stahl B. and Kankeleit E. (1997) A high luminosity UHV orange type magnetic spectrometer developed for depth selective Mössbauer spectroscopy, *Nucl. Instrum. Methods Phys. Res., Sect. B* **122**, 149-161.

CONVERSION ELECTRON MÖSSBAUER SPECTROSCOPY STUDY OF LANGMUIR-BLODGETT FILMS

D. HANŽEL[1] AND B. STAHL[2]

[1] *J. Stefan Institute, P.O.B. 3000, SI-1001 Ljubljana, Slovenia*
[2] *Fachbereich Materialwissenschaft, Fachgebiet Dünne Schichten, Petersenstrasse 23, TU Darmstadt, D-64287, Germany*

Abstract

We report on characterization and magnetic properties of very thin films of Fe-oxides formed by thermal desorption of iron stearate Langmuir-Blodgett films using low temperature conversion electron Mössbauer spectroscopy (CEMS). Langmuir-Blodgett films of iron stearate (23 and 27 monolayers) have been formed on silicon wafers and afterwards thermally treated in air at 260°C for 45 minutes. The resulting, very well defined oxide films exhibited very smooth and homogeneous surfaces. These oxide films have been investigated by ultra high vacuum low temperature CEMS. The very thin oxide film, supposedly γ-FeOOH, contains only 6 monolayers of iron corresponding to only three crystallographic unit cells (or 1.5 magnetic unit cells) and thus enables the study of non-equivalent sites in the oxide film due to surface and interface effects. The magnetic ordering temperature of the 27 layers sample corresponds almost to the bulk value, while the 23 layers sample orders at lower temperature.

1. Introduction

A reduction in size of particles may alter the magnetic behavior compared to a bulk material. Intermediate steps in the reduction of the spatial dimensions are studies of thin films, where the film thickness is systematically reduced below the range of the changed near-surface behavior. This may be accomplished by varying the temperature while using a film of constant thickness. Due to the Larmor precession of the nuclear moment, a time scale for observing relaxation phenomena is given in the order of 10^{-8} s.

Very smooth and homogeneous thin metal-oxide films can be fabricated using Langmuir-Blodgett (LB) multilayer films as precursors [1,2,3]. The LB film forming molecules were metal salts of fatty acids. Heating or exposure to UV light can remove the organic part of the film. The resulting continuous films or droplets of metal oxides remain at the substrate surface. We report on characterization of Fe-oxides formed by thermal desorption of iron stearate LB films using low temperature conversion electron Mössbauer spectroscopy (CEMS).

53

M. Mashlan et al. (eds.), Material Research in Atomic Scale by Mössbauer Spectroscopy, 53–58.
© *2003 Kluwer Academic Publishers. Printed in the Netherlands.*

2. Experimental

Langmuir-Blodgett films with 23 (LB23) and 27 (LB27) monolayers of ^{57}Fe stearate have been prepared by use of a commercial film balance. Naturally oxidized silicon wafers served as substrates. A very high purity of substances is required. Further preparation details and characterization by photoelectron spectroscopy (XPS) and atomic absorption spectroscopy (AAS) are given in references [4,5,6,7,8]. Thermal desorption, that removed the organic part of the layers, has been performed at 530K in air for 45 minutes. The resulting films have been investigated by atomic force microscopy (AFM) and showed a very smooth and homogeneous surface [8] with a remarkable resistance against corrosion [9] in aggressive environment.

CEMS spectra have been recorded in UHV in two spectrometers: (i) in an energy integral one and (ii) in an energy dispersive Orange spectrometer [10]. The spectra in the Orange spectrometer have been recorded at the K-edge of the ^{57}Fe transition.

A relatively simple low temperature CEMS apparatus has been constructed to enable measurements down to 13K. It consists of a He-flow Cryostat (Cryovac) mounted inside the ultra high vacuum chamber. A special heat shield with an opening for the emerging conversion electrons has been developed. UHV is necessary to avoid condensation of water and other residual vapors which otherwise disturb the measurement drastically already after 1-2 hours. The detection of electrons is by means of a low noise single channel electron multiplier (Philips X959) biased at 10-100V to increase the collection of conversion electrons. A lead collimator perpendicular to the sample has collimated the incident gamma rays from a 25 mCi $^{57}Co/Rh$ source, while the electron multiplier is oriented at $60°$ to the sample surface. The chamber is pumped by a magnetic bearing turbo pump connected via bellows. An additional weight of 50 kg is attached to the pump in order to minimize the transmitted vibrations. The base pressure in the chamber is below 10^{-8} mbar, which allowed measuring times of one to two days before the condensation of residual gases prevented the resonant electrons from escaping from the sample. The spectrum recording time has been three days or longer due to small spatial angle of detection covered by the single channel electron multiplier.

3. Results and Discussion

Spectra recorded in the channeltron apparatus between room temperature and 13 K are shown in Fig. 1. The room temperature CEM spectrum shows mainly Fe^{3+} with a quadrupole splitting distribution. On cooling down, the sample orders magnetically at 60 K. The ordering temperature T_N of well crystallized γ-FeOOH is about 73 K with a magnetic hyperfine field of about 44 to 46 T at 4 K, and the transition range is often larger than 10 K [12,13,14]. The difference in the transition temperature compared to the bulk sample can be attributed to a difference between the sample surface temperature and the cold finger temperature due to radiative heat transfer and the thermal contact. On further cooling down, a paramagnetic contribution within a temperature range of 15 K can still be observed. There seems to be no evidence of superparamagnetic effects, which might be suppressed due to the very homogeneous oxide surface as confirmed by AFM. Mössbauer spectrum recorded at 13 K shows a hyperfine field of 46 T confirming that the oxide formed by thermal desorption is similar to γ-FeOOH [12,13,14]. The spectrum at 13 K shows a small quadrupole shift of -0.24 mm/s, which is slightly larger than the value reported for synthetic lepidocrocite. The oxide film contains only 6 layers of iron [5,7]corresponding to only three crystallographic unit cells (or 1.5 magnetic unit cells), thus the line broadening can be explained by non-equivalent sites in the oxide film due to surface and interface effects. Unfortunately, the signal to noise ratio in this experiment was not good enough to allow for a more detailed evaluation of Mössbauer spectra. Therefore further experiments were performed in the energy dispersive Orange spectrometer.

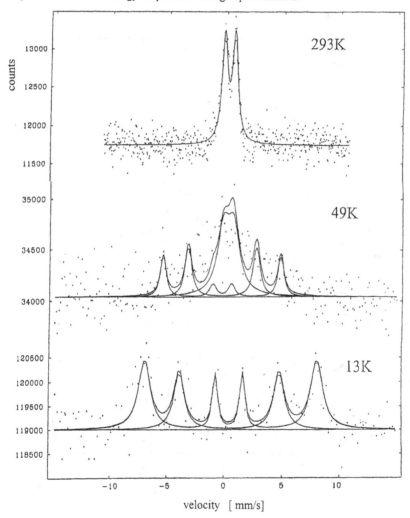

Figure 1. Mössbauer spectra of thermally desorbed ^{57}Fe stearate Langmuir-Blodgett films (27 monolayers), recorded in energy integral spectrometer.

CEM spectra of sample LB27 recorded between 10 K and room temperature in Orange spectrometer are shown in Fig.2. On cooling down the sample LB27 orders magnetically at 65 K, while the sample LB23 orders at 40 K. On further cooling down a paramagnetic contribution within a temperature range of 25K can still be observed in LB27, while in LB 23 it can be observed down to 3K (Fig.3). The outermost sextet fits well into the behavior of bulk reference samples of γ-FeOOH. They show a decomposition into two fractions: the bulk component and the near surface phase. Figure 4 shows the dependence of the magnetic hyperfine field on temperature. In sample LB27, the same saturation hyperfine field at low temperatures and roughly the same Néel temperature is observed as for the bulk γ-FeOOH.

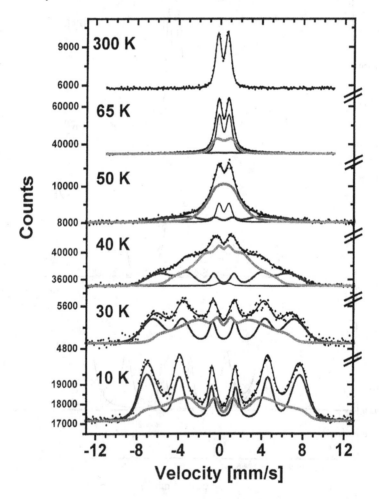

Figure 2. Mössbauer spectra of of thermally desorbed ^{57}Fe stearate Langmuir-Blodgett films (27 monolayers), recorded in the Orange spectrometer.

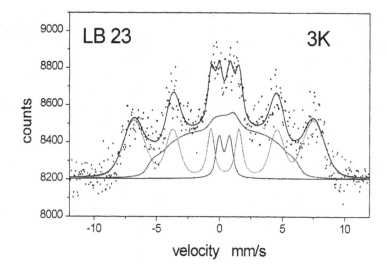

Figure 3. Mössbauer spectrum of of thermally desorbed ^{57}Fe stearate Langmuir-Blodgett films (23 monolayers) at 3 K recorded in the Orange spectrometer.

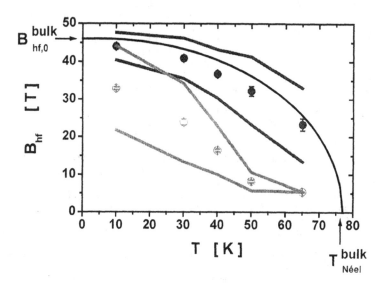

Figure 4. Dependence of the hyperfine field in the LB27 sample, solid curves indicate the distribution width. The magnetization curve of the bulk reference sample.

58

The second fraction in the Mössbauer spectra has a significantly different dependence of the magnetic hyperfine field as a function of temperature. It decreases steeper than the bulk component. The steeper decrease is attributed to the surface component [11] as in the work on FeBO$_3$ crystal. The width of the distribution has the same signature of collective relaxation. By raising the temperature, the bulk fraction diminishes gradually to zero. Within the present state of data collection it cannot be decided, whether this fraction vanishes at or below the Néel temperature. The attempts to fit the spectra to the superparamgnetic relaxation model have not been successful.

Acknowledgements

We are grateful to W. Exel and W. Meisel for providing the samples. We acknowledge the financial support of Ministry of Science and Technology of Slovenia, BMBF, DAAD cooperation programme.

References:

1. P. Tippmann-Krayer, W. Meisel, U. Höhne, H. Möhwald (1991) Ultrathin Metal-Films and Inorganic Clusters via Thermodesorption of LB Films, Makromol. Chem. Macromol. Symp. **46**, 241-246.
2. W. Meisel, P. Tippmann-Krayer, H. Möhwald, P. Gütlich (1991) CEMS XPS Study of Iron Stearate Langmuir-Blodgett Layers, Fresenius J. Anal. Chem. **341**, 289-291.
3. D.T. Amm, D.J. Johnson, T. Laursen, S.K.Gupta (1992) Fabrication of Ultrathin Metal-Oxide Films Using Langmuir-Blodgett Deposition, Appl. Phys. Lett **61**, 522-524.
4. T. Faldum, W. Meisel, P. Gütlich (1994) Formation and Characterization of Oxidic and Metallic Fe/Ni Multilayers Prepared from Langmuir-Blodgett-Films, Hyperfine Interact. **92**, 1263-1269.
5. T. Faldum (1995) Doctoral thesis, Fachbereich Physik der Universität Mainz .
6. W. Meisel, T. Faldum, D. Sprenger, P. Gütlich (1993) Phase-Composition of Fe-Containing Langmuir-Blodgett Layers After Thermal Treatment in a Reactive atmosphere, Fresenius J. Anal. Chem. **346** 110-113.
7. T. Faldum, W. Meisel, P. Gütlich (1996) Determination of the absolute density of Fe^{3+} and Ni^{2+} ions in Langmuir-Blodgett films, Surf. Interface Anal. **24**, 68-73.
8. W. Exel (1998) Doctoral thesis, Fachbereich Chemie/Pharmazie der Universität Mainz .
9. J.F. Marco, M.Gracia, J.R. Gancedo, A. Agudelo, W. Exel, W. Meisel, D. Hanzel (2001) Corrosion of Ultrathin Iron Layers and Titanium Nitride Coated Iron, Hyperfine Interact. **134** 37-51.
10. B. Stahl, E. Kankeleit (1997) A high luminosity UHV orange type magnetic spectrometer developed for depth selective Mössbauer spectroscopy, Nucl. Inst. Meth. **B 122**, 149-161.
11. B. Stahl, M. Ghafari, H. Hahn, A. Kamzin, D. Hanzel (2001) Depth Selective Mössbauer Spectroscopy as a Bridge between Surface and Small Particle Phenomena in Magnetic Materials, Mat. Res. Soc. Symp. Proc. Vol. **676**, Y7.6.1-Y7.6.10.
12. C.E. Johnson (1969) Antiferromagnetism of γ FeOOH: a Mossbauer effect study, J. Phys. C: Solid State Phys. **2** 1996.
13. E. Murad and U. Schwertmann (1984), The Influence of Cristallinity on the Mössbauer Spectrum of Lepidocrocite,Min. Mag. **48**, 507-511.
14. E. De Grave, R.M. Persoons, D.G. Chambaere, R.E. Vandenberghe, L.H. Bowen (1986), An Fe-57 Mössbauer-Effect Study of Poorly Crystalline γ-FeOOH, Phys. Chem. Minerals **13**, 61-67.

MODIFICATION OF STEEL SURFACES FOLLOWING PLASMA AND ION BEAM IMPLANTATION INVESTIGATED BY MEANS OF CEMS

A.L. KHOLMETSKII, V.V. UGLOV, V.V. KHODASEVICH,
J.A. FEDOTOVA, V.M. ANISCHIK, V.V. PONARYADOV,
D.P. RUSALSKII AND A.K. KULESHOV
Department of Physics, Belarus State University, Skorina Ave 8, Minsk, Belarus

1. Introduction

The method of ion treatment of steel surfaces is well documented [1-5]. It has been ascertained that the practical significance of this topic could pave the way for finding methods permitting the creation of modified surface layers with comparably increased thickness. To this end, we focused our research on methods that make it possible to create modified layers ranging in thickness from one to ten micrometers: high current implantation, plasma immersion implantation and compression plasma flow (CPF) method.

The present work is aimed at investigating the phase and structural surface changes in high-speed M2 steel and Y8A carbon steel after the above type of implantation and, to reveal possible correlation of such changes with the mechanical properties of the modified layers. The main research methods were conversion electron Mössbauer spectroscopy (CEMS), conversion X-rays Mössbauer spectroscopy (CXMS) combined with tri-biological investigations and micro-hardness measurements.

2. Materials

We used AISI M2 steel samples comprising martensite Fe(C, M) and carbides M_6C. The samples were subjected to thermal processing typical for this class of steels. Further, they were mechanically polished to a mirror finish on one face by means of diamond slurry. The roughness of the samples was $R_a=0.01$ μm. The hardness of the steel after thermal treatment was 64 HRC. The regimes for high current are presented in the table 1.

Plasma-immersion implantation was carried out by nitrogen ions under the following conditions: ion energy 40 keV; current density 3.5 мА/см^2; dose $8 \cdot 10^{18}$ ions/cm^2; temperature 380°C.

CPF treatment of Y8A steel was carried out for the accelerators energy accumulator ($W_0=10$ kJ, $U_0=4$ kV) enabling maximum value of a discharge current (80 kA); discharge duration was 140 μs. CPF diameter was 1 cm, length $d=12$ cm, 6 cm. Charge

M. Mashlan et al. (eds.), Material Research in Atomic Scale by Mössbauer Spectroscopy, 59–68.

particle concentration in CPF was $7 \cdot 10^{17}$ cm^{-3}; characteristic velocity of plasma formations was $4 \cdot 10^6$ cm/s. Nitrogen pressure in working volume was 400 Pa.

TABLE 1. High current implantation (HCII).

Ion	Energy, keV	Current density, mA/cm^2	Irradiation time, min	Dose, ion/cm^2	Temperature, °C
N	1	3.5	30	$7 \cdot 10^{19}$	500
B	20	0.53	10	$2 \cdot 10^{18}$	500
B+N	20+1	0.53+3.5	10+30	$2 \cdot 10^{18} + 7 \cdot 10^{19}$	500
N+B	1+20	3.5+0.53	30+10	$7 \cdot 10^{19} + 2 \cdot 10^{18}$	500

3. Results and Discussion

Analysis of M2 steel surface structure after implantation was carried out by means of CEMS and CXMS. All measurements were made under room temperature with ^{57}Co source in Rh matrix.

The concentration profiles of elements were measured by Auger electron spectroscopy, as well as by Rutherford backscattering. Röntgen measurements were carried out in glancing geometry. The cross-sectional patterns were investigated using optical microscopy. We also applied the hardness measurements and tribological investigations under conditions mentioned in [6].

We will now consider the experimental results obtained consistently for three methods of implantation. We begin from the high current implantation.

The photographs of the cross-sectional patterns of steel implanted with nitrogen and boron indicate that the thickness of modified layers after N, B and B+N treatment as being approximately 40 μm, while after N+B as 20 μm. Fig. 1 shows the results of Auger electron spectroscopy of steel surface after nitrogen implantation. The nitrogen concentration is maximum on the surface (approx. 17%), then decreases to ~9% and remains approximately constant up to a depth of 14 μm. Boron is present inside the steel only in a thin surface layer of about 0.4 μm. This did not penetrate into the steel but condensed onto the surface. After B+N irradiation a segregation of boron is observed on a surface of sleet, while in the layer up to 20 μm we revealed 5-6% of nitrogen concentration.

Our research of mechanical properties of steel shows that microhardness of the implanted samples decreases with increasing depth, reaching the magnitude of microhardness of non-implanted samples. The maximum extent of hardness, exceeding two times the value for non-irradiated sample, was ascertained for the sample implanted with nitrogen and nitrogen plus boron. Our tribological research indicated that for the sample implanted with nitrogen plus boron, a decrease of the friction coefficient takes place even in the continuous wear regime.

So, one may conclude that high-current implantation with nitrogen and boron improves the mechanical properties of M2 steel. The best results (increase of microhardness and decrease of friction coefficient by two times) have been obtained in the variant of implantation with nitrogen plus boron. The structural and phase changes responsible for enhancement of mechanical properties, were investigated by CEMS and XMS.

Fig. 2 shows CEMS spectrum of the original sample after standard thermal treatment, and Fig. 3 presents p(H) factor for this spectrum. This distribution is typical for the martensite-carbide mixture of M2 steel. The concentration of austenite is of about 5%.

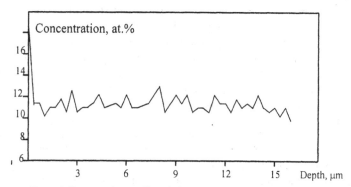

Figure 1. Concentration profiles of nitrogen in M2 steel after HCII.

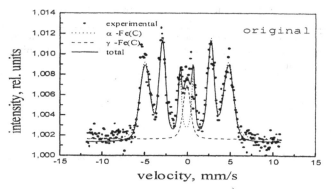

Figure 2. CEMS spectrum of the original sample.

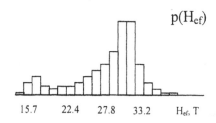

Figure 3. p(H) distribution for the original sample of M2 steel.

Table 2 shows the results of phase analysis of steel surface after different variants of high-current implantation. The obtained results reveal that a formation of the nitrides ε-Fe_xN (with x≈3) and $Fe_{16}N_2$ takes place in all scanning depths (about 10 µm). Observe that a formation of ε-Fe_3N nitrides is confirmed by the Roentgen diffraction. The borides FeB and Fe_2B are revealed only in CEMS measurements, that is in agreement with the results of Auger spectroscopy, indicating an absence of boron atoms at a depth of more than 0.4 µm. CEMS measurements of surface of steel after combined variants of implantation with boron plus nitrogen and nitrogen plus boron do not allow to distinguish the lines of ε-Fe_3N and Fe_2B owing to similar hyperfine parameters. Their total concentration is equal to about 10%. We also observed that a phase composition of thin surface layers (0.1 µm) of steel after N+B and B+N variants of implantation is different, for example, by different concentration of austenite, while under CXMS measurements (the scanning depth is about 10 µm) the phase composition is similar. Nevertheless, the mechanical properties of steels in both cases are essentially different. From this we can hypothesize that the compounds formed by nitrogen, boron and carbon with doped elements of M2 steel essentially influence the properties of steel. Indirect confirmation of high probability of formation of such compounds represents an effect of great narrowing of resonance lines in the spectra of irradiated samples. However, a full interpretation of this effect remains to be established.

TABLE 2. Phase composition of AISI M2 steel after HCII by nitrogen and boron.

Variant of implantation	Method of measurement	Concentration of martensite Fe(C,M) and carbides M_6C	Concentration of austenite	Concentration of nitrides and borides (%)
Original sample	CEMS	(82±2) %	(2.7±0.5) %	-
M2←N	CEMS	(88±2) %	(4.5±0.6) %	$Fe_{16}N_2$- (5±1) % ε-Fe_xN - (2.±1) %
M2←N	CXMS	(78±6) %	(5.3±0.5) %	$Fe_{16}N_2$- (10±1) % ε-Fe_xN - (4.±1) %
M2←B	CEMS	(80±1) %	(4.0±0.3) %	Fe_2B - (7.6±0.5) % FeB - (8.5±0.7) %
M2←B	CXMS	(91±2) %	(9±1) %	-
M2←B←N	CEMS	(77±2) %	(4.1±0.3) %	$Fe_{16}N_2$- (3±1) % ε-Fe_xN+Fe_2B - (10±1) % FeB - (5.1±0.3) %
M2←B←N	CXMS	(79±4) %	(6±1) %	$Fe_{16}N_2$- (5±2) % ε-Fe_xN - (10±2) %
M2←N←B	CEMS	(71±3) %	(11±1) %	$Fe_{16}N_2$- (5±2) % ε-Fe_xN +Fe_2B - (9±1) % FeB - (4.4±1) %
M2←N←B	CXMS	(73±2) %	(5.8±0.6) %	$Fe_{16}N_2$- (7±1) % ε-Fe_xN - (10.±1) %

Now let us consider the results of investigation of M2 steel after plasma-immersion implantation by nitrogen. This method is one of the new methods of treatment of surface. It allows for improving the properties of comparably thick surface layers of about 10 μm in thickness. The positive influence of PIII on the mechanical properties of steel is related to the formation of S-phase (expanded austenite). Its lattice expands due to the additionally doped nitrogen.

The surface nitrogen concentration profile of M2 steel after plasma-immersion implantation essentially differs from the corresponding profile after high-current implantation, fig. 4. Under plasma-immersion implantation nitrogen penetrates to the depth of about 2 μm. Its concentration is maximal on the surface and equals 23 %. Therefore, one may expect an essential difference in the structure of steel at depths 0.1 μm (CEMS measurements) and 10 μm (CXMS measurements). The spectra obtained in these cases are shown in fig. 5. The CEMS spectrum does not reveal any magnetic phases of iron with the effective magnetic field comparable with α-iron. We suppose that a thin surface layer of 0.1 μm fully consists of S-phase that is described by a superposition of paramagnetic and magnetic expanded austenite with their relative intensities of 25% and 75% respectively.

p(H) distribution for CXMS spectrum of M2 steel after plasma-immersion implantation is shown in fig. 6. It is interesting that this distribution essentially differs from the corresponding distribution for the original sample in spite of a low penetration depth of nitrogen atoms (2 μm) in comparison with the scanning depth (10 μm). Such a difference of p(H) factors is not fully understandable. Perhaps, a heating up of the sample

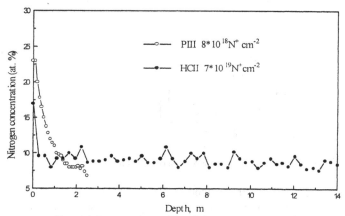

Figure 4. Concentration profiles of nitrogen after HCII and PIII.

under plasma-immersion implantation stimulates some transitions in the systems Fe-C and Fe-M. Fig. 5 also shows the positions of lines γ-Fe, ε-Fe₃N and α-Fe, which were revealed under Röntgen measurements.

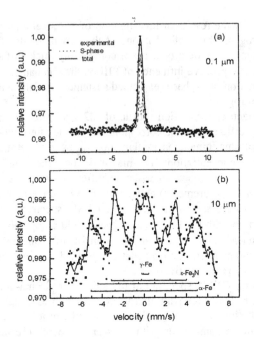

Figure 5. Mössbauer spectra of M2 steel: a – CEMS; b – CXMS.

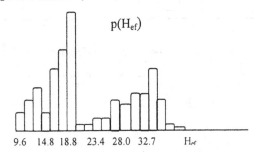

Figure 6. p(H) distribution for XMS spectrum of M2 steel after PIII.

Thus, our results show that the method of plasma-immersion implantation leads to essential alterations in the structure of thin surface layer of steel. In particular, a full transformation of martensite-carbide mixture into S-phase takes place in the surface layer of 0.1 μm; in the surface layer of 10 μm an essential transformation of p(H) factor in comparison with the original sample is observed. These effects lead to improvement in the mechanical properties of steel: we found that the hardness of implanted steel increases two fold in the very least, and the friction coefficient decreases in the same extent when compared with the original sample.

Now let us consider the results of CPF treatment of Y8A steel. The CPF method creates a powerful discharge during a short interval of about 100 microseconds. Concentration of the electrons in the plasma reaches the value up to 10^{18} cm^{-3}, and the temperature of plasma is 2÷3 eV. As a result, the sample surface acquires an increased amount of energy (approximately 10 Joules per square cm) over a short period. It leads to local and intensive heating of the sample surface and even to its partial evaporation. In our research we implemented CPF treatment of Y8A steel for the distances between sample and electrode d=12 cm and 6 cm in nitrogen and oxygen atmospheres. Fig. 7 shows cross section patterns of the sample for the distance d = 12 cm. One can clearly see two surface layers: a white layer with uneven profile and green layer. The thickness of the white layer varies from 5 to 15 micrometers, whereas thickness of the green layer is about 7 micrometers. Deeper we see non-modified structure of Y8A steel. Uneven profile of the surface can be explained by the intensive evaporation processes on the surface under CPF. Measuring the dependence of hardness on the depth of the surface layer, we revealed that the microhardness is maximal on the surface and more than three times exceeds the micro hardness of the initial sample. In addition, the friction coefficient is equal to 0.28. At the same time, the friction coefficient for the original sample was 0.92. Thus, we definitely observe an enhancement of mechanical properties of the steel. The problem is to understand this phenomenon at microscopic level, taking into account that the Auger electron measurements indicate a presence of carbon and oxygen on the surface of the sample.

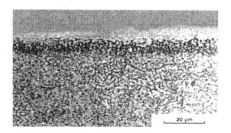

Figure 7. Cross-sectional patterns of Y8A steel after CPF treatment for *d*=12 cm.

CEMS spectrum of the original sample is shown in the Fig. 8, and a corresponding p(H) distribution is depicted in Fig. 9. This distribution is typical for Y8A steel, which contains ferrite in a concentration of more than 95%. CEMS spectrum of Y8A steel after CPF for the distance between the electrodes 12 cm is shown in Fig. 10a. Corresponding p(H) distribution is shown in Fig. 10b This spectrum reflects a structure of white modified layers. According to this measurement, this layer consists of γ-Fe in a concentration of more than 70% and a mixture of martensite and ferrite in a concentration of about 30 %. Such phase transformations occur during the quenching of steels. We propose that similar mechanism takes place in our case due for fast heating of thin surface layers followed by its fast cooling via bulk layers of the sample. In order to investigate the central group of the spectrum in more detail, we repeated the measurement in the velocity range ±2 mm/s, Fig. 10, c. Processing of this spectrum with a discrete approach reveals the presence of γ-Fe(C,M) in concentration of about 50%, as well as non-stoichiometrical iron oxide FeO in concentration of about 20%. Here we did not observe

66

any compounds of iron with nitrogen. We propose that all these compounds get evaporated during CPF treatment.

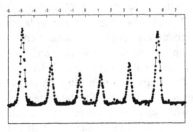

Figure 8. CEMS spectrum of the original sample Y8A steel.

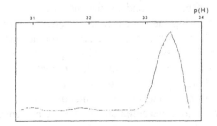

Figure 9. p(H) distribution for the original sample of Y8A steel.

In addition to CEMS measurements we implemented CXMS measurements of the sample; the corresponding spectrum is shown in Fig. 11. The obtained p(H) distribution is depicted in fig. 12. Concentration of austenite decreases to approximately 15%, that allows concluding that only white layer contains γ-Fe. p(H) distribution is similar to corresponding distribution for the original sample, and according to the Mössbauer measurements we did not reveal a cementite. Perhaps, its concentration is small, and the obtained statistical quality of the spectrum does not allow exactly determining the concentration of cementite.

In addition, we also performed similar measurements for the distance of 6 cm between the electrodes in nitrogen atmosphere. We again reveal an intensive $\alpha \rightarrow \gamma$ transformation on the sample surface, as well as ferrite-martensite transformation for magnetic iron. It again is in favor of the quenching mechanism. The central group of the spectrum represents the Mössbauer spectrum of austenite with no iron phases with nitrogen and with no iron oxides. That is why we may conclude that CPF treatment in nitrogen atmosphere, in fact, is equivalent to selective quenching of a thin surface layer of steel. This result explains the increase in micro hardness of the surface layer and the decrease of the friction coefficient.

4. Conclusion

The Mössbauer and tribological investigations of the surface of high-speed M2 steel after different kinds of implantation allow us to conclude that the methods of high-current, plasma-immersion implantation and CPF lead to the formation of a multi-layer multiphase system, which enhances the mechanical properties of steel. After high-current implantation with nitrogen and boron the formation of thin surface layers with nitrides α''-$Fe_{16}N_2$, ε-Fe_xN and borides FeB, Fe_2B takes place, and the formation of a layer with a thickness up to 10 μm containing only nitrides. In addition, high-current implantation leads to some ordering of the martensite structure. A presence of double-layer system with borides and nitrides increases the micro hardness of surface and decreases the friction coefficient by approximately two-fold. Plasma-immersion implantation leads to the formation of S-phase (expanded austenite) in a thin surface layer of about 0.1 μm, that enhances the mechanical properties of steel. CPF treatment creates a double layer system with increased microhardness and decreased friction coefficient.

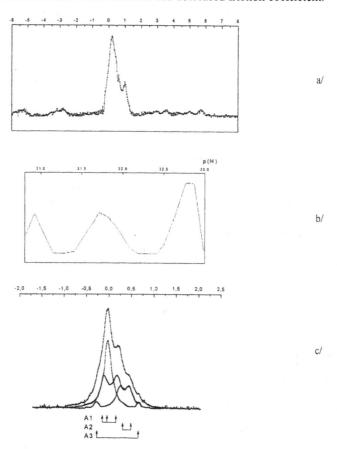

Figure 10. CEMS spectra of Y8A steel after CPF (d=12 cm): a – velocity range ±6 mm/c; b – distribution of the effective magnetic field; c – velocity range ±2 mm/c. A1 - γ-Fe(C,M), A2 – $Fe_{0.95}O$, A3 – 3, 4 lines of magnetic iron.

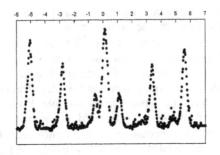

Figure 11. CXMS spectrum of the sample Y8A steel after CPF (*d*=12 cm).

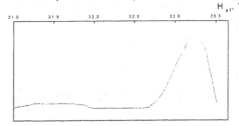

Figure 12. p(H) distribution for the sample of Y8A steel after CPF (*d*=12 cm).

Mössbauer research allows concluding that CPF is equivalent to local quenching (hardening) of the surface layer with the thickness about of 10 micrometers. According to metallographic and Röntgen researches the deeper layer represents grained pearlite $\alpha+(\alpha+Fe_3C)$. At the same time, in our research we failed to develop a consistent mathematical model of the local environment of the iron atoms after the considered above methods for surface treatment. Partially it is explained by complex structure of the chosen samples. However, the results obtained with these samples have a direct practical significance. Nevertheless, now we plan similar measurements with the samples of a simple structure (for example, pure ferrite) in order to consistently understand the enhanced mechanical properties of steel at microscopic level.

References

1. Uglov, V.V., Rusalskii, D.P., Khodasevich, V.V., et al. (1997) *Physics and Chemistry of Materials* 6, 33-38 (in Russian).
2. Kholmetskii, A.L., Misevich, O.V., Mashlan, M., Chudakov, V.A., Anashkevich, A.F. and Gurachevskii, V.L. (1997) Air scintillation detector for conversion electrons Mössbauer spectroscopy (CEMS), *Nucl. Instrum. & Meth.* **B129**, 110-116.
3. Uglov, V.V., Rusalskii, D.P., and Kholmetskii, A.L. (1999) *Advanced Materials*, № 4, 83 (in Russian).
4. Uglov, V.V., Rusalskii, D.P., Khodosevich, V.V., Kholmetskii, A.L., Wei, R., Vajo, J.J., Rumyanceva, I.N. and Wilbur, P.J. (1998) Modified layer formation by means of high current density nitrogen and boron implantation, *Surface and Coating Technology* **103-104,** 317-322.
5. Uglov, V.V., Kholmetskii, A.L., Kuleshov, A.K., Fedotova, J.A., Rusalsky, D.P., Khodasevich, V.V., Ruebenbauer, K., Richter, E., Geunzel, R. and Parascandola, S. (1999) CEMS-investigations of AISI M2 steel after nitrogen plasma immersion ion implantation, *Nucl. Instrum. & Meth.* **B148,** 841-845.
6. Kholmetskii, A.L., Uglov, V.V., Fedotova, J.A., et al. (2000) In: *Current Advances in Materials and Processes,* The Iron and Steel Institute of Japan, Nagoya, **13,** pp. 1417-1425.

CEMS STUDIES OF LASER TREATED DISORDERED AMORPHOUS AND NANOCRYSTALLINE ALLOYS

M. MIGLIERINI[1], K. SEDLAČKOVÁ[1], E. CARPENE[2]
AND P. SCHAAF[2]
[1]Department of Nuclear Physics and Technology, Slovak University of
Technology, Ilkovičova 3, 812 19 Bratislava, Slovakia
[2]Universität Göttingen, II. Physikalisches Institut, Bunsenstrasse 7/9,
D-37073 Göttingen, Germany

Abstract

Conversion electrons provide information from a depth of about 150 nm for most iron rich alloys. Thus, surface effects can be effectively studied by means of Conversion Electron Mössbauer Spectroscopy (CEMS). Our study adopted this technique as the principal method for the investigation of surface modifications induced by pulsed excimer laser irradiation. The effects of laser beam on the magnetic microstructure of amorphous materials with Curie temperatures close-to-room temperature are presented. Such systems do not show the six-line Mössbauer patterns and represent an analytical challenge because the line-intensity ratios, which are correlated to magnetic texture, cannot be used as a measure of laser-induced modifications. In addition, we have chosen such materials, which after suitable temperature annealing exhibit nanocrystalline behavior. This enables the effects of laser treatment on both the amorphous and the nanocrystalline disordered structures to be investigated and discussed. Laser treatment performed under nitrogen and argon atmosphere affects the short-range order arrangement in a different manner. Through surface melting and subsequent rapid quenching, internal stresses are generated. The crystalline phase is eventually removed completely from the surface of nanocrystalline samples when exposed to laser fluence higher than 0.75 Jcm^{-2}.

1. Introduction

Disordered nature of structural arrangement in amorphous and nanocrystalline alloys gave rise to advantageous (from a practical application point of view) magnetic properties. Especially the latter have attracted a lot of scientific interest because, contrary to their amorphous counterparts, their magnetic parameters do not substantially deteriorate at elevated temperature during the process of their practical exploitation. In order to understand the correlation between structural arrangement and the resulting magnetic properties, both amorphous and nanocrystalline systems are widely studied.

69

M. Mashlan et al. (eds.), Material Research in Atomic Scale by Mössbauer Spectroscopy, 69–78.
© 2003 *Kluwer Academic Publishers. Printed in the Netherlands.*

They are exposed to various external conditions (temperature, magnetic fields) comprising also extreme impacts on structure (neutron irradiation, ion bombardment) aiming to unveil how the modification of local atomic arrangement and/or formation of crystalline phases influence the resulting macroscopic properties. Invaluable results are obtained from microscopic investigations that can probe the immediate vicinity of atoms. Among them, Mössbauer spectrometry plays an important and an indisputable role.

Modification of surfaces can help in improving such properties as corrosion resistance or hardness on one hand but at the same time, it might affect macroscopic magnetic parameters. Especially promising and a relatively inexpensive technique for surface modification is the use of lasers. A process of nitriding, which when performed conventionally requires high temperatures in order to incorporate nitrogen into the materials' surfaces, can be effectively and relatively simply accomplished by means of lasers – the so-called laser nitriding [1].

During surface modification, magnetic properties of the treated material are also affected. Consequently, magnetically soft alloys such as amorphous metallic glasses and their nanocrystalline parallels are very sensitive to laser processing. On the other hand, we can benefit from a combined effect acting both upon structural arrangement and magnetic microstructure thus tailoring not only mechanical but also magnetic behavior of the treated system. Mössbauer spectrometry is a technique which provides us with simultaneous information namely on the structural arrangement and magnetic microstructure [2-5].

Recent studies performed on $Fe_{78}B_{13}Si_9$ [2], Fe-B-Si-C [3], and Fe-B-Si-C and Fe-Ni-Mo-B [4] metallic glasses have revealed the effects of laser beams on the structure of magnetic domains. Depending on magnetostriction of amorphous alloys, the magnetic moments adopt positions in an out-of-plane direction [2, 3] or of random orientation [4]. Seemingly opposite effects of crystalline phase formation and removal are illustrated in [5] when high-energy lasers caused surface as well as bulk crystallization of metallic glasses and in [2], respectively, where surface amorphization of originally fully crystallized alloy is reported. These examples illustrate the complexity of laser treatment procedures as well as its flexibility when, using appropriate irradiation parameters, the resulting effects can be varied.

This study reports on the magnetic microstructure of amorphous materials with Curie temperature close-to-room temperature exposed to laser treatment. Such systems do not show a six-line Mössbauer pattern and represent an analytical challenge because of the line-intensity ratios, which are correlated to magnetic texture, but cannot be used as a measure of laser-induced modifications in the present case as they were employed in previous studies [2-5]. In addition, we have chosen such a material which after suitable temperature annealing exhibits nanocrystalline behavior. This enables the effects of laser treatment on both the amorphous and nanocrystalline disordered structures to be investigated and discussed.

Conversion Electron Mössbauer Spectrometry (CEMS) was chosen as the principal method of investigation. Complementary results were obtained from conventional transmission Mössbauer Spectrometry (TMS). Here, we present the results which are a follow-up to the recently reported laser treatment of $Fe_{76}Mo_8Cu_1B_{15}$ alloy [6]. We will concentrate on the effects of atmosphere in which the laser treatment was instituted in order to afford a discussion on the influence of surface modification on disordered nanocrystalline system of this alloy. For comparison, some results taken from [6] have been illustrated.

2. Experimental Details

The amorphous alloy of the nominal composition $Fe_{76}Mo_8Cu_1B_{15}$ was produced by the method of planar-flow casting at the Institute of Physics, Slovak Academy of Sciences in Bratislava in the form of 6 mm wide and 20 μm thick ribbons. The amorphicity of the specimen was verified by X-ray diffraction (XRD) along with Mössbauer spectrometry. No traces of crystallites were revealed in the as-quenched state. Nanocrystalline parallels were obtained by isothermal annealing at temperatures specified with respect to the results of differential scanning calorimetry (DSC) during one hour in a vacuum. The annealing temperatures were chosen so as to cover the structural relaxation and the primary crystallization regions. XRD and Mössbauer effect measurements confirmed that the crystalline phase present is bcc-Fe.

Crystallites that emerge from the amorphous precursor alter not only the structure but also the magnetic structure [7] and their amount can be easily controlled by the annealing temperature at a specific composition of the master alloy [8].

Room temperature ^{57}Fe Mössbauer spectra were collected with an apparatus adopted for simultaneous recording of transmitted photons and conversion electrons. The escape depth of the latter is about 150 nm. A $^{57}Co/Rh$ source was mounted on a constant acceleration drive. Spectral parameters were refined by the NORMOS DIST program [9]. The amorphous samples were fitted with the help of distributions of quadrupole splitting and distributions of hyperfine magnetic field. Details on the fitting model employed to the nanocrystalline samples can be found elsewhere [10, 11].

Laser treatments were performed with a XeCl excimer laser (Siemens XP2020) operating in a pulsed mode. Pulses of 55 ns duration, 308 nm wavelength were focused through a fly-eye homogenizer in order to obtain a uniform intensity distribution over a spot of 5 x 5 mm^2. Details on the laser beam characteristics are published in [1]. Amorphous and nanocrystalline samples were treated by varying the number of pulses with a frequency of 1 Hz and with an average laser fluence H up to 3 Jcm^{-2} on the dull side of the ribbons. Samples were placed in an irradiation chamber which was first evacuated to a residual pressure lower than 10^{-3} Pa then filled with pure nitrogen and/or argon to the pressure of 0.1 Pa.

3. Results and Discussion

3.1. AMORPHOUS STRUCTURE

3.1.1. *Nitrogen atmosphere*
Examples of TMS and CEMS spectra of untreated as-quenched specimen and of that treated with 64 laser pulses with a fluence of 3 Jcm^{-2} are shown in Fig. 1 together with their spectral components. Broadened patterns confirm the amorphous nature of the

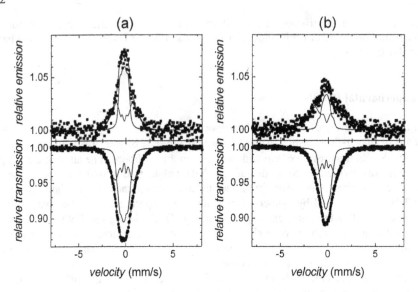

Figure 1. CEM (top) and TM (bottom) spectra including spectral components of the as-quenched $Fe_{76}Mo_8Cu_1B_{15}$ alloy: (a) untreated, and (b) after laser treatment in nitrogen atmosphere with 64 pulses with a fluence of 3 Jcm^{-2}.

samples. They exhibit combined electric quadrupole and magnetic dipole interactions characteristic for magnetic structures in the vicinity of the magnetic ordering transition. Consequently, the spectra were evaluated with the help of distributions of quadrupole splitting $P(\Delta)$ which modeled the non-magnetic regions in the samples and distributions of hyperfine magnetic fields $P(B)$ assigned to magnetically active resonant atoms. To account for the spectrum asymmetry, a linear coupling between the respective hyperfine parameters and the isomer shift was introduced.

In the as-quenched state, the relative areas of the magnetic components are of 47% and 52% in the TMS and CEMS, respectively. This indicates a homogeneous occurrence of magnetic and non-magnetic structural positions of the resonant atoms in the amorphous $Fe_{76}Mo_8Cu_1B_{15}$ alloy at room temperature both in the bulk and on the surface. The error in the determination of spectral line areas is estimated to be of ±2.5%.

Pronounced changes are detected in the magnetic microstructure of the surface regions after the laser treatment with 64 pulses with the fluence of 3 Jcm^{-2}. The corresponding CEMS in Fig. 1b displays notable broadening in comparison with the untreated sample. Contribution of the magnetic component was increased to 65%. The same tendency was observed also in TMS where the amount of magnetic active regions was raised to 54%.

The detected alternations of the spectral parameters in the bulk (TMS) are more delicate than those related to the surface (CEMS) as expected because the laser treatment extends to less than 1 μm in depth [1]. In this respect, the CEMS technique is more appropriate for investigation of laser treatment effects. Nevertheless, a proper choice of the system's composition yielding close-to-room Curie temperature makes this model of amorphous alloy suitable and sensitive also to TMS. Bulk properties are affected by the arrangement of magnetic regions which are, in turn, influenced by the reorientation of magnetic domains due to internal stresses [2-5] along with modifications of chemical and/or topological short-range order. The laser beam acts as a source of intense heat which create a plasma above the irradiated surface thus causing the two above mentioned effects.

We have studied systematically the mutual relation between the number of laser pulses delivered to one irradiation spot and a given energy density (fluence). For the sake of further discussion, we repeat the main conclusions [6]: The average value of the hyperfine magnetic field distribution $<B>$ increases with the number of laser pulses applied per irradiation spot as well as with the energy density of the laser beam. For a low-energy density of 1 Jcm^{-2}, the increase in $<B>$ as a function of the number of pulses is not accompanied by an increase in the relative fraction of the magnetic component, as observed for irradiation with $H > 1$ Jcm^{-2}. This result suggests that the sole 'quantitative' increase (i.e., relative fraction) of regions with magnetic dipole interactions cannot explain this seemingly controversial behavior.

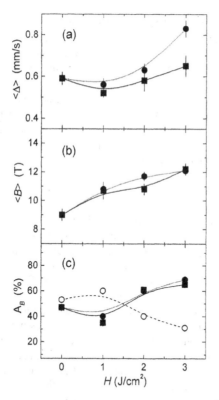

Figure 2. (a) Average value of P(Δ) distribution, $<\Delta>$, (b) average value of P(B) distribution, $<B>$ and (c) relative area of magnetic component, A_B plotted against the fluence of the laser beam H, applying 64 pulses under nitrogen (solid squares) and argon (solid circles) atmosphere. Open circles in (c) represent the relative area of the non-magnetic component under argon atmosphere. Connecting lines act only as guides.

Laser beam introduces internal stresses generated by the excess heat. At the same time, the excess heat created by plasma on the surface propagates further into the bulk thus causing structural relaxation accompanied by rearrangement of the constituent elements in the disordered structure. This effect can be hardly observed in ordered crystalline materials. Thus, in addition to surface stresses it is also the structural rearrangement that influences the bulk. Nevertheless, the alternation of the short-range order is particularly strong in surface regions because we have to take into consideration incorporation of atoms from the surrounding atmosphere (nitrogen) into the material.

3.1.2. *Argon atmosphere*

Laser treatment in argon atmosphere resulted in a further increase in the content of magnetic regions localized on the surface of amorphous specimens as derived from CEMS (up to 69% for the fluence of 3 Jcm^{-2}). Average value of quadrupole splitting distribution $<\Delta>$, average value of hyperfine magnetic field distribution $<B>$, and the relative area of magnetic component A_B derived from CEMS are plotted in Fig. 2 against the laser fluence H for irradiations performed in nitrogen and in argon atmosphere with 64 pulses.

It is noteworthy that even though the changes in the relative contributions of respective spectral components as well as in $<B>$ are rather moderate, a significant increase in $<\Delta>$ is observed. Moreover, the deviations between samples irradiated in nitrogen and in argon progressively increase with rising laser fluence. TMS did not reveal any considerable change in any of the above mentioned spectral parameters with respect to the nitrogen atmosphere. Thus, we can conclude that laser treatment affects predominantly the surface regions of amorphous samples. As regards the atmosphere in which the laser treatment was performed, the incorporation of argon has introduced a pronounced increase in $<\Delta>$ even though the relative content of paramagnetic regions was decreased (see the open circles in Fig. 2c).

3.2. NANOCRYSTALLINE STRUCTURE

Annealing of the as-quenched amorphous alloy at temperatures above the first crystallization peak resulted in the formation of bcc-Fe crystalline phase. Its identification was confirmed by Mössbauer as well as by XRD measurements. From the latter, the size of crystallites was estimated to be of about 6 nm [6]. CEMS and TMS Mössbauer spectra of the sample annealed for 1 hour at 600°C are illustrated in Fig. 3.

The presence of nanocrystals is documented by narrow lines. They are superimposed on broad spectral features ascribed to the residual amorphous matrix. The prominent central line indicates that the amorphous remainder contains non-magnetic regions that are, however, smaller in amount than in the as-quenched sample. This is a consequence of segregation of Fe atoms into the bcc grains with subsequent modification of the chemical short-range order of the amorphous rest. At the same time, a relative portion of the magnetic fraction increases partially due to penetrating exchange interactions from ferromagnetic bcc grains into the retained amorphous phase [12, 13].

Both CEMS and TMS exhibit the same features except for the difference in relative line intensities of the crystalline phase. Formation of closure domains on the surface [2] forces the magnetic moments to turn into the plane of the ribbon-shaped sample as

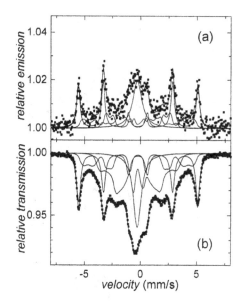

Figure 3. (a) CEM and (b) TM spectra including spectral components of the untreated nanocrystalline (annealed for 1 hour @ 600° C) $Fe_{76}Mo_8Cu_1B_{15}$ alloy.

indicated by a higher contribution of the second and the fifth CEMS lines in Fig. 3a.

The influence of laser treatment on nanocrystalline samples with a different content in the crystalline phase is documented by Table I. Average value of the hyperfine magnetic field of the residual amorphous phase $<B>$, its relative content A_B, and the relative content of non-magnetic amorphous regions A_{QS} are listed for untreated nanocrystalline samples and for the samples treated with one laser pulse with the fluence $H = 0.2$ Jcm^{-2} in nitrogen atmosphere. The nanocrystalline samples were prepared by annealing for one hour at annealing temperatures $T_a = 490$, 510, and 600° C giving rise to 26, 39, and 60% of crystalline phase before laser irradiation, respectively.

TABLE I. Average value of the hyperfine magnetic field of the residual amorphous phase $<B>$, its relative content A_B, and the relative content of non-magnetic amorphous regions A_{QS} derived from CEMS of nanocrystalline $Fe_{76}Mo_8Cu_1B_{15}$ heat treated for 1 hour at annealing temperature T_a and irradiated with laser fluence H. The numbers in brackets represent errors in the determination of the last digit.

T_a (° C)	H (Jcm^{-2})	$<B>$ (T)	A_B (%)	A_{QS} (%)
490	0	12.7(5)	22	52
	0.2	9.2(6)	50	20
510	0	14.4(8)	21	40
	0.2	12.0(7)	29	28
600	0	14.8(6)	17	23
	0.2	14.6(7)	18	22

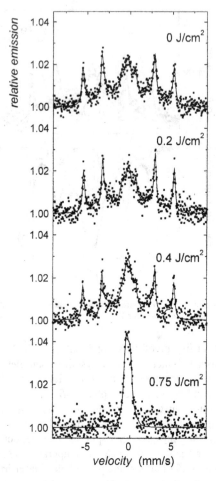

Figure 4. CEM spectra of the nanocrystalline $Fe_{76}Mo_8Cu_1B_{15}$ alloy after laser treatments with one pulse in the indicated fluences.

CEMS Mössbauer spectra of the 600°C annealed alloy after laser treatment with one pulse in the indicated fluences are shown in Fig. 4. Laser treatment was performed in nitrogen atmosphere. Corresponding TMS do not show any qualitative changes with increasing laser fluence and that is why they are not provided on a separate figure. Surface melting caused by the laser beam and subsequent rapid quenching due to thermal contact with the neighbouring regions of the sample's material have lead to complete re-amorphization after irradiation with a fluence of 0.75 Jcm^{-2}. As a result, the corresponding Mössbauer spectrum shows only broadened central doublet which was fitted with one distribution on quadrupole splitting values. So, the magnetic regions of the retained amorphous phase (comprising in this case 100% of the surface) have vanished entirely, too.

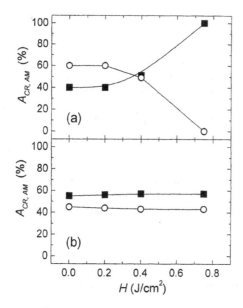

Figure 5. Relative area of crystalline component, A_{CR} (open circles) and amorphous component, A_{AM} (solid squares) of the: (a) surface regions (CEMS) and (b) bulk (TMS) plotted against the fluence of the laser beam H, applying one pulse to the nanocrystalline $Fe_{76}Mo_8Cu_1B_{15}$ alloy.

Quantitative features of the content of the crystalline A_{CR} and the amorphous A_{AM} phases on the surface (CEMS) and in the bulk (TMS) of the laser treated nanocrystalline (600° C/1 h) alloy are presented in Fig. 5 as a function of the laser fluence. A gradual decrease of the crystalline component is observed on the surface (Fig. 5a). No changes in the relative content of the TMS components were revealed within the frame of the experimental error (Fig. 5b).

4. Conclusions

We have investigated the effect of laser treatment on disordered amorphous and nanocrystalline $Fe_{76}Mo_8Cu_1B_{15}$ alloy. A close-to-room Curie temperature of this alloy provides particular shape of the Mössbauer spectra corresponding to the as-quenched specimens. Consequently, pronounced changes in the spectral parameters are observed when the samples are either heat treated by annealing (partial nanocrystallization occurs) and/or subjected to surface modification by laser. Due to the latter, mutual ratio of the non-magnetic and magnetic regions both on the surface and in the bulk of the amorphous samples varies. These changes are associated also with variations of the average values of the respective hyperfine parameters (quadrupole splitting, hyperfine magnetic field).

Laser treatment of the amorphous specimens performed under two different atmospheres (nitrogen and argon) has unveiled the sensitivity of especially the non-magnetic spectral component to changes in the short-range order arrangement of the resonant atoms located in the surface regions. CEMS technique proved to be suitable for studying such phenomena. Their understanding, however, deserves closer attention and further experiments with other types of filling gasses.

Nanocrystalline $Fe_{76}Mo_8Cu_1B_{15}$ alloy has exhibited surface re-amorphization after laser treatment with sufficiently high fluence (0.75 Jcm^{-2}). Lower irradiation fluences have affected only the relative ratio of the crystalline-to-amorphous phase content of the surface regions. Practically no changes were observed in the bulk. It should be noted, however, that the above mentioned effects were obtained after laser treatment with a single pulse per irradiation spot.

Laser treatment of amorphous and nanocrystalline alloys can be effectively studied by Mössbauer effect techniques. In particular CEMS proved to be quite suitable. With the expectations of practical applications of these materials, the impact of laser treatment on the surface properties comprising mechanical hardness and corrosion resistance will be subjected to further investigations.

Acknowledgement

This work was supported by the projects DAAD 17/2002 and SGA 1/8305/01.

References

1. Schaaf, P. (2002) *Prog. Mater. Sci.* **47** 1-161.
2. Gonser, U. and Schaaf, P. (1991) Surface phase analysis by Conversion X-ray and Conversion Electron Mössbauer spectroscopy *Fresenius J. Anal. Chem.* **341** 131-135.
3. Sorescu, M. (2000) Magnetic moment distributions in laser-ablated metallic glass thin films *J Appl. Phys.* **87** 5855-5857.
4. Sorescu, M. (2000) Direct evidence of laser-induced magnetic domain structures in metallic glasses *Phys. Rev.* **B61** 14338-14341.
5. Sorescu, M. (2000) Magnetic properties of metallic glasses using the laser-Mössbauer method *J. Magn. Magn. Mater.* **218** 211-220.
6. Miglierini, M., Schaaf, P., Škorvánek, I., Janičkovič, D., Carpene, E. and Wagner, S. (2001) Laser-induced structural modification of FeMoCuB metallic glasses before and after transformation into a nanocrystalline state *J. Phys.: Condens. Matter* **13** 10359-10369.
7. Suzuki, K., Makino, A., Inoue, A. and Masumoto, T. (1991) Soft magnetic properties of nanocrystalline bcc Fe-Zr-B and Fe-M-B-Cu (M=transition metal) alloys with high saturation magnetization *J. Appl. Phys.* **70** 6232-6237.
8. Suzuki, K. (1999) Nanocrystalline soft magnetic materials: A decade of alloy development *Mater. Sci. Forum.* **312-314** 521-530.
9. Brand, R.A., Lauer, J. and Herlach, D.M. (1983) The evaluation of hyperfine field distributions in overlapping and asymmetric Mössbauer spectra: a study of the amorphous alloy $Pd_{77.5-x}Cu_6Si_{16.5}Fe_x$ *J. Phys. F: Met. Phys.* **13** 675-683.
10. Miglierini, M. and Grenèche, J.-M. (1997) Mössbauer Spectrometry of Fe(Cu)MB-Type Nanocrystalline Alloys: I. The Fitting Model for the Mössbauer Spectra *J. Phys.: Condens. Matter* **9** 2303-2319.
11. Miglierini, M., Škorvánek, I. and Grenèche, J.-M. (1998) Microstructure and Hyperfine Interactions of the $Fe_{73.5}Nb_{4.5}Cr_5CuB_{16}$ Nanocrystalline Alloys: Mössbauer Effect Temperature Measurements *J. Phys.: Condens Matter* **10** 3159-3176.
12. Hernando, A., Navarro, I. and Gorría, P. (1995) Iron exchange-field penetration into the amorphous interphase of nanocrystalline materials *Phys. Rev.* **B51** 3281-3284.
13. Suzuki, K. and Cadogan, J.M. (2000) Effect of Fe-exchange-field penetration on the residual amorphous phase in nanocrystalline $Fe_{92}Zr_8$ *J. Appl. Phys.* **87** 7097-7099.

CEMS MEASUREMENT OF SURFACE HARDENING
OF TOOTHED WHEELS

M. SEBERÍNI, J. LIPKA AND I. TÓTH

Department of Nuclear Physics and Technology, Slovak University of Technology, Ilkovičova 3, 81219 Bratislava, Slovakia

Abstract

Surface hardening is one of the final steps in production of toothed wheels. It is done by bombardment of the tooth surface in two steps: 1) by glass balls and 2) by steel balls. A search goes on after a reliable method, which could detect the quality and the intensity of the hardening process. Three samples in the form of steel plates were measured in common Mössbauer scattering geometry and by CEMS method. The samples were: 1) unprocessed, 2) shot for shorter process time and 3) shot for full process time. After CEMS measurement, the surface layer of the unprocessed sample was ground off and measured again in order to obtain a reference on bulk material. While little significant changes were found in the results of the common Mössbauer backscattering measurement, the effect in the change of CEMS spectra is striking. A distinct paramagnetic component, with the area comparable to the spectrum of the basic material, appeared in the middle of the two spectra of the processed samples and, moreover, it seems to be sensitive to the time (or dose) of bombardment of the material surface.

1. Introduction

A requirement was conveyed to our Mössbauer laboratory by the VUSTAM research insitute whether the Mössbauer spectroscopy could provide a measurable proof of surface hardening of toothed wheels. Originally, there was a call for such a method by a car components producing factory. Such a method should serve as a comparative tool in final optimization of production.

As samples, three steel plates were provided after different process stages: 1) none, 2) shorter time of processing and 3) full time of processing. Under processing, the final operation of bombardment the surface by glass and steel balls is understood after the regular operations of case hardening, quenching and tempering the toothed wheels. These three samples were measured in common Mössbauer backscattering geometry and by the CEMS method.

The basic material was ZF 1A alloyed steel, annealed isothermically, case-hardened to 0.4 - 0.7 mm and HRC 61 quenched in the endogas atmosphere (in a continual furnace

M. Mashlan et al. (eds.), Material Research in Atomic Scale by Mössbauer Spectroscopy, 79–82.

Aichelin). Time of bombardment by steel balls was 0.83 min; the speed of the launching wheel was 1800 rpm. No additional X-ray measurements were performed.

The measurement of dose is carried out in rather a spectacular way. A normalized steel plate of the same material is fitted in a grip and shot simultaneously with the product. For any particular product, the bending angle of the sample plate is prescribed, i.e. the bombardment process is finished when the sample plate bends to the prescribed angle.

2. Experimental Details

In the backscattering geometry, the samples were measured at a distance 4.5 cm from the 20 mCi ^{57}Co(Rh) source; the source-detector angle was 80°. The measurement time was about 2 days per sample. The CEMS measurements were carried out with a gas-flow detector of thickness 1+1 mm with the He-13 % CH$_4$ mixture. The average measurement time was about 3 days per sample.

3. Experimental Results

As can be seen from the Fig. 1 and from the table 1, no significant changes were found among the samples above the statistical error. It was, however, possible to fit the spectra with three sextets and to confirm the martensitic structure of the bulk, which is clearly visible in these spectra.

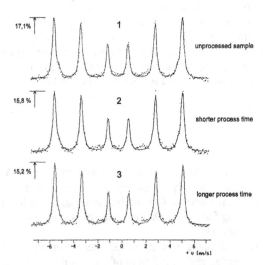

Figure 1. Spectra of the Mössbauer backscattering experiment.

Different situation appeared with the CEMS results (Fig. 2, table 2). In the unprocessed sample, magnetite, Fe$_3$O$_4$ was found in the surface layer as a result of thermal processes. (It is very likely that Fe$_2$O$_3$ was present as well; this should be confirmed by a measurement with slightly higher velocity range. However, at this stage of preliminary measurement, this question was of minor importance). To obtain information on the bulk, the surface layer was then ground off and the sample was measured again.

In the spectra of the samples processed by bombardment, a distinct paramagnetic phase appeared which could be fitted by two doublets with rather broad lines. Moreover, the measurement seems to be sensitive to different dose of bombardment. It can be supposed that some ferromagnetic phase arose too; however, we were not able to fit it due to insufficient statistics. (This also, was the reason why the ferromagnetic phase was fitted by an only sextet).

TABLE 1. Parameters of the fitting procedure of the backscattering measurement.

sample	sextet 1				sextet 2			sextet 3			
	$\Gamma_1=\Gamma_2$ mm/s	H_1 T	IS^*_1 mm/s	$A_{rel\,1}$ %	H_2 T	IS_2^* mm/s	$A_{rel\,2}$ %	Γ_3 mm/s	H_3 T	IS_3^* mm/s	$A_{rel\,3}$ %
1	0,24	33,7	-0,109	25,7	33,0	-0,115	56,4	0,49	31,0	-0,121	17,9
2	0,27	33,5	-0,113	31,6	33,0	-0,114	51,5	0,46	31,0	-0,101	16,9
3	0,27	33,6	-0,112	37,2	33,0	-0,115	47,1	0,37	30,8	-0,098	15,7

TABLE 2. Parameters of the CEMS spectra.

sample	sextet 1				sextet 2			sextet 3			
	Γ_1 mm/s	H_1 T	IS_1^* mm/s	$A_{rel\,1}$ %	H_2 T	IS_2^* mm/s	$A_{rel\,2}$ %	Γ_3 mm/s	H_3 T	IS_3^* mm/s	$A_{rel\,3}$ %
un-processed	0,28	48,74	0,079	8,15	45,97	0,496	30,85	0,394	33,00	-0,228	61,0
bulk	0,255	32,83	-0,227	100							

sample	sextet				doublet A				doublet B			
	Γ_1 mm/s	H_1 T	IS_1^* mm/s	$A_{rel\,1}$ %	Γ_A mm/s	IS_A^* mm/s	QS_A mm/s	$A_{rel\,A}$ %	Γ_B mm/s	IS_B^* mm/s	QS_B mm/s	$A_{rel\,B}$ %
shorter process time	0,414	32,80	-0,227	74,26	0,724	0,028	0,755	13,30	1,010	0,719	1,143	12,44
longer process time	0,265	32,82	-,0237	53,17	0,647	0,089	0,708	22,00	0,889	0,747	0,993	24,83

* All isomer shifts are relative to Rh matrix

4. Conclusions

Bombardment by steel balls introduces an additional energy into the thin surface layer and influences a creation of new paramagnetic phase/s. Creation of a new ferro-magnetic phase is very likely too; however, for a more detailed phase analysis, a selective separation of surface layer and its measurement in transmission geometry would be necessary, which would mean introducing chemical methods of separation.

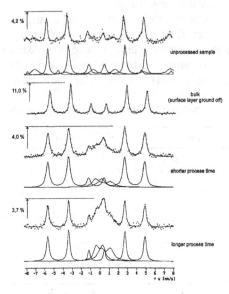

Figure 2. CEMS spectra.

Since the growth of the new phase is dose-dependent and the CEMS measurement seems to be sensitive to its thickness, it can be concluded that the thickness of the layer does not exceed the effective penetration depth of the 7.4 keV electrons, which is about 200 - 300 nm. Thus, from this point of view, the CEMS meets the requirements almost ideally.

There are drawbacks, however. Apart from long measurement time (2-3 days at least), it would be somewhat more requesting to design a detector to fit to a curved surface for a real *in-situ* measurement, for example.

If the newly created phase proved stable, a measurement in the constant velocity mode seems to be an ideal solution for quick checks in the process of industrial production.

Acknowledgment

This work was partly supported by the grant Vega No. 1/7631/20

References

1. Thomas, J.M., Tricker, M.J. and Winterbottom, A.P. (1975) Conversion Electron Mössbauer Spectroscopic Study of Iron Containing Surfaces, *J. Chem. Soc. - Faraday Trans.* II **71**, 1708.
2. Sawicki, J.A. and Sawicka, B.D. (1983) Experimental Techniques for Conversion Electron Mössbauer Spectroscopy, *Hyperfine Interactions* **13**, 199.
3. Camara, A.S. and Keune, W. (1975) Oxidation Study of Iron by Mössbauer Conversion Electron and Gamma Ray Scattering, *Corrosion Science* **15**, 441.
4. Volkswagen Internal Regulatory Guide VW 500 19.

MAGNETIC IRON-BASED OXIDES INVESTIGATED BY ^{57}Fe MÖSSBAUER SPECTROMETRY

JEAN-MARC GRENÈCHE
Laboratoire de Physique de l'Etat Condensé, UMR CNRS 6087, Université du Maine, Faculté des Sciences, 72085 Le Mans Cedex 9, France
E-Mail : greneche@univ-lemans.fr

Abstract

Oxides exhibit different structural types giving rise, especially, to a large variety of magnetic and transport properties. New synthesis routes allow obtaining new oxides in microcrystalline, nanocrystalline and nanostructured states. The understanding of their physical properties requires a structural characterization at micrometer and nanometer scales, using different spatial and time scales techniques. In the case of iron-based oxides, both zero-field and in-field ^{57}Fe Mössbauer spectrometry are suitable to investigate the local environments of iron and to probe dynamic effects. Indeed, the high efficiency is essentially due to the local probe and relaxation phenomena sensitivity. We report thus several examples to illustrate the good ability of Mössbauer spectrometry to investigate iron-based oxides.

1. Introduction

Iron oxides are very common in nature but numerous kinds of oxide materials can be also synthesized. They are important industrial products which can be used as magnetic recording media, as hard magnets, as magnetoresistive sensors, etc. and can also constitute the outcome of iron corrosion and can be found in rocks, colored pigments or proteins. Iron oxides exhibit different crystalline structural types among which one can distinguish ferrites, garnets, hexaferrites, manganites and double perovskites. They become very fascinating materials because of their physical properties which are strongly dependent on the chemical content (oxygen stoichiometry, metallic cationic species), on the synthesis methods leading to crystalline phases, amorphous phases, nanocrystalline particles, nanostructured powders, thin films and ferrofluids. Oxides received thus great attention within the last decades, especially the research of new synthesis methods to get new phases and to create new architectures, and of unusual properties in the case of magnetic materials.

Oxides are traditionally prepared by ceramic technique which consists of alternating high temperature treatments and low temperature grinding in order to improve the

M. Mashlan et al. (eds.), Material Research in Atomic Scale by Mössbauer Spectroscopy, 83–92.
© 2003 *Kluwer Academic Publishers. Printed in the Netherlands.*

microstructural properties but these procedures are time and energy consuming. Consequently, low temperature chemical and soft sol-gel based processes were developed to synthesize microstructured and nanoparticle oxides while mechanochemistry and mechanical grinding were used to obtain nanostructured powders.

The atomic structures of oxides are usually based on packing of corner-sharing octahedral and tetrahedral (and sometimes dodecahedral) units centered on a transition metal (TM) and linked each other through oxygen atoms (O). These bonds originate the occurrence of superexchange and direct interactions favoring either antiferromagnetic or ferromagnetic coupling between magnetic moments according to the Kanamori Goodenough rules [1], the sign and the amplitude of the exchange coupling being strongly dependent on the superexchange angle TM-O-TM and on the degree of covalency of the cationic sites. Let us mention that the supersuperexchange term TM-O-O-TM might significantly contribute to the magnetic ordering for some local configurations. Consequently, iron oxides usually behave as ferrimagnets or antiferromagnets in the case of crystalline structures while the topological disorder in amorphous phases gives rise to non-collinear magnetic arrangements as speromagnetism or derived structures. But both the presence of cationic substituting or atomic doping species and the non homogeneous chemical order do induce some local canted magnetic structures. Otherwise superparamagnetic fluctuations are expected to occur in confined nanocrystalline and nanostructured systems when the contribution of surface or interfaces becomes extremely higher than that of microcrystalline materials. Indeed, superexchange bonds can be broken when oxygen atoms are missed either in the case of non stoichiometric systems or at the surface of nanoparticles.

To investigate the structural and magnetic properties of iron oxides, diffraction X-ray and neutron techniques associated to static magnetic measurements and to electron transmission microscopy provide useful and complementary information.

Nevertheless, some questions remain opened or unclearly solved as (i) the content of transition metal atoms located in tetrahedral and octahedral units (ii) the valence states of the iron species, (iii) the magnetic arrangement of iron network, (iv) the presence of relaxation phenomena due to clustering chemical effects and (v) the surface magnetic effects and magnetic grain boundaries in the case of nanoparticles and nanostructured powders, respectively.

Both zero-field ^{57}Fe Mössbauer spectrometry and in-field ^{57}Fe Mössbauer spectrometry are appropriate tools to investigate some aspects of iron oxides. Indeed, local techniques offer a great advantage because they are atomic scale sensitive to short-range chemical and topological (dis)orders. Consequently, a local analysis of magnetic hyperfine field characteristic of atoms located either at interfaces, at grain boundaries or at surfaces can be achieved, providing consequently information about the local structure and the composition of immediate probe surroundings. In addition, in-field Mössbauer spectrometry allows determining the in-field magnetic arrangement of magnetic iron moments. It is important to emphasize first that that the magnetization is antiparallel to the hyperfine field B_{hf} in the case of metallic iron and iron oxides, resulting from the large and negative Fermi term contribution, and then the hyperfine field of octahedral iron sites are expected to be larger than that of tetrahedral iron sites. Let us also remember that when B_{hf} is inclined at an angle ϑ from the γ–ray direction, the line areas of Mössbauer sextets are in the ratio 3:x:1:1:x:3 with x = $4\sin^2\vartheta/(1+\cos^2\vartheta)$. By normalizing the total area $\Sigma A_{i,7-i} = 1$, the area of each of the lines 2,5 is $A_{2,5} = \frac{1}{4}\sin^2\vartheta$.

Finally, Mössbauer spectrometry is sensitive to relaxation phenomena: because the time scale given by the Larmor precession time is about 10^{-8} s in the case of ^{57}Fe, one expects thus the observation of the blocked state to superparamagnetic state transition and to paramagnetic state in most iron-based particle systems using Mössbauer technique within the usual temperature range (0 – 400 K). Indeed, fast fluctuations of magnetic iron yields to paramagnetic like Mössbauer spectra (single line or quadrupole doublet) while at lower temperatures, the fluctuation frequency decreases and a magnetic hyperfine splitting occurs for frequencies smaller than the Larmor frequency of the ^{57}Fe magnetic moment (5.10^8 s^{-1}). Besides, static magnetic measurements (characteristic time: ~10 s), ac-susceptibility at various frequencies are convenient for investigating the dynamic properties while neutron scattering experiments are shorter times sensitive (10^{-10} - 10^{-11}s).

The next sections report different examples illustrating how both zero-field and in-field Mössbauer spectrometry bring relevant information to understand some unusual magnetic behaviors.

2. Substituted, Doped and Mixed Spinel Ferrites

The ideal ferrites AB_2O_4 (A and B divalent and trivalent atoms) which exhibit a spinel structure (the same as the mineral spinel $MgAl_2O_4$), can be divided into two families: direct and inverse type according to the location of the metallic ions into tetrahedral and octahedral units, giving rise to either antiferromagnetic or ferrimagnetic order. The magnetic structures were first described by Néel's theory of ferrimagnetism [2] on the basis of two non equivalent magnetic sublattices, assuming that the magnetic moments are either parallel or antiparallel throughout the whole temperature range. But non collinear magnetic structures with lower energy were proposed by Yafet and Kittel,

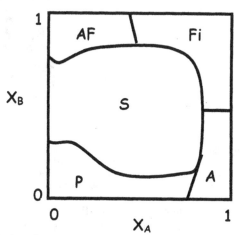

Figure 1. Magnetic phase diagram showing different magnetic behaviors in the case of a mixed spinel system.

giving rise to triangular spin structures [3], or to more complex helical or screw magnetic structures, as observed in many spinel compounds. Indeed, magnetic frustration originates from the competing A-A, A-B and B-B interactions. By mixing several kinds of metallic and diamagnetic cations, very complex magnetic structures can be observed, as spin glass, reentrance, resulting from the short-range order competing interactions or clustering effects, as well as the occurrence of reverse spins. Such a complexity is illustrated by the schematic magnetic phase diagram proposed by Poole and Farach [4] in Figure 1.

The main questions are thus where iron cations are located, the proportions of octahedral and tetrahedral iron sites and the average canting angle of both kinds of magnetic moments. In opposite to zero-field Mössbauer spectra which exhibit overlapped and unresolved lines preventing a clear and accurate distinction between tetrahedral and octahedral components, in-field spectra show split outermost lines and intermediate lines with a certain intensity, both allowing accurate estimation of tetrahedral and octahedral iron sites and canting angle for the two types of magnetic iron moments, respectively. Examples of theoretical spectra are reported in Figure 2.

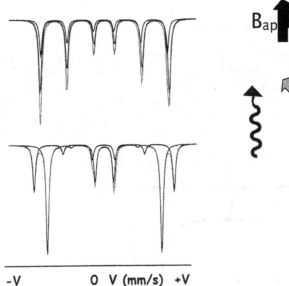

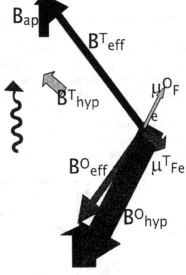

Figure 2. Typical zero-field (Top) and in-field (Bottom) Mössbauer spectra in the case of a ferrimagnet with both octahedral and tetrahedral iron sites.

Figure 3. Schematic representation of magnetic moments, hyperfine field and effective field in presence of an external magnetic field with octahedral and tetrahedral iron sites.

Thus, the presence of an external magnetic field leads to the schematic representation of iron moments, hyperfine field and effective field B_{eff} which results from the vector sum of the hyperfine field B_{hf} and the applied field B_{app}

$\left(\vec{B}_{eff} = \vec{B}_{hf} + \vec{B}_{app}\right)$, as it is shown in Figure 3. The values of the mean hyperfine field of the octahedral and tetrahedral sites can be estimated from those of the in-field effective field using the following relationship

$$<B_{hf}^2> = <B_{eff}^2> + B_{app}^2 - 2*<B_{eff}> *B_{app}*\cos<\vartheta> \qquad (1)$$

where the angle $<\vartheta>$ defines the direction of the mean effective field with respect to the gamma ray direction, oriented parallel to the external field.

It is important to emphasize that the fitting procedure requires great attention, especially when the magnetic lines are broadened and overlapped, due to chemical disorder. The effective field distribution does result from a distribution of hyperfine field combined to that of angles ϑ, while the quadrupolar shift can be assumed equal to zero and the isomer shift values are similar for each type of iron site. It is consequently important to check if the resulting hyperfine field distribution is consistent with that obtained from zero-field Mössbauer study. The low and high limits of hyperfine field distributions assigned to octahedral and tetrahedral iron sites, respectively, have to be carefully analyzed, in order to evidence for the existence of reverse spins.

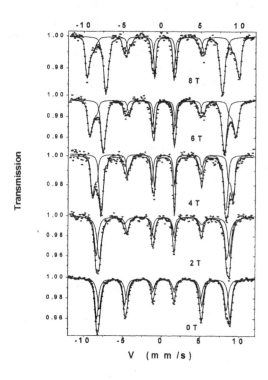

Figure 4. ^{57}Fe Mössbauer spectra recorded at 10 K from 8% Sn-doped γ –Fe$_2$O$_3$ in an external magnetic field of 2–8 T parallel to the γ-radiation see [5].

An example which concerns Sn-doped maghemite (γ-Fe_2O_3) [5] is illustrated in Figure 4. It shows the evolution of the hyperfine structure, especially the splitting of the outermost lines and the intensity of intermediate lines, as a function of the external field which is oriented parallel to the γ-beam. It is important to emphasize that low applied fields are unable to resolve the hyperfine structure, while applied fields higher than 6 T allow splitting outermost lines into asymmetrical lines. It is finally concluded in the present study that Sn prefers to be located in octahedral sites, in agreement with EXAFS measurements [5].

3. Clustering Effects in Manganites

When the chemical disorder becomes inhomogeneous within a crystalline matrix, magnetic clusters might occur, giving rise to superparamagnetic behavior when the intracluster exchange interactions are larger than those of the matrix. It is important to emphasize the role of both superexchange and supersuperexchange interactions. Such a situation was particularly observed in iron doped manganite $La_{0.75}Ca_{0.25}MnO_3$ [6] and $La_{0.7}Sr_{0.3}MnO_3$ [7]. At low temperature, the zero-field Mössbauer spectra consist of resolved sextet with narrow lines but at higher temperature, they spread into a magnetic component with asymmetric broadened lines and a broadened quadrupolar doublet, which progressively collapse into a paramagnetic spectrum. In the intermediate temperature range, the external magnetic field allows to resolve clearly the hyperfine structure. It exhibits only a broad distribution of hyperfine fields with a shift towards higher fields and with a low field tail, i.e. the disappearance of the quadrupolar feature.

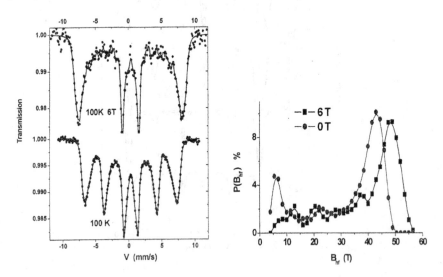

Figure 5. Mössbauer spectra of iron doped $La_{0.75}Ca_{0.25}MnO_3$ (Fe content 0.045) recorded at 100K in zero and in a 6T external field applied parallel to the γ-beam and corresponding hyperfine field distributions [6].

It is thus clearly established that the occurrence of paramagnetic-like clusters deep into the so-called ferromagnetic metallic regime provides a clear evidence of the two-phase character of the metallic state in the mixed valence manganite matrix. Indeed the Mn partial substitution by Fe gives rise to a clear competition between AF Fe-Mn and F Mn-Mn exchange interactions while the ferromagnetic character of the unsubstituted manganites is mainly due to double exchange between Mn^{3+}-Mn^{4+} ions [6].

4. Anti-site and Anti Phase Boundaries

The ideal double perovskites $A_2BB'O_6$ (A alkaline earth cation = Sr, Ba, Ca) exhibit a NaCl-type superlattice structure, in which the two cations B (Cr,Fe) and B'(Mo,W,Re) are expected to be ordered. This series was recently reinvestigated because Kobayashi et al [8] reported that Sr_2 FeMoO$_6$ is a half-metallic ferromagnet with a Curie temperature of about 415 K, significantly higher than for any mixed valence manganite. In addition it exhibits appreciable negative magnetoresistance at room temperature.

The Mössbauer technique is very useful for distinguishing localized and delocalized iron states, and for probing the influence of the cation environment around Fe nuclei. Thus, the presence of two iron components in both magnetic and paramagnetic states allows to estimate the cation disorder relative to the ideal NaCl-type order of Fe and B' in the double perovskite structure [9-10]. In addition, the isomer shift values allow the electronic configuration of iron to be estimated: it differs from $3d^5$ and the value is dependent on the alkaline earth cation and probably on the cationic disorder. Finally, we used a model to explain both the magnetization and zero-field and in-field Mössbauer data on the basis of structural defects with anti-site and antiphase boundaries [10]. Such defects are illustrated in Figure 6. Antiphase boundaries arise between two coherent crystallites with different starting sites for the NaCl-type order. They lead to planes of Fe-O-Fe bonds, which are strongly antiferromagnetic. These bonds are absent in the defect-free NaCl structure. The number of iron moments located at antiphase boundaries can be estimated from in field Mössbauer spectra because the antiferromagnetic coupling can be observed from the intermediate lines, in opposite to the iron moments which are ferromagnetically coupled to B' moments. Such a point is discussed in [10] in the case of Ca$_2$FeMoO$_6$.

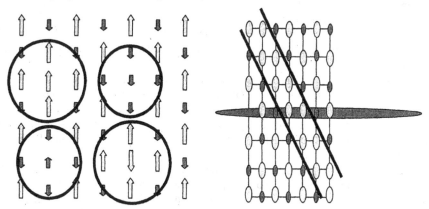

Figure 6. Schematic representation of antisite defects (left) and antiphase boundary (right).

5. Nanostructured Powders

The high energy milling procedure applied to micrometric particles of oxides lead to microstructural transformation with the appearance of nanocrystalline grains welded each other by grain boundaries, the thickness of which is strongly dependent of milling conditions. The magnetic properties are thus strongly dependent on first the structural nature of both the crystalline grains and of the grain boundaries. By analogy to ferric fluoride for which a complete description of the nanostructured phase has been described from both zero-field and in-field Mössbauer spectrometry on the basis of two components, antiferromagnetic crystalline grains and speromagnetic grain boundaries [11], a microstructural modeling of ground ferrites was attempted on $CdFe_2O_4$, $ZnFe_2O_4$ and $NiFe_2O_4$. The nanocrystalline grains exhibit usually sizes ranged from 5 to 20 nm. It is first clearly observed than the milling treatment favors the cationic redistribution. In the case of $CdFe_2O_4$, $ZnFe_2O_4$, it favors a transformation from AF (microcrystalline state) to ferrimagnetic behavior (nanostructured state), as clearly observed by in-field Mössbauer spectrometry [12]. To understand the temperature dependency of magnetic properties, a modeling of grain boundaries suggested that they rather consist of a random packing of octahedral units which behave a speromagnetic behavior [12]. Such approach might explain the superparamagnetic relaxation phenomena which are dependent on grinding conditions, giving rise to different blocking temperatures. Indeed the magnetic coupling between single domains grains is as effective as the thickness of grain boundaries is small, compared to the magnetic exchange length which is less than 1nm in those insulating systems [13]. Further experiments are in progress.

6. Magnetic Nanoparticles

Nanomagnetism became an important topic during the last decade: indeed, the magnetic properties of ultrafine particles are dominated by finite-size effects and interaction effects between particles, giving rise to either superparamagnetic effects or blocked magnetic states. The main characteristics of nanoparticle are based on the large surface-to-volume ratio which favors a significant increase of surface effects which result basically from the breaking of symmetry of the lattice. Consequently the magnetic arrangement originates from the competition between surface and core magnetic properties.

Numerous studies were devoted to maghemite (γ-Fe_2O_3) prepared by sol-gel process, in order to follow the evolution of magnetic behavior as a function of size, morphology and dispersion. Different time scale measurement techniques are used to investigate from an assembly of nanoparticles, the surface anisotropy, intraparticle exchange coupling between surface and core, and interparticle interactions. Reviews of experimental and theoretical aspects are given in [14-15].

Zero-field Mössbauer spectrometry allows to follow the magnetically blocked state into the superparamagnetic state and to paramagnetic state and to estimate the blocking temperature, to be thus compared to that deduced from magnetic measurements. Then, in-field Mössbauer spectrometry is an excellent tool to evidence for the average direction of the net magnetization of each particle: smaller the particle, higher the canting is, as it is shown in Figure 7. On the basis of a core-shell model, it allows to quantify the proportions of moments contained in the core and the external shell, for which the iron moments are collinear and oriented parallel to the applied field and are randomly oriented, respectively [16-18].

A schematic representation is given in Figure 8. The thickness of the magnetic surface layer can be estimated from the relative areas of intermediate lines. One gets $A_{2,5} = 1/2~((1-q)~\sin^2\vartheta_{core} + q~\sin^2\vartheta_{shell}) = 1/2~\sin^2\vartheta$ (Mössbauer) with $q = 1 - (1-e/r)^3$ where q and (1-q) represent the proportions of magnetic moments located in the shell and core, respectively, e the thickness of the external shell and r the radius of the particle. Assuming a perfect alignment of core iron moments and a random distribution of canted iron moments, $\vartheta_{core} = 0$ and $\vartheta_{shell} = 54.7°$, leading finally to $e = r/2~\sin^2\vartheta$. It is generally found a thickness of about 2 atomic layers, as expected.

Such an analysis provides relevant local information, allowing thus to perform Monte Carlo simulation, in order to follow the magnetic configurations, both in the core and in the outer shell, as a function of the size of the particle and of surface anisotropy [19-21].

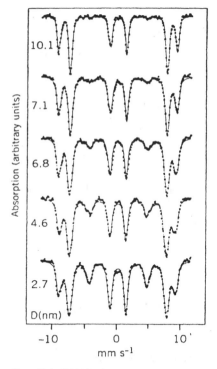

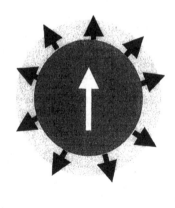

Figure 7. In-field Mössbauer spectra recorded at 10K on maghemite nanoparticles as a function of size (the 6 T field is applied parallel to the γ-radiation [18].

Figure 8. 2D representation of the core-shell model showing a collinear magnetic structure and a random orientation of moments for core and shell, respective.

7. Conclusions

This brief review demonstrates how the combination of zero-field and in-field Mössbauer spectrometry is powerful and provides relevant information which essentially concerns local magnetic aspects. It is also important to emphasize that further measurements are needful and that the fitting procedure remains an important stage in interpreting the hyperfine structure, especially when it is not resolved.

References

1. Goodenough, J.B. (1963) Magnetism and the Chemical Bond, Interscience, New York
2. Néel L. (1948) Théorie du Ferrimagnétisme, applications aux ferrites, *Annales de Physique* **3**, 137.
3. Yafet, Y. and Kittel, C. (1952) Triangular ferrimagnetic structures, *Phys Rev* **87**, 290.
4. Poole, C.P. Jr and Farach, H.A. (1982) Magnetic phase diagram of spinel spin-glasses, *Zeitschrift für Physik B-Condensed Matter* **47**, 55-7.
5. Helgason, Ö. , Greneche, J.-M., Berry, F. J., Mørup, S. and Mosselmans, F. (2001) Tin- and Titanium-doped γ-Fe$_2$O$_3$ (Maghemite) *J. Phys.: Condens. Matter* **13**, 10785-10797.
6. Hannoyer, B., Marest, G., Greneche, J.M., Ogale, S.B., Bathe, R. and Patil S.I. (2000) Colossal magnetoresistance and hyperfine interactions in iron doped La$_{0.75}$Ca$_{0.25}$MnO$_3$, *Phys. Rev. B* **61**, 9613-9620.
7. Barandiarán, J.M., Greneche, J.M., Hernández, T., Plazaola, F. and Rojo, T. (2002) Non conventional magnetic order in Fe substituted Ln$_{0.7}$Sr$_{0.3}$MnO$_3$ giant magnetoresistance manganites, *J. Phys Condensed Matter*, submitted.
8. Kobayashi, K. I., Kimura, T., Sawada, H., Terakura K., and Tokura, Y. (1998) Nature (London) **395**, 677
9. Pinsard-Gaudart, L., Suryanarayanan, R. , Revcolevschi, A., Rodriguez-Carvajal, J., Greneche, J.M., Smith, P. A. I., Tho-mas, R. M., Borges, R. P. and Coey, J. M. D. (2000) Ferrimagnetic order in Ca$_2$FeMoO$_6$, *J. Appl. Phys.* **87**, 7118.
10. Greneche, J.M., Venkatesan, M., Suryanarayanan, R. and Coey J.M.D. (2001) Mössbauer spectrometry of A$_2$FeMoO$_6$ (A = Ca, Sr, Ba): Search for antiphase domains, *Phys. Rev. B* **63**, 174403(5).
11. Guérault, H. and Greneche, J.M. (2000) Microstructural modelling of iron-based nanostructured fluoride powders prepared by mechanical milling, *J. Physics: Condensed Matter* **12**, 4791-4798.
12. Chinnasamy, C. N., Narayanasamy, A., Ponpandian, N., Chattopadhyay, K., Guérault H. and Greneche J.M. (2000) Magnetic properties of nanostructured ferrimagnetic zinc ferrite, *J. Physics: Condensed Matter* **12**, 7795-7805.
13. Greneche, J.M. (2002) Grain boundaries in magnetic Nanostructured Systems, *Czechoslovak Journal of Physics* **52** 139-144.
14. Dormann, J.L., Fiorani, D. and Tronc, E. (1997) *Adv. Chem. Phys.* **98**, 283.
15. Tronc, E., Ezzir, A., Cherkaoui, R., Chanéac, C., Noguès, M., Kachkachi, H., Fiorani, D., Testa, A.M.., Grenèche, J.M. and Jolivet, J.P. (2000) Surface-related properties of γ-Fe$_2$O$_3$ nanoparticles, *J. Magn. Magn. Mater.* **221**, 63-79.
16. Coey, J. M. D. (1971) *Phys. Rev. Lett.* **27**, 1140.
17. Kodama, R.H. and Berkowitz, A.E., McNiff, E.J. Jr. and Foner S. (1996) Surface Spin Disorder in NiFe2O4 Nanoparticles *Phys. Rev. Lett.* **77**, 394-397.
18. Tronc, E., Prené, P., Jolivet, J.P., Dormann, J.L. and Greneche J.M. (1997) Spin-canting in γ-Fe$_2$O$_3$ Nanoparticles, *Hyper. Inter.* **112**, 97-100.
19. Kodama, R.H. and Berkowitz, A.E. (1999) Atomic-scale magnetic modeling of oxide nanoparticles, *Phys. Rev. B* **59**, 6321-6336.
20. Kachkachi, H., Ezzir, A., Noguès, M. and Tronc, E. (1998) Eur. Phys. J. B 14 (2000) **81** 3976.
21. Labaye, Y., Crisan, O., Berger, L., Grenèche, J.M. and Coey, J. M. D. *(2002)* Surface anisotropy in ferromagnetic nanoparticles, *J. Appl. Physics* **91** 8715-8717.

MATTIS' MAGNETICS AND DISORDERED SYSTEMS

Theoretical Considerations, Experimental Detection, Local Magnetic Structure, Teperature Behavior

E.P.YELSUKOV[1], E.V.VORONINA[1], A.V.KOROLYOV[2] AND
G.N.KONYGIN[1]
[1] *Physical-Technical Institute UrB RAS, 426001, Izhevsk, Russia*
[2] *Institute of Metal Physics, UrB RAS, 620219, Ekaterinburg, Russia*

1. Introduction

For non-ordered Fe alloys with concentration x>45 at.% of Al and x_{Sn}>50 at.% of Sn the dependence of magnetic parameters on the concentration, temperature and applied magnetic field are not characteristic of ferromagnets. Data [1-4] on the magnetic behavior of thin amorphous Fe-Sn films, being the closest in properties to the disordered nanocrystalline alloys, indicates that at x_{Sn}> 60 at.% Sn the ground magnetic state is of spin glass (SG), and in the interval of x=55-60 at.% Sn one may assume the coexistence of ferromagnetism (F) and SG. As far as the Fe-Al system is concerned, it is impossible to draw a conclusion on the peculiarities of the magnetic structure of higher Al concentration disordered alloys from magnetic measurements data alone where only the average magnetic moment per Fe atom [5,6] is registered. The magnetic state [7] of the disordered $Fe_{60}Al_{40}$ alloy was also determined as spin-glass. A considerable difference in the metalloid concentration of the vanishing average magnetic moment and the average hyperfine field, and the failure of saturation in the magnetization curves in the high external magnetic fields up to 5T (for Fe-Al alloys according to [6] even up to 15 T) can be regarded as the experimental signs of non-ferromagnetic behavior in high-concentration Fe-Al and Fe-Sn alloys. Concurrently, for these alloys there are no observed conventional signs for the spin-glass state such as the shifted about H_{appl}=0 hysteresis loop or true temperature hysteresis in the magnetization curves. Thus, the magnetic structure of disordered nanocrystalline Fe alloys with Al and Sn needs to be specified.

For the systems with atomic and spin disorders, especially for the non-ordered metal systems with RKKY interaction, Mattis [8] considered the model in which spin-spin interactions were random in sign but did not result in the frustrations experienced with the Ising spin glass model supposing randomness of competitive exchange interaction - the main cause of frustrations. Mattis' theoretical model describes the structure of the ground state in which the magnetic moments are directed opposite to each other, randomly distributed over the lattice sites and the total magnetic moment $\overline{m} = 0$. Further

93

M. Mashlan et al. (eds.), Material Research in Atomic Scale by Mössbauer Spectroscopy, 93–104.
© 2003 *Kluwer Academic Publishers. Printed in the Netherlands.*

on we will refer to this magnetic structure as Mattis' spin glass. Besides, from the theoretical calculations [9] for non-ordered Fe alloys with metalloids it followed that the magnetic moment of the Fe atom with a certain number of metalloid atoms in the nearest neighborhood can orient in the direction opposite to that of the local magnetization. But Mattis' model does not include the dynamics of spin variances and the thermodynamic properties of such a system are characterized from itinerant magnetism. In particular, in [10] it was shown that in such a non-ordered system the temperature dynamics of the magnetic moments directed along (+) and opposite to the magnetization (-) was different due to heightening an effect of the Stoner type excitations. The average local magnetic moments (-) vanish at temperatures lower than the magnetic ordering temperature yielding a maximum on the temperature dependence of magnetization (Fig.1).

This attempt aims to characterize the ground state magnetic structure of disordered non-equilibrium Fe alloys with Al, Sn having a high metalloid content in terms of the Mattis' model, to trace and interpret their temperature dependency as an itinerant magnetism phenomenon proceeding from Mössbauer and magnetic measurements data versus temperature. For comparison purposes disordered Fe-Si alloys with ferromagnetic type of magnetic moments ordering [11-13] are considered.

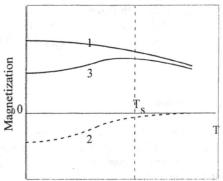

Figure1. The theoretical prediction of the temperature dependence of magnetization for Mattis' spin glass : 1 - $\overline{m}_+ = f(T)$, 2 - $\overline{m}_- = f(T)$, 3 - $\overline{m} = \overline{m}_+ + \overline{m}_- = f(T)$.

2. Experimental Details

The Fe-Al, Fe-Si and Fe-Sn ingots were alloyed from high-purity components (99.99 % Fe and 99.99 % sp-element) in an induction furnace in Ar atmosphere. They were homogenized in a vacuum furnace at 1400K over a period of 6h. The chemical analysis showed that after melting and homogenization the Al concentration in the alloys was 35, 46, 52 and 60 at.%, the Sn content 46, 55 and 62 at.% and Si 33,42 and 50 at.%. Nanocrystalline powders were obtained by mechanical grinding in a planetary ball mill. The method and the conditions of preparation, phase composition and structure are described in detail in [13-15]. The average size of the powder grains was 2-6 μm for all alloys.

Magnetic measurements were carried out using a SQUID-magnetometer in external magnetic fields up to 50 kOe at temperatures ranging from 5 to 300 K. Magnetization curves and hysteresis loops were measured, FC and ZFC experiments were carried out. From the extrapolation of the high-field section of the magnetization curve to $H_{ext} = 0$ we derived the value of specific magnetization σ_0, from which the average magnetic

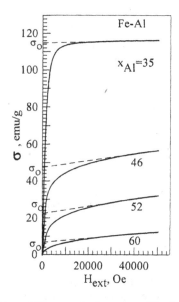

Figure 2. The magnetization curves of the disordered Fe-Al alloys.

moment per Fe atom $\overline{m}_{Fe}^{mm}$ ("mm" is for "magnetic measurement") was determined. The ^{57}Fe Mössbauer spectra were taken with a constant acceleration spectrometer, using ^{57}Co in the Cr matrix source in the temperature range of 6 to 380 K. The regularization method with parameters correction was used [16] to calculate the distribution P(H) of the hyperfine magnetic field (HFMF), for some spectra the optimal parameter regularization and distribution error were estimated. A slight linear dependence of isomer shift on HFMF $\delta=\delta(H)$ was taken into account. Calculating $\overline{H} = \int_0^{H_{max}} HP(H)dH$, we assigned H=0 to the interval of the HFMF from 0 to 35

kOe, in which P(H) actually corresponds to the part of the spectrum without hyperfine magnetic splitting. The limit of 35 kOe was chosen from the HFMF distribution of MS of the alloys that are paramagnetic. The average magnetic moments per Fe atom $\overline{m}_{Fe}^{ms}$ ("ms" is for "Mössbauer measurement") were estimated using the conversion factor $\overline{H}_{Fe}/\overline{m}_{Fe}$=120-130 kOe/$\mu_B$. From Mössbauer measurements the value representing the fraction of the non-magnetic part of the spectrum was also calculated. Additionally, the Mössbauer measurements assisted us in the estimation of the temperature of magnetic ordering T_C, at which the hyperfine magnetic splitting in the MS collapses.

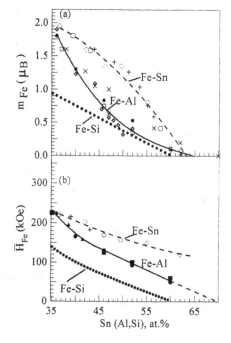

Figure 3. The average magnetic moment per Fe atom ($\overline{m}_{Fe}^{mm}$) from the magnetic measurements -(a) and the average HFMF $\overline{H}$ - (b) in the disordered nanocrystalline Fe-Al, Fe-Sn and Fe-Si alloys: □ (5K), ■ (6K) - the present paper; ▲ - [18]; ◊ - [6]; ● - [15]; x- [17], +, □-[1,2], O –[14].

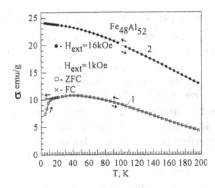

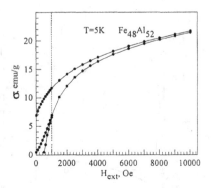

Figure 4. ZFC and FC temperature dependences of the magnetization of the $Fe_{48}Al_{52}$ alloy at $H_{ext} = 1$ kOe (1) and $H_{ext} = 16$ кOe (2).

Figure 5. The magnetization curve and part of the hysteresis loop of the disordered $Fe_{48}Al_{52}$ alloy; $T = 5$K.

3. Magnetic characterization

The magnetization curves measured at $T=5$K for some milled samples are shown in Fig.2. All of the Fe-Al and Fe-Sn samples, with the exception of $Fe_{65}Al_{35}$ and Fe-Si alloys, have a considerable slope in the high-field segment of the magnetization curves. The contribution of the process providing the slope of the magnetization curve into the total magnetization can be estimated from the ratio σ_0/σ_{50} (σ_{50} is the magnetization under $H_{ext} = 50$ kOe) (Tables 1, 2).

The calculated values of average magnetic moments per Fe atom $\overline{m}_{Fe}^{mm}$ and the data from [1,2,6,14,15,17] are presented in Fig. 3. The same figure presents the low-temperature values of the average HFMF $\overline{H}$ from the present work and refs. [1,2,14,15]. The difference in the Al, Sn concentrations at which $\overline{m}_{Fe}^{mm}$ and $\overline{H}$ vanish, is one of the signs of the non-ferromagnetic type of the magnetic structure at x > 40 at. %Al, Sn. The same can be inferred from the values of $\overline{H} / \overline{m}_{Fe}^{mm}$, which are 122 kOe/$\mu_B$ for the $Fe_{65}Al_{35}$ and 450 kOe/μ_B for the $Fe_{40}Al_{60}$ alloys or 1700 kOe/μ_B for $Fe_{38}Sn_{62}$. Note also two more important features of the magnetic properties of the alloys: independent of the magnetic pre-history of the sample, hysteresis loops are symmetric with respect to $H_{ext} = 0$. According to Beck [19] and Hurd [20], one of the signs of the cluster spin glass (CSG) or ideal spin glass (SG) is a shifted hysteresis loop, which was not observed in our experiments.

For all samples Fe-Al and Fe-Sn samples the FC and ZFC experiments were performed. The measurements of Fe-Sn alloys using $H_{ext} > H_C$, where H_c is the coercive force, showed the absence of hysteresis phenomena. In Fig.4 curves 1 show the hysteresis on heating and cooling in $H_{ext} = 1$kOe for the $Fe_{48}Al_{52}$ alloy. This fact seems to be evidence in favor of the spin-glass behaviour of the sample, which agrees with the study of the ball-milled $Fe_{60}Al_{40}$ [7]. However, note that the value of magnetization $\sigma = 10.2$ emu/g under $H_{ext} = 1$ kOe on the FC curve at $T=5$K agrees well with the value of $\sigma = 11.6$

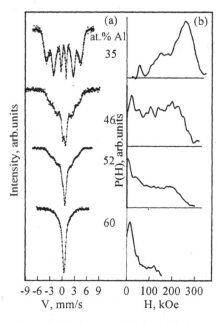

Figure 6. Mössbauer spectra (a) and distribution functions of HFMF P(H) (b) of Fe-Al alloys: $Fe_{54}Al_{46}$, $Fe_{48}Al_{52}$ at T = 6K and $Fe_{65}Al_{35}$, $Fe_{40}Al_{60}$ at 11K.

4. Mössbauer Spectroscopy Measurements

The Mössbauer spectra and the HFMF distributions P(H) for the temperatures T=6 K ($Fe_{54}Al_{46}$, $Fe_{48}Al_{52}$) and 11K ($Fe_{65}Al_{35}$, $Fe_{40}Al_{60}$) are presented in Fig.6. It is important to note that the shape of the MS of the disordered crystalline alloys agrees well with that of the thin amorphous films [3,8,12,17] except for the $Fe_{38}Sn_{62}$ alloy, which can be characterized as a partially disordered one.

Therefore, we conclude that under the given conditions of milling we obtain completely disordered nanocrystalline alloys. The non-magnetic component P(H=0) in MS is rising as the metalloid content in the alloys increases. The average magnetic moment per Fe atom

emu/g on the descending branch of the hysteresis loop (Fig.5). Analogously, the value of σ=7.2 emu/g on the ZFC curve (Fig.4) corresponds to the value σ=7.0 emu/g at H_{ext} =1 kOe of the magnetization curve and of the ascending branch of hysteresis loop (Fig.5). If FC and ZFC experiments are carried out at H_{ext} > 8 кOe, no temperature hysteresis is found as it is seen from Fig.4, curves 2 for the $Fe_{48}Al_{52}$ alloy under H_{ext} = 16 kOe. From these results we deduce that the real reason for the temperature hysteresis at H_{ext} < 8 кOe is the shape of the hysteresis loop but not the spin-glass state. All the data given above do not characterize the Fe-Al and Fe-Sn alloys with a metalloid content of more than 40 at.% either as a ferromagnetic or spin glass or cluster spin glass.

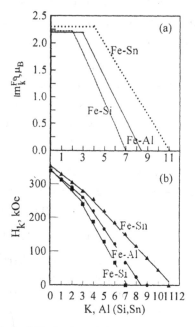

Figure 7. The local magnetic mo-ments models of Fe atom (m_k^{Fe}) and local HFMF at ^{57}Fe nucleus H_k.

$\overline{m}_{Fe}^{ms}$ estimated using the conversion factor $\overline{H}_{Fe}/\overline{m}_{Fe} = 120\text{-}130$ kOe/μ_B is given in Tables 1 and 2. This factor was derived from the low - temperature magnetic, neutron diffraction and Mössbauer spectroscopy data for the stoichiometric compounds of the Fe-Al, Fe-Si, Fe-Sn systems (100-160 kOe/μ_B).

5. The Local Atomic and Magnetic Structure

The analysis of the magnetic structures is supposed to be on the basis of a comparison of the experimental values of the average magnetic moment of the Fe atoms $\overline{m}_{Fe}^{mm}$ and $\overline{m}_{Fe}^{ms}$, obtained from the magnetic and Mössbauer measurements, with the values of $\overline{m}_{Fe}$ calculated in terms of the characteristics of the local atomic environment. In our earlier papers [13-15] we proposed the models of local atomic arrangement for the disordered metastable Fe-Al, Fe-Si and Fe-Sn alloys with the metalloid concentration of up to 60 at.%. These models describe quantitatively the concentration dependences of such magnetic characteristics as the $\overline{m}_{Fe}$ and $\overline{H}_{Fe}$. The models of local magnetic moments of Fe atoms built on the data for inter-metallic compounds and data on local hyperfine magnetic field for disordered alloys are given in Fig.7. These models differ from each other in the different number of metalloid atoms k_{cr} in the nearest neighborhood of an iron atom at which $m_k^{Fe} = 0$. It is evident from Fig.7 that k_{cr} differs in various models changing from 7 to 11. The values of the average magnetic moment per Fe atom $\overline{m}_{Fe}^{mm}$ calculated from σ_0 are presented in (Tabs. 1,2). We suppose that the values of $\overline{m}_{Fe}^{mm}$ obtained in such a way do not correspond to the real average magnetic moment per Fe atom in the alloys where σ_0/σ_{50} are lower than 1. Indeed, the values of the average magnetic moments per Fe atom according to the Mössbauer spectroscopy $\overline{m}_{Fe}^{ms}$ are much higher than those of $\overline{m}_{Fe}^{mm}$. The values of the magnetic moment of Fe atom calculated for the ferromagnetic ordering $\overline{m}_{Fe}(F)$, using the data on local magnetic moments and local atomic arrangement within the measurement error, correlate well with the data of the Mössbauer spectroscopy. But, both $\overline{m}_{Fe}(F)$ and $\overline{m}_{Fe}^{ms}$ are higher than $\overline{m}_{Fe}^{mm}$ at x > 40 at. % Al, Sn. For the Fe-Si alloys there is a quantitative agreement of $\overline{m}_{Fe}(F)$ and $\overline{m}_{Fe}^{mm}$ values, which allows considering the Fe-Si alloys as ferromagnetic materials. In [9] within the double-band Hubbard's model it was shown for non-ordered transition metal - metalloid systems, that the magnetic moment of the Fe atom surrounded by a certain number k of metalloid atoms orients opposite to the local magnetization. We calculated the average magnetic moment per Fe atom $\overline{m}_{Fe}$ for the opposite orientation of the magnetic moment of Fe atom with a certain number of Al and Sn atoms in the nearest surrounding k^-. Comparing them with the data obtained from the magnetic

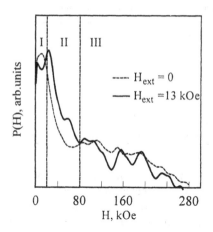

Figure 8. The distributions of HFMF P(H) for the Fe$_{54}$Al$_{46}$ alloy at T=80K, H$_{ext}$= 0 (a dashed line) and H$_{ext}$=13 kOe (a solid line).

measurements, we draw the conclusion that the best agreement is achieved with the opposite orientation of Fe magnetic moments in the local configurations with k$^-$=9 and more Sn atoms and with k$^-$=7 and more Al atoms in the nearest environment (Tabs). Some difference for the Fe$_{38}$Sn$_{62}$ alloy can be accounted for by an incomplete chemical disorder in the sample.

The Mössbauer measurements data in applied magnetic field for the Fe$_{54}$Al$_{46}$ alloy can be considered as proof of these results (Fig.8). The whole range of HFMF can be divided into 3 intervals: 0-20 (I), 20-80 (II) and higher than 80 kOe (III). From Fig. 9 it is evident that P(H) under H$_{ext}$≠ 0 is located below P(H) for H$_{ext}$ =0 in interval I from 0 to 20 kOe and higher in interval II from 20 to 80 kOe. That is, the external magnetic field decreases the fraction of the non-magnetic component in the MS and considerably increases the fraction of the magnetic component in the interval of HFMF from 20 to 80 kOe. It should be noted that this particular interval corresponds to the local HFMF of the Fe atoms having 7 and 8 Al atoms in the nearest environment (Fig.7) [15] and opposite directed magnetic moments are attributed exactly to these very Fe atoms. The magnetic structure with randomly arranged in the lattice and contra-oriented magnetic moments with zero total magnetization $\overline{m}_{Fe}^{mm}$ = 0 is Mattis' spin glass [8]. In our case the Fe$_{40}$Al$_{60}$ and Fe$_{38}$Sn$_{62}$ alloys are closest to the MSG. We suggest designating the other alloys with $\overline{m}_{Fe}^{mm}$ ≠0 (except for Fe$_{65}$Al$_{35}$) as the non-compensated MSG (NCMSG).

TABLE 1. Magnetic parameters of the Fe-Al alloys.

x, at.%	Fe$_{65}$Al$_{35}$	Fe$_{54}$Al$_{46}$	Fe$_{48}$Al$_{52}$	Fe$_{40}$Al$_{60}$
σ_0/σ_{50} (5K)	0.97	0.84	0.68	0.43
$\overline{m}_{Fe}^{mm}$, μ_B	1.60±0.03	0.70±0.03	0.34±0.03	0.11±0.03
$\overline{m}_{Fe}^{ms}$, μ_B	1.8±0.2	1.0±0.2	0.8±0.2	0.5±0.2
$\overline{m}_{Fe}$ (F), μ_B	1.8	1.1	0.7	0.3
$\overline{m}_{Fe}$, $(-m_{k\geq7}^-)\mu_B$	1.8	0.9	0.4	0.1
P^{exp}(H=0)/P^{calc}(H=0)	0.01/0.02	0.12/0.14	0.23/0.25	0.48/0.49

TABLE 2. Magnetic parameters of the Fe-Sn and Fe-Si alloys.

x, at.%	$Fe_{54}Sn_{46}$	$Fe_{45}Sn_{55}$	$Fe_{38}Sn_{62}$	$Fe_{67}Si_{33}$	$Fe_{58}Si_{42}$	$Fe_{50}Si_{50}$
σ_0/σ_{50} (5K)	0.95	0.88	0.59	0.99	0.97	0.86
$\overline{m}_{Fe}^{mm}$, μ_B	1.2	0.8	0.2	1.2	0.8	0.2
$\overline{m}_{Fe}^{ins}$, μ_B	1.4	1.3	1.0	1.4	0.9	0.2
$\overline{m}_{Fe}(F)$, μ_B	1.6	1.15	0.85	1.1	0.7	0.3
$\overline{m}_{Fe}(-m_{k\geq 9}^{Fe})$, μ_B	1.4	0.8	0.35			
$P^{exp}(H=0)/P^{calc}(H=0)$	0.04/0.02	0.07/0.06	0.06/0.16	0.04/0.07	0.2/0.34	0.9/0.61

In all the alloys beginning with a certain metalloid concentration the non-magnetic Fe atoms (NMFe) (their local magnetic moment is equal to 0) occur, which follows from the local magnetic moments models. It should be noted that the calculated values of fraction of non-magnetic Fe atoms are qualitatively consistent with the experimental ones estimated from Mössbauer spectra (Tabs). Using the values of k^- and k_{CR} we determined the ground state magnetic structures of the studied alloys of different metalloid concentration. The results of the calculation can be presented by the following scheme:

$$F \xrightarrow{35} NCMSG \xrightarrow{40} NCMSG + NMFe \xrightarrow{62} MSG + NMFe \xrightarrow{80} P , x, at.\% \, Al$$

$$F \xrightarrow{35} NCMSG \xrightarrow{45} NCMSG + NMFe \xrightarrow{72} MSG + NMFe \xrightarrow{90} P , x, at.\% \, Sn$$

$$F \xrightarrow{25} F + NMFe \xrightarrow{70} P , x, at.\% \, Si .$$

6. The temperature behavior of the magnetic structure

To obtain information on both the magnetic structure and its evolution with temperature the magnetic and Mössbauer measurements for all alloys were carried out. For some Fe-Al and Fe-Sn alloys curves the maximum is found in the $\sigma(T)$. It is clearly evident for the $Fe_{48}Al_{52}$ alloy presented in Fig.5 at T = 40K for H_{ext} = 1 kOe but it is absent for H_{ext} =16 kOe. For the $Fe_{54}Al_{46}$ alloy this behavior is illustrated by the FC curves (Fig. 9(a)) for H_{ext} ranging from 0.6 to 10 kOe. For the ferromagnetic disordered $Fe_{65}Al_{35}$ alloy no maximum were found (Fig. 9(b)).

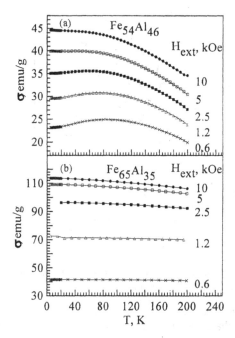

Figure 9. Temperature dependences of the magnetization of $Fe_{54}Al_{46}$ (a) and $Fe_{65}Al_{35}$ (b) alloys for the different values of the external magnetic field.

As an example, Fig. 10(a) presents the Mössbauer spectra and calculated P(H) distributions for the $Fe_{54}Al_{46}$ alloy at different temperatures. First of all, it should be noted that the non-magnetic component in the MS arises at rather low temperatures long before the temperature at which MS transforms into a single line. Thus, the decrease of $\bar{H}$ with temperature is not only the result of the monotonic shift of P(H) to the lower H values but it is also due to the growth of the non-magnetic component of MS. For the ferromagnetic disordered $Fe_{65}Al_{35}$ alloy (Fig.10,b) (at all temperatures of measurements) the non-magnetic component in the MS is not found and consequently, the decrease of $\bar{H}$ with temperature is caused only by the P(H) shift into the low-values range. In Fig.11 the dependences of the reduced average HFMF $\bar{H}$ (T)/H(0) and the fraction of the non-magnetic component in the MS P(H=0) on the reduced temperature T/T_c for Fe-Al and Fe-Sn systems are shown in comparison with the same dependence for the ferromagnetic Fe_3Al [21], ferromagnetic Fe_3Sn_2 [22] and anti-ferromagnetic $FeSn_2$ intermetallic compounds [23] (by a solid line) We have not show the temperature behavior of these two parameters for the Fe-Si alloys because it is analogous to the $Fe_{65}Al_{35}$ alloy The given experimental data $\bar{H}$ (T)/H(0) for all alloys, except for the $Fe_{65}Al_{35}$ alloy, are much lower in comparison with the solid line in the range $0.07 < T/T_c < 1$ for the Fe-Al and in $0.3 < T/T_c < 1$ - Fe-Sn alloys. This difference is due to the increase of the fraction of the non-magnetic component in the MS P(H=0)-P_0 with the temperature rise. P_0 - is the fraction of the non-magnetic component in the MS obtained by extrapolation of P(H = 0) to 0 K. On the face of it, the growth of the non-magnetic component at $T \ll T_c$ points to a superparamagnetic behavior of the alloys studied. But, as it was mentioned above, the powders have an average particle size of 2-6 μm with a minimum value of 1 μm and thus, the particles of this size can hardly behave like superparamagnetic ones. In [24], the emergence of the non-magnetic component in the MS of the milled intermetalloid $FeSn_2$ at room temperature was accounted for by unison spin fluctuations that are due to the nanometric grain size. Nevertheless, in the $Fe_{65}Al_{35}$ alloy, with an average grain size of 6 nm, the non-magnetic component in MS is not revealed up to $T/T_c = 0.72$ (Fig. 10(b)).

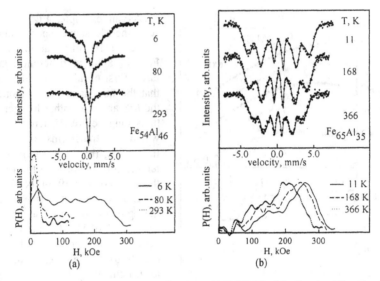

Figure 10. Mössbauer spectra and distribution functions of HFMF P(H) of the disordered $Fe_{54}Al_{46}$ alloy (a) and $Fe_{65}Al_{35}$ (b) at different temperatures.

On the basis of theoretical considerations [25], we believe that the temperature behavior of the Mössbauer and magnetic characteristics is conditioned by short-wave spin excitations (SWSE) mainly of the Stoner type rather than by superparamagnetism. If we suppose the SWSE to be primarily connected with the Fe atoms surrounded by a great number ($k \geq k^-$) of metalloid atoms [10], which, consequently, have magnetic moments aligned conversely to the magnetization, we can account for the maximum in the dependences $\sigma(T)$ under $H_{ext} < 5$ kOe (Fig. 4 and 9). The experimental points in Fig.11 for the $Fe_{54}Al_{46}$, $Fe_{48}Al_{52}$ and $Fe_{40}Al_{60}$ alloys can be approximated by a complex curve with the inflection point $T_S = 0.3\ T_c$. For the $Fe_{54}Al_{46}$ alloy we obtained the value $T_S = 108$ K, for the $Fe_{48}Al_{52}$ alloy $T_S = 33$K. The values obtained are in good agreement with the position of maxima in the dependences $\sigma(T)$ in Figs.9 and 4, accordingly. It implies that at the temperature T_S the average value of the local magnetic moments goes down to zero $\left(m_{k \geq k^-} \right)_{T_S} = 0$ because of SWSE of the Fe atoms with the magnetic moments directed conversely to the magnetization. Then, T_S can be regarded as the temperature of transition from NCMSG state to ferromagnetic or paramagnetic ones.

7. Conclusions

The analysis of the Mössbauer and magnetic measurements data allows classifying the ground state of the disordered Fe-Al and Fe-Sn alloys with a high metalloid content as Mattis' spin glass type magnets, with Fe-Si alloys being non-ordered ferromagnets for all the alloys studied. The main feature of this structure is the magnetic moments randomly distributed over the lattice, different in magnitude and aligned conversely to each

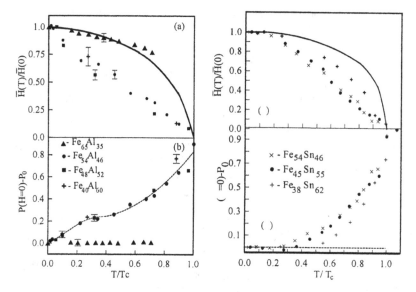

Figure 11. The dependence of the reduced average HFMF ($\overline{H}$ (T)/H(0)) - (a) and the reduced fraction of non-magnetic component in MS (P(H = 0) - P_0) - (b) depending on the reduced temperature (T/T_C): solid lines - ferromagnetic ordered Fe_3Al alloy [21]; ferromagnetic Fe_3Sn_2 [22] and antiferromagnetic $FeSn_2$ intermetallic compounds [23].

other. The temperature behavior of the magnetic characteristics of these systems is conditioned by short- wave spin excitations, which are most significant for the Fe atoms with the great number of metalloid neighbors.

Acknowledgement

The authors express their gratitude to Prof. S.K.Godovikov for the help in low-temperature Mössbauer experiment and to Prof. A.K. Arzhnikov for useful arguments during the discussions of the results of this work.

References

1. Mangin,Ph., Piecuch, M., Marchal, G., Janot, Chr. (1978) About the magnetic behaviour $Fe_x Me_{1-x}$ (Me= Si,Ge,Sn) amorphous alloys, *J.Phys.F: Metal Phys.* **8**, 2085-2092.
2. Rodmacq, B., Piecuch, M., Janot, Chr., Marchal, G., Mangin, Ph. (1980) Structure and magnetic properties of amorphous Fe_xSn_{1-x} alloys, *Phys.Rev.B* **21**, 1911-1923.
3. Tierlinck, D., Piecuch, M., Geny, J.F., Marchal, G., Mangin, Ph., Janot, Chr. (1981) Magnetic phase diagram in Fe_xSn_{1-x} amorphous alloys , *IEEE Trans. Magn.* **17**, 3079-3081.
4. Piecuch, M., Janot, Chr., Marchal, G., Vergnat , M. (1983) Magnetic behaviour of Fe_xSn_{1-x} amorphous alloys near the critical composition, *Phys.Rev.B* **28**, 1480-1489.
5. Gengnagel, H., Besnus M.J., Danan, H. (1972) Temperature and field dependence of magnetization of Fe-Al powders in cold-worked and annealed state, *Phys.Stat.Sol.(a)* **13**, 499-503.
6. Besnus, M.J., Herr, A., Meyer, A.J.P.(1975) Magnetization of disordered cold-worked Fe-Al alloys up to 51 at.% Al, *J.Phys.F: Metal Phys.* **5**, 2138-2147.

7. Amils, X., Nogués, J., Suriañch, S., Baró, M.D., Muñoz, J.S. (1998) Magnetic properties of ball milled Fe-40 Al at.% alloys, *IEEE Trans.Magn.* **34,** 1129-1131.
8. Mattis, D.C. (1976) Solvable spin systems with random interactions, *Phys.Lett* **56A,** 421-422.
9. Arzhnikov, A.K., Dobysheva, L.V. (1992) The formation of the magnetic moments in disordered binary alloys of metal-metalloid type, *J.Magn.Magn.Mater.* **117,** 87-92.
10. Arzhnikov, A.K., Dobysheva, L.V. (1994) The Stoner excitations in disordered metal-metalloid alloys, *Phys.Letters* **A195,** 176-180.
11. Mangin, Ph., Marchal, G. (1978) Structure and magnetic properties in Fe_xSi_{1-x} amorphous alloys, *J.Appl. Phys.* **49** 1709-1711.
12. Bansal, C., Campbell, S.J., Stewart, A.M.(1982) Mössbauer and magnetic resonance experiments on amorphous iron-silicon films, *J.Magn.Magn.Mater.* **27,** 195-201.
13. Elsukov, E.P., Konygin, G.N., Barinov, V.A., Voronina, E.V. (1992) Local atomic environment parameters and magnetic properties of disordered crystalline and amorphous iron-silicon alloys, *J.Phys.: Condens. Matter.* **4,** 7597-7606.
14. Yelsukov, E.P., Voronina, E.V., Konygin, G.N., Barinov, V.A., Godovikov, S.K., Dorofeev, G.A., Zagainov, A.V. (1997) Structure and magnetic properties of $Fe_{100-x}Sn_x(3.2<x<62)$alloys obtained by machanical milling, *J.Magn.Magn.Mater.* **166,** 334-348.
15. Yelsukov, E.P., Voronina, E.V., Barinov, V.A. (1992) Mössbauer study of magnetic properties formation in disordered Fe-Al alloys, *J.Magn.Magn.Mater.* **115,** 271-280.
16. Nemtsova, O.M., Ageev, A.L., Voronina, E.V. (2002) The estimation of the error of the hyperfine interaction parameter distribution from Mössbauer spectra, *NIM B* **187,** 132-136.
17. Shiga, M., Kikawa, T., Sumijama,K., Nakamura, Y. (1985) Magnetic properties of metastable Fe-Al alloys produced by vapor quenching, *J.Magn.Soc.Jap.* **9,** 187-190.
18. Huffman, G.P., Fisher, R.M. (1967) Mössbauer studies of ordered and cold-worked Fe-Al alloys containing 30 to 50 at.% aluminum, *J.Appl.Phys.* **38,** 735-742.
19. Beck, P.A. (1971) Some recent results on magnetism in alloys, *Met.Trans.* **2,** 2015-2024.
20. Hurd, C.M. (1982) Varieties of magnetic order in solids, *Contemp.Phys.* **23,** 469-493.
21. Stearns, M.B. (1968) Internal field variations with temperature for the two sublattice ordered Fe_3Al and Fe_3Si, *Phys.Rev.* **168,** 588-592.
22. LeCaër, G., Malaman, B., Roques, B. (1978) Mössbauer effect study of Fe_3Sn_2 *J.Phys.F: Met.Phys.* **8** 323-326.
23. Le Caër, G., Malaman, B., Venturini, G., Fruchart, D., Roques, B. (1985) A Mössbauer study of $FeSn_2$, *J. Phys.F: Met.Phys.* **15,**1813-1827.
24. Sánchez, F.H., Socolovsky, L., Cabrera, A.F., Mendoza-Zelis, L. (1996) Magnetic relaxations in mechanically ground $FeSn_2$, *Mater.Sci.Forum* **225-227,** 713-718.
25. Arzhnikov, A.K., Vedyaev, A.V. (1988) Coherent potential approximation for Heisenberg ferromagnet diluted by nonmagnetic impurities, *Theor.Math.Phys.* (Russia) **77** 440-449.

HYPERFINE MAGNETIC STRUCTURE AND MAGNETIC PROPERTIES OF INVAR Fe-Ni-C ALLOYS

V.M. NADUTOV, YE.O. SVYSTUNOV, T.V. YEFIMOVA AND A.V. GORBATOV

G.V. Kurdyumov Institute for Metal Physics of the N.A.S. of Ukraine, Kyiv, Ukraine, nadvl@imp.kiev.ua

Abstract

Fcc Fe-Ni-C alloys in comparison with binary Fe-Ni alloys were studied by the Mössbauer spectroscopy and magnetic methods. The saturation magnetization and the magnetic susceptibility were measured for the alloys within a temperature range of 77 – 450 K. The martens tic points and Curie temperatures were obtained. A supplement to the Window method for the fitting of asymmetrical Mössbauer spectra of invar alloys is proposed. The probability distribution $P(H)$ was determined and isomer shifts were estimated.

1. Introduction

Invar is an alloy of 65at% Fe and 35at% Ni, which was discovered and named by Guillaume in 1897 [1]. The well known distinctive property of Invar is its low thermal expansion coefficient (TEC) in a wide temperature range [2]. Invar is of technical importance because of zero thermal expansion near to room temperature.

The small TEC of this alloy is due to the large positive spontaneous volume-magnetostriction below the Curie temperature [2,3]. The magnetically induced expansion balances the normal lattice contraction with decreasing temperature giving almost zero net thermal expansion. One of the mechanisms is based on the assumption that some iron atoms in configurations described in [4] may loose their localized effect with increasing temperature and become non-magnetic above the Curie temperature and this collapse may be accompanied by a decrease of the atomic volume.

A number of models for the Invar anomalies are based on the analysis of inhomogeneities in the magnetic structure of the Fe-36%Ni alloy. The Mössbauer effect with ^{57}Fe has been used to study the distribution of the internal magnetic field in Invar [5,6], which has been correlated to the distribution of the exchange field [5]. The lower run of reduced magnetization vs. reduced temperature compared to the Brillouin type curve has been discussed on the basis of local fluctuations of the iron moment, which reduces from 2.8 μ_B to lower values keeping the ferromagnetic state. Then the

M. Mashlan et al. (eds.), Material Research in Atomic Scale by Mössbauer Spectroscopy, 105–116.

antiferromagnetic state (or paramagnetic state) is found at places with Fe richer concentrations where, however, the major part of iron moments is reduced to smaller values, rather than the appearance of the antiferromagnetic clusters [5].

The Mössbauer effect has been studied for $(Fe_{1-x}Ni_x)_{97}C_3$ with $0.28 \leq x \leq 0.36$ and $Fe_{65}(Ni_{1-x}Mn_x)_{35}$ with $0.05 \leq x \leq 0.3$ alloys at liquid He temperatures in order to know the mechanism of the abrupt decrease of the spontaneous magnetization near the Invar region [4]. In order to stabilize the fcc phase at low temperatures, alloying with C and Mn was used. The distribution of the internal field is characterized by two distinguishable contributions, one is a larger internal field and the other is the smaller one, which may originate from Fe atoms having about 2.8 μ_B atomic moment and other with zero moment, respectively. The Mössbauer spectra under an external field of 50 kOe revealed that the Fe atoms have two electronic states ($\mu = \sim3$ μ_B and $\mu = \sim0$) as proposed by Weiss [7]. One assumed that part of Fe atoms becomes to the non-magnetic state depending upon their nearest neighborhoods [4].

Alloying with carbon is an effective factor to control magnetic structure in the Fe-Ni alloys. The Mössbauer study of Fe-Ni-C invar has shown the essential dependence of the distribution of internal magnetic field on the extent of carbon dissolution [8.9]. The distribution of internal fields shows the narrow low-field component and the broadened high-field contributions, which are attributable to antiferromagnetic and ferromagnetic interactions respectively.

Mössbauer spectra of the Fe-28%Ni-3%C alloy has been measured for the alloy at 22 K with longitudinal external field, $H_{ext} = 50$ kOe [10].. To improve the fitting of these spectra by Window's model the authors [10] assumed that some fraction of the Fe atoms exhibiting large internal field are ferromagnetically coupled and align collinear to the external magnetic field and another fraction of the Fe atoms exhibiting small internal fields tend to align perpendicular to the H_{ext}.

It should be noted that the distribution of internal field was obtained in Refs [4,8-10] within models assuming that other parameters are the same for all spectra components and the shape of a spectrum is symmetrical. Contrary to these assumptions, the spectra of Fe-Ni-C and even Fe-Ni alloys exhibit a distinct asymmetry, which should be taken into account. One of the reasons for this asymmetry is the different isomer shift of the subspectra caused by different nearest neighborhoods.

Thus the main goal of the work reported here is to study the effect of C on the Curie point, isomer shift and distribution of internal magnetic fields in fcc Fe-Ni-C invar alloys using magnetic methods and Mössbauer spectroscopy. For this purpose we studied the alloys with reduced Ni concentration (30%) in comparison with the base Fe-36%Ni invar.

2. Experimental Details

The Fe-25.3%Ni-0.78%C, Fe-30.3%Ni, Fe-36.0%Ni, Fe-36.1%Ni-0.55%C, Fe-29.7%Ni-0.97%C, Fe-30.5%Ni-1.5%C, Fe-30.2%Ni-0.8%Mn-1.15%C (wt.%) alloys were melted in vacuum induction furnace in protective argon atmosphere. The ingots were aged at 1000°C during 3 hours. The metal plates of 80-100 μm thick solution treated at 1100°C in silica slag and subsequently quenched in water. The samples were 20-25 μm foils obtained by chemical aching. The carbon concentration was determined by chemical analysis and the content of nickel was obtained by means of X-ray fluorescence analysis. The phase content of the alloys was controlled by X-ray analysis.

Mössbauer spectra of the alloys were obtained at room temperature on NP255 spectrometer (KFKI production, Hungary) and MS1101E spectrometer (MosTech Company, Russia). The isotope ^{57}Co in Cr matrix with activity of 25÷50 mCi was used as gamma quantum source.

The spectra were stored in a multichannel scaler with 512 channels. Velocity calibration was performed at room temperature with a α-Fe foil and sodium nitroprusside. The line width for sodium nitroprusside was 0.2 mm/s in the velocity range ±10 mm/s. The isomer shifts are given with respect to α-Fe.

Low field AC magnetic susceptibility was measured by induction method on samples of 1.8x1.8x10.5 mm in size. The magnetic field amplitude was 400 A/m and frequency was 1 kHz. The temperature of samples was varied within the 77 – 450 K interval with 3-5 K/min. Thermocouple was contacted with a sample.

The saturation magnetization was measured by means of ballistic magnetometer in magnetic field H = 800 kA/m within the temperature range 77 – 450 K on samples 1.8x1.8x10.5 mm in size.

3. Approximation of Spectra

Using the standard discrete fitting procedure, some Mössbauer spectra with resolved components were fitted. In order to fit the Mössbauer spectra with an asymmetrical distribution of hyperfine parameters, the Window method [11,12] and supplementary the program complex **"FUN-DENS"** is used here.

The superposition of simple Zeeman sextets, determined by $f(H, \delta, v)$ function, is written as:

$$I(v) = \int_H p(H)f(H,\delta,v)dH \quad , \tag{1}$$

where $I(v)$ is the intensity of spectrum in channel v; $p(H)$ is the probability of existence of a sextet with magnetic field H; δ is the isomer shift.

Window proposed to present the distribution function $p(H)$ as follows:

$$p(H) = \sum_{i=1}^{n} a_i(\cos(i\pi H / H_{max}) + (-1)^i) \quad , \tag{2}$$

where the number of components of sum n has meaning the number of "elementary" quasi-sextets which are substituted into experimental spectrum. The unknown coefficients a_i are calculated using the minimum of sum:

$$S = \sum_v \left(I(v) - I^*(v)\right)^2 \rightarrow min \tag{3}$$

where $I^*(v)$ is the intensity of experimental spectrum in channel v. The least-square method is used to find minimum of this function. We obtain a system of n linear equations:

$$\frac{dS}{da_i} = 0 \tag{4}$$

To solve this system the Gauss method with defining principle elements was applied [13].

For this step of fitting procedure the background I_{cont} was estimated as an arithmetic mean of 100 points (50 first points from each side of an experimental spectrum). At first, the isomer shift δ is taken as being the same for all sextets and its value is taken from calibration. We assume that Γ_i is the same for all sextets and is taken as the natural lines width or from calibrating values for a given spectrometer. Moreover the Γ_i value could be different between the lines in a sextet. The intensity of line ε_i is calculated from the relation: $\Gamma_1\varepsilon_1 : \Gamma_2\varepsilon_2 : \Gamma_3\varepsilon_3 = 3 : 2 : 1$.

After finding solutions (coefficients a_i) we calculated the function of density $p(H)$ by formula (2) and simulate spectrum using (1). The Simpson method was used for integration in (1). The quality of fitting is estimated by calculating the standard χ^2 value:

$$\chi^2 = \frac{\sum_v \left(I(v) - I^*(v)\right)^2 / I(v)}{R - n}, \tag{5}$$

where R is the number of experimental points. If $\chi^2 \gg 1$ we should repeat this program with greater n. If $\chi^2 < 1$ then the smoothness of the simulated spectrum is too small and it is follows statistical fluctuations, so that we have to repeat this program once again with smaller n. We found that for all our spectra consisting of 512 points and for $H_{max} = 36$ T the optimal value $n = 35$ - 36.

The Window method normally fits well with the symmetrical distribution functions. The spectra obtained from carbon-containing Fe-Ni-C-based alloys are usually asymmetrical [8-10] and after fitting the value $\chi^2 > 1$. In order to restore values of parameters on asymmetrical experimental spectra we supplemented the $P(H)$ method by the procedure of isomer shifts variation. This is because carbon considerably changes the isomer shift of the subspectra.

The scheme of this procedure is as follows. The $p(H)$ distributions function obtained in the first step is fitted by Gaussians using the least-square method. In our case it was the Origin program. The isomer shifts of "elementary" quasi-sextets within one Gaussian group are accepted to be the same. The number of these groups is denoted N.

In order to find out the optimal set of δ the exhaustive search method was used. We vary the set of δ in definite limits with the given step. For every variant we simulate the spectrum with a certain set of δ and find χ^2. A criterion for optimal set of δ is the closeness of χ^2 to 1 ($\chi^2 - 1 \longrightarrow 0$). It is a time consuming process due a number of variants requiring search. The number of variants is determined using formula obtained on the basis of the methods of combinatory [14]:

$$QV = \left(\frac{2 * \delta_{dev}}{\delta_{step}} + 1 \right)^N \tag{6}$$

where δ_{dev} is the maximal deviation of δ, δ_{step} is the step on δ. These values were chosen for physical reasons.

In order to find a better fitting spectrum and to find out the final result on $p(H)$ distribution the standard Window procedure described above using the set of optimal values of δ is repeated.

4. Results and Discussion

4.1. Fe-25.3%Ni-0.78%C

The Fe-25%Ni alloy has high martensitic point M_s that prevents remaining austenitic state at room temperature. Alloying with carbon decreases M_s. Mössbauer spectra of the Fe-25.3%Ni-0.78%C alloy taken at room temperature after solution treatment at 1100°C and ageing at 500°C are shown in Fig. 1.

The conventional approximation shows that the spectra consist of two subspectra: the doublet and broadened central line. The doublet is attributed to Fe atoms with one C atom or atom pair as the first nearest neighbors (n.n.) (Fe_C). The central line is related to Fe atoms with no C n.n. (Fe_0) and described by low field sextet since its broadening results from magnetic ordering [15]. The isomer shift of doublet δ_{Fec} is positive as compared to that of sextet δ_{Fe0} and the difference $\delta_{Fec} - \delta_{Fe0} = \Delta\delta = 0.031$ mm/s, that points to decreasing the s-electron density at the Fe nuclei by C n.n..

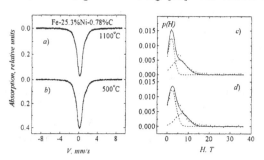

Figure 1. Mossbauer spectra of the Fe-25.3%Ni-0.78%C alloy after solution treatment at 1100°C (a) and ageing at 500°C (b). The hyperfine field distribution curves $p(H)$ (c,d).

We treated the asymmetrical spectra of the Fe-25.3%Ni-0.78%C alloy by the proposed program and $p(H)$ distribution was obtained (Fig. 1 c,d). Two sets of isomer shifts were found. The low field components in the spectrum have higher δ than high field subspectra. The fitting was good enough when the difference $\Delta\delta = 0.018$ mm/s. This is qualitatively consistent with that derived from the discrete approximation model mentioned above. This shows the validity of restoration of δ parameters from asymmetrical spectrum using the developed Window method.

The doublet with the larger quadruple splitting, which could be attributed to ordered structure of $(FeNi)_4C$ type (($FeAl)_4C$ in case of Fe-Al-C austenite [16]) was not revealed in the spectrum using both approximation approaches. The experimental and theoretical

110

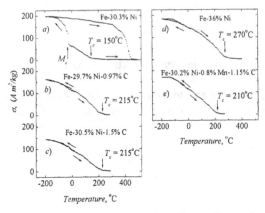

Figure 2. Temperature dependence of saturation magnetization of the alloys.

analysis has shown that the distribution of C atoms in the Fe-25.3%Ni-0.49%C austenite is not random and is characterized by existence of C-C clusters [15]. The shape of the spectrum was not changed even after ageing at 500°C (Fig. 1 b). The intensity of the doublet from Fe_C decreases under ageing at 800°C [17]. This is accounted for by clustering of carbon and/or the formation of graphite [8,17].

Clustering in Fe-Ni-C austenite occurs due to soft C-C interaction in the first and second coordination shells (the energies $w_1 \approx -0.015 \div 0.045$ eV, $w_2 > 0.08$ eV) [18,19] and a stronger repulsion of C atoms by Ni compared to Fe, characterized by the difference of energies $V_{Ni-C} - V_{Fe-C} \approx 0.07 \div 0.12$ eV [15,20] which is consistent with the increase of the thermodynamical carbon activity by Ni in f.c.c. Fe [21].

4.2. Fe-36.0%Ni

The magnetic transition in the Fe-36.0%Ni invar derived from temperature dependence of saturation magnetization is close to 270°C (Fig. 2 d), which is higher by 8-10% than Curie point (250°C, [2]) and explained by effect of external magnetic field [22].

Mössbauer spectrum of the alloy at room temperature shows the smeared and broadened lines (Fig. 3 a). The Window and/or Hesse-Rübartsch's methods are usually used to obtain the distribution of internal magnetic field [11,12]. Since the obtained spectra of the Fe-36.0%Ni alloy show small asymmetry we analyzed them within the proposed approximation model described above in order to obtain data on p(H) distribution and isomer shifts δ.

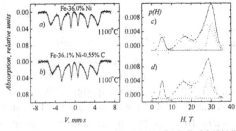

Figure 3. Mossbauer spectra of Fe-36.0% Ni (a), Fe-36.1% Ni-0.55% C (b) alloys after solution treatment. The hyperfine field distribution curves (c, d).

As seen from Fig. 3c, the distribution of the internal field can be divided in to four contributions: the low-, middle- and two high-field ones. However, unlike previous investigations, we obtained the δ values in addition. Three sets of isomer shift, δ = 0.009 mm/s, δ = -0.003 mm/s and δ = 0.061 mm/s were found for low-, middle- and high-field subspectra (Table 1). It was suggested that two high-field groups (Fig. 3 c) have the same isomer shift $\delta = 0.061$ mm/s.

Thus, there are at least three neighborhoods in invar. The low-field and middle-field groups with small δ are attributed to iron atoms in configurations with Fe-Fe bonds or in configurations with less than three Ni n.n. [4]. The high-field components belong to atomic configurations where the Fe-Ni bonds are favored to Fe-Fe ones. As it was considered in [4,5] a Fe atom becomes magnetic with high magnetic moment $\mu = 2.8$ μ_B when it has 3 Ni n.n., and at least 3 magnetic Fe atoms [4] or it is surrounded by less than 9 n.n. iron atoms [5]. Since the high-field subspectra have positive isomer shift ($\delta = 0.061$ mm/s) with respect to that of low- and middle-field ones ($\delta = 0.009$ mm/s and

TABLE 1. The isomer shift of subspectra with respect to α-Fe.

Alloys (wt.%)	IS (mm/s)
Fe-36.0%Ni	0.009
	-0.003
	0.061
Fe-36.1%Ni-0.55%C	0.058
	0.035
	0.055
Fe-30.3%Ni	-0.029
	0.0
Fe-29.7%Ni-0.97%C (1100°C)	0.040
	0.081
(500°C)	-0.032
	0.003

$\delta = -0.003$ mm/s), it means that Ni n.n. decreases the s-electron density at the nuclei of magnetic Fe atoms in configurations enriched by nickel. This suggests redistribution of s-electrons from Fe to Ni n.n.

Thus, if the description of the $p(H)$ distribution is correct (Fig. 3 c) the addition of C into the Fe-36%Ni invar should change the isomer shift of the low- and middle-field components.

4.3. Fe-36.1%Ni-0.55%C

Alloying with C of invar did not change in general the structure of the $p(H)$ distribution. It increases the low and middle field contributions (Fig. 3 d). The isomeric shifts of these components became positive with respect to pure iron, $\delta = 0.058$ mm/s, $\delta = 0.035$ mm/s, the value of δ for large fields is changed small (Table 1). This result means that carbon atoms occupy positions in Fe neighborhoods or in atomic configurations when an iron atom is surrounded by less than three Ni n.n. [4]. The inhomogeneous distribution of C atoms in Fe-Ni invar is qualitatively consistent with increasing carbon activity in austenite by Ni [21] and with the sign of difference of energies $V_{Ni-C} - V_{Fe-C} \approx 0.07 \div 0.12$ eV [15,20].

The values of δ show, that carbon decreases the s-electron density at the nuclei of n.n. Fe atoms in occupied atomic configurations that is consistent with the literature (see

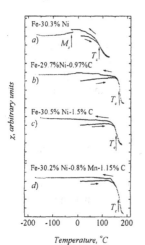

Figure 4. Temperature dependence of magnetic susceptibility of the alloys.

e.q. [15] and references). It does not affect the density of *s*-electrons at iron nuclei surrounded by Ni n.n.

4.4. Fe-30.3%Ni

The Fe-30.3%Ni alloy heated at 1100°C and water quenched has stable f.c.c. structure at room temperatures. This is due to low martensitic point $M_s \approx$ -30°C (Fig. 4). The saturation magnetization shows higher $M_s \approx$ -25°C (Fig. 2). Increasing the martensitic point in the magnetic field is consistent qualitatively with the results [23]. The Curie point of the alloy obtained from the magnetic susceptibility data is 85°C (Fig. 2 *a*) and the average magnetic transition temperature derived from magnetization curve is 150°C (Fig. 3 *a*). Such a difference is explained by the influence of a magnetic field on the magnetic transition temperature [22].

Mössbauer spectrum of the Fe-30.3%Ni alloy at room temperature shows rather symmetrically broadened single line with very weak satellite lines (Fig. 5 *a*). Such a shape of the spectrum results from weak magnetic ordering at room temperature, which differs from developed magnetic structure of Fe-36.0%Ni invar (Fig. 3 *a*).

We observed the narrow distribution of internal magnetic field (Fig. 5*b*). As seen in the figure, there are two distinguishable components. The most probable internal magnetic fields lie in the 2 ÷ 10 T interval and the fields of 10 ÷ 22 T have weak intensity. There are no fields with larger values observed in the spectra of Fe-36.0%Ni invar (Fig. 3 *a*). We estimated the isomer shift relating to considered subspectra. The low-field components have δ = -0.029 mm/s and for high-field components $\delta \approx$ 0 (Table 1).

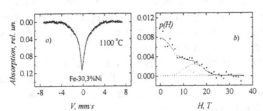

Figure 5. Mossbauer spectrum of the Fe-30.3%Ni alloy after 1100°C (*a*) and the hyperfine field distribution curve *p(H)* (*b*).

Thus in the Fe-30.3%Ni alloy the *s*-electron density at the nuclei of iron atoms with Fe n.n. is larger than in α-iron and in configurations enriched by Ni it is not affected. We assumed that carbon should change this balance.

4.5. Fe-29.7%Ni-0.97%C and Fe-30.5%Ni-1.5%C

Alloying with C of the alloy with low Ni concentration (29.7%Ni and 30.5%Ni) essentially affects the magnetic properties as compared to those of binary Fe-30.3%Ni austenite (Fig. 2, Fig. 4). Carbon stabilizes the f.c.c. phase by shifting the martensitic point lower the temperature of liquid nitrogen. This provides observations of reversibility of σ(T) curve in wide temperature range (Fig. 2 *b,c*). As seen from the χ(T) curve (Fig. 4 *a,b,c*) the Curie point growths from 85°C in Fe-30.3%Ni alloy to T_c = 165 - 175°C in the alloys containing carbon.

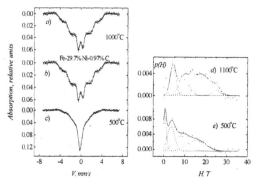

Figure 6. Mossbauer spectra of the Fe-29.7%Ni-0.97%C alloy after solution treatment at 1100°C (a,b) and ageing at 500°C (c) and the hyperfine field distribution curves. The fitting by Window method with no variation (a) and with variation of δ (b).

We observe distinct changes in Mössbauer spectra as a result of alloying with carbon (Fig. 6 a,b, Fig. 7 a). After 1100°C the spectra show smeared magnetic structure with broadened lines and are asymmetrical. Such a shape of spectra was observed also in [4,8-10].

Unlike previous studies we have fitted the spectra taking into consideration their asymmetry. In order to show the quality of the fitting we did not vary (Fig. 6 a) and varied isomer shifts (Fig. 6 b). The p(H) distribution obtained by fitting including the δ variation is shown in Fig. 6 d and Fig. 7 b. The low-field and high-field components with corresponding δ (δ = 0.040 mm/s and 0.081 mm/s, Table 1) were found. Notice that isomer shift of high-field components in interval 10–30 T was accepted to be the same.

The high-field contribution essentially increases comparing to that of C-free Fe-30.3%Ni alloy (Fig. 5 b). The growth of carbon content from 0.97%C to 1.5%C increases this distinction (Fig. 6 d and Fig. 7 b). This is connected with atomic redistribution under dissolution of carbon resulting in inhomogeneous short-range order of the Fe and Ni atoms [8]. For example, carbon may provide configurations in which an Fe atom has 3 Ni n.n. and at least 3 magnetic Fe atoms as it was considered in [4]. Since atomic order becomes inhomogeneous from one to another site we observe p(h) distribution of internal field in wide range. The magnetic inhomogeneities in the Fe-30.5%Ni-1.5%C alloy were indicated by means of small-angle neutron scattering [24].

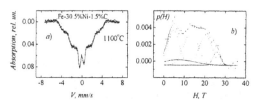

Figure 7. Mossbauer spectrum of Fe-30.5%Ni-1.5%C alloy after heating at 1100°C (a). The distribution of internal magnetic field (b).

The second point concerns redistribution of electrons between substitution atoms and doped carbon impurities that results in variation of spin density at the Fe nuclei and magnetic moment of iron atom.

In order to show that carbon induces such a magnetic structure we analyzed spectra derived from sample of the Fe-29.7%Ni-0.97%C alloy after ageing at 500°C. It is clearly seen from Fig. 6 b that after ageing when clustering and graphitization take place the p(H) function becomes narrow and tends to be similar to that for binary Fe-30.3%Ni alloy (Fig. 5 b). Clustering of carbon or its exit from solid solution under graphitization should increase the s-electron density. In fact, the isomer shift decreases from δ =0.040 mm/s and 0.081 mm/s

$(1100°C)$ to $\delta = -0.032$ mm/s and 0.003 mm/s respectively (500°C) and became close to that for binary Fe-30.3%Ni alloy (Table 1).

4.6. Fe-30.2%Ni-0.8%Mn-1.15%C

This alloy exhibits stable austenitic structure at room temperature. The Curie point (170°C) is close to that for Fe-29.7%Ni-0.97%C and Fe-30.5%Ni-1.5%C alloys ($T_c =$ 165, Fig. 4). The spectrum of this alloy shows asymmetrical shape with satellite lines (Fig. 8) characterized by magnetic fields smaller than those observed in the Fe-29.7%Ni-0.97%C and Fe-30.5%Ni-1.5%C alloys. This means that Mn affects the magnetic structure. Decreasing the internal magnetic fields at the nuclei of Fe n.n.. results from a

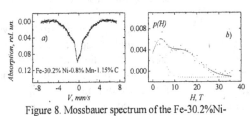

Figure 8. Mossbauer spectrum of the Fe-30.2%Ni-0.8%Mn-1.15%C alloy after 1100°C (*a*). The distribution of internal magnetic field (*b*).

redistribution of atoms in carbon austenite containing Mn, which accelerates clustering of carbon [20]. The reason lies in decreasing thermo dynamical carbon activity by Mn [21] and in relatively soft inter-atomic Mn-C repulsion than Fe-C one ($V_{Mn-C} - V_{Fe-C} = -0.28$ eV) [20].

There are two main peaks on *p(H)* curve which are located near $2 \div 10$ T and $10 \div 24$ T. The isomer shifts $\delta = -0.017$ mm/s and $\delta = 0.038$ mm/s were estimated for two groups of subspectra (Table 1). This result shows that Mn increases the *s*-electron density on the iron nuclei decreasing δ with respect to isomer shift attributed to the Fe-Ni-C alloy and partially weakening the effect of carbon (Table 1). Such effect of Mn was considered in Ref. [25].

5. Conclusion

The magnetic properties of the f.c.c Fe-Ni alloys with Ni concentration close to 30% are considerably changed with the addition of 0.97 and 1.5%C.carbon. The Curie point of the Fe-Ni-C alloys is higher on 80-90°C than that of binary Fe-30.3%Ni alloy. It should be noted that the Curie point derived from χ(T) curve was lower by 10 - 65°C than the temperature of the ferromagnetic-paramagnetic transition obtained from σ(T) dependence that results from effect of external magnetic field.

The hyperfine magnetic fields increasing under alloying with carbon are distributed in wide $2 \div 27$ T range and remained narrow as compared to those of Fe-36.0%Ni invar ($2 \div 32$ T). Addition of Mn narrows the *p(H)* distribution interval (2-24 T).

The isomer shifts and their transformation under alloying of the alloys were estimated on the basis of the Window method developed for approximation of asymmetrical spectra. It was found that C decreases and Mn increases the *s*-electron density at the iron nuclei in f.c.c. Fe-Ni alloy. The effect of Ni on the density of *s*-

electrons depends on its concentration. In the Fe-30.3%Ni alloy the s-electron density at the iron nuclei is larger than in Fe-36.0%Ni invar.

Acknowledgement

This work was completed with the support of the Science and Technology Center in Ukraine (project #2412). The authors thank Dr. Vladimir Chernenko for providing measurements of magnetic susceptibility of the alloys.

References

1. Guillaume, C.E. (1897) C.R. Acad. Sci. (Paris) 124 176.
2. Zakharov, A.I. (1986) *Physics of precession alloys with special thermal properties*. Moscow, Metallurgija (in Russian).
3. Precession alloys (1983) in B.V. Molotilov (eds.), *The reference-book*, Moscow, Metallurgija, p. 438 (in Russian).
4. Shiga, M., Maeda, Y., Nakamura, Y., (1974) Mössbauer Effect of Invar Type Fe-Ni-C and Fe-Ni-Mn Alloys in the Critical Concentration, *J. Phys. Soc. Jap.* **37**, No2, 363-370.
5. Tomiyoshi, S., Yamamoto, H., Watanabe, H. (1971) Temperature-Dependent Distribution of Internal Magnetic Fields at Fe57 Nuclei in fcc Iron-Nickel Alloys, *J. Phys. Soc. Jap.* **30**, No6, 1605-1613.
6. Window, B. (1973) Invar anomalies, *J. Appl. Phys.* **44**, No6, 2853-2865.
7. Weiss, R.J., (1963) *Peroc. Phys. Soc.*, **82**, 281.
8. Gavrilyuk, V. G.. Nadutov, V. M. (1983) Effect of Carbon on Magnetic and Atomic Ordering in Iron-Nickel Alloys, *Fizika Metallov Metallovedenie* **56**, No3, 555-563.
9. Gavriljuk, V., Nadutov, V. (1994) Hyperfine Structure and Properties of New Invar Alloy Fe-Ni-C at Low Temperatures, *Cryogenics* **34**, 485-488.
10. Gonser, U., Nasu, S., Keune, W., Weis, O. (1975) Fe^{57} Hyperfine Field Distributions in Fe-28%Ni-3%C Invar Alloy, *Solid State Comm.* **17**, 233-236.
11. Window, B. (1971) Hyperfine field distributions from Mössbauer spectra. *J. Phys. E: Scientific Instruments* **4**, 402 -.402.
12. Hesse, J., Rübartsch, R. (1974) Model independent evaluation of overlapped Mössbauer spectra, *J. Phys. E: Scientific Instruments* **7**, 526-532.
13. Bakhvalov, N.S., Zhydkov, N.P., Kobelkov, G.M. (1987) Numerical methods, Moscow, Nauka, The main publisher of physical and mathematical literature, 600 p.
14. Vilenkina, N.Ya., (1969) *Combinatorics*, Moscow, Nauka..
15. Bugaev, V.A., Gavrilyuk, V.G., Nadutov, V.M., Tatarenko, V.A., (1983) Mössbauer Study of Carbon Distribution in Fe-Ni-C Austenite, *Acta Met.* **31**, No3, 407-418.
16. Kozlova, O.S., Makarov, V.A. (1979) Study of austenite of the Fe-Al-C alloy by means of Mossbauer spectroscopy. *Fizika Metallov Metallovedenie* **48**, No5, 974-978 (in Russian).
17. Gridnev, V.N., Gavrilyuk, V.G., Nadutov, V.M. (1984) *DAN SSSR*, **277**, No2 360-362 (in Russian).
18. Nadutov, V.M., Tatarenko, V.A., Tsynman, K.L. in V.G. (1993) Interatomic Interaction and Structure of Phases in F.C.C. Fe-N and Fe-C Alloys, Gavrilyuk and V.M. Nadutov (Eds.) *Proc. 3rd Int. Conf on High Nitrogen Steels*, Kiev, Ukraine, 14-16 September, 1993, Institute for Metal Physics, Ukraine, pp. 106-113.
19. Balanyuk, A.G., Bugaev, V.N., Nadutov, V.M., Sozinov, A.L. (1998) Estimation of the Energies of N-N and C-C Interactions in F.C.C. Fe-N and Fe-C Alloys on the Basis of Mössbauer Spectroscopy Data, *Phys. Stat. Sol.* (b) **207** 3-12.
20. Bugaev, V.A., Gavrilyuk, V.G., Nadutov, V.M., Tatarenko, V.A. (1989) Distribution of Carbon in Fe-Ni-C and Fe-Mn-C Alloys with F.C.C. Lattice, *Fizika Metallov Metallovedenie*, **68**, No5 931-940 (in Russian).

21. Mogutnov, B.M., Tomilin, I.A., Shwartsman, L.A. (1984) *Thermodynamics of Iron Alloys*, Metallurgy, Moscow.
22. Chernenko, V. A., Kokorin, V. V., Minkov, A. V. (1992) Peculiarities of the Magnetic State in Aging Iron-Nickel Invars, *Phys. Stat. Sol.* **134**, 193 - 199.
23. Shimizu, K., Kakeshita, T., (1989) Effect of Magnetic Fields on Martensitic Transforrmations in ferrous Alloys and Steels, *ISIJ International* **29**, No2, 97 - 116.
24. Nadutov, V.M. Garamus, V.M., Willumeit, R., Svystunov, Ye.O. (2002) Small-Angle Neutron Scattering in Iron-Based Austenite. P. II, The Effect of Carbon on SANS in F.C.C. Fe-Ni- Alloy, *Metallofizika i novejshie technologii*, **5** (in press).
25. Nadutov, V.M. (1998) Mössbauer Analysis of the Effect of Substitutional Atoms on the Electronic Charge Distribution in Nitrogen and Carbon Austenite, *Mater. Sci .Eng.*, **A254**, 234-241.

INVESTIGATION OF MAGNETIC PROPERTIES IN IRON-BASED NANOCRYSTALLINE ALLOYS BY MÖSSBAUER EFFECT AND MAGNETIZATION MEASUREMENTS

J. HESSE[1], O. HUPE[1], C. E. HOFMEISTER[1], H. BREMERS[1],
M. A. CHUEV[2] AND A. M. AFANAS'EV[2]

[1] *Institut für Metallphysik und Nukleare Festkörperphysik, Technische Universität, Mendelssohnstrasse 3, D-38106 Braunschweig, Germany*
[2] *Institute of Physics and Technology, Russian Academy of Sciences, Nakhimovskii pr. 34, 117218 Moscow, Russia*

1. Introduction

Nanostructured magnetic materials are challenging scientists and engineers. Considering magnetic properties in the modern applications it becomes possible to achieve extremally soft magnetic materials [1] (with applications for loss free transformers) and extremly hard magnetic materials [2] (with application for permanent magnets) in both cases applying properly chosen nanostructured materials. The Mössbauer effect as a local nuclear probe has contributed a lot to the deeper physical understanding of such complex alloy systems. Two examples, the paper of Mørup [3] concerning the superparamagnetic behaviour of nanosized particles and the exciting contribution by Miglierini and Greneche [4, 5] concerning the information content of hyperfine field distributions may evidence the power of this method.

In this contribution we focus on nanostructured alloys exhibiting very soft magnetic properties if choosen a suitable composition and, what is very important, a suitable thermal treatment producing nanograins from a former amorphous alloy. One question concerns the magnetic behaviour of the nanograins when the sample temperature exceeds the (nominal) Curie temperature of the residual amorphous matrix. A similar question has been discussed in the work of Kemény et al. [6] applying Mössbauer spectroscopy. With magnetization measurements and Mössbauer effect experiments performed on samples of the Fe-Cu-Nb-B alloy after different heat treatments we tried to answer the following questions: Is it possible to find conditions where a superparamagnetic behaviour of the nanograins is observed? Is there experimental evidence about the interaction between the nanograins and what about the influence of the nanograins on the behaviour of the whole alloy system? Is there super-ferromagnetism, what means a spontaneous magnetic order due to particle-particle interactions?

M. Mashlan et al. (eds.), Material Research in Atomic Scale by Mössbauer Spectroscopy, 117–126.
© 2003 *Kluwer Academic Publishers. Printed in the Netherlands.*

2. The Samples

We performed experiments on ferromagnetic $Fe_{(86-x)}Cu_1Nb_xB_{13}$ (x=4,5,7) nanostructured alloys studied by Mössbauer effect on ^{57}Fe and magnetization measurements in a wide temperature range, below and above the Curie temperature of the amorphous matrix. Here we will focus on the alloys with x=7, i.e. with 7 at% Nb. The alloys have been produced by the Vacuumschmelze GmbH, Hanau, Germany and provided by Dr. G. Herzer. The thickness of the long ribbons is about 24 μm and their width 14.8 mm. One very important feature of the system in question is that the nano-crystallites appearing after suitable annealing consist of pure bcc-Fe of about 5.6 nm in diameter. This makes the application of the Mössbauer effect favourable because the evaluation of the spectra becomes in some sense simple.

Smaller parts of the Mössbauer absorbers were used as samples for magnetization measurements. To reduce the demagnetizing factor the external magnetic field is applied parallel to the foil plane. Due to different annealing temperatures for the as quenched amorphous samples it becomes possible to produce different amounts of crystallites. These crystallites are nearly of the same size and exhibit a rather narrow particle size distribution. So it becomes possible to use the relative amount of the spectral component for bcc-Fe in the Mössbauer spectra as a measure for the number of nanograins. A higher amount of nanograins affects the inter-granular coupling and so the total magnetic behaviour of the sample below and above the Curie temperature of the residual amorphous matrix.

3. Results

In Figure 1 a sequence of room temperature Mössbauer spectra for differently annealed $Fe_{79}Cu_1Nb_7B_{13}$ samples is plotted. The indicated parameter is the annealing temperature. On top the spectrum of the as quenched sample is shown. On bottom a spectrum collected on a foil of pure bcc iron is added for comparison. All samples were annealed for one hour in a He-gas atmosphere.

In Figure 2 the increase of the bcc-iron content versus annealing temperature as determined from the areas of the room temperature Mössbauer spectra from Figure 1 is plotted. The dramatic influence of the nanograin content on the magnetic properties is shown in Figure 3.

Magnetization measurements have been performed on all samples presented in Figure 1. The magnetization curves measured versus temperature in different external magnetic fields were fitted with a model based on an extended mean-field-theory, which provides information about the particle-particle coupling.

4. The Evaluation of Magnetization Measurements

To describe the magnetization of an amorphous ferromagnetic matrix with embedded nanograins in the whole temperature range and different applied external magnetic fields some basic assumptions should be done. In this description the molecular field approximation is used.

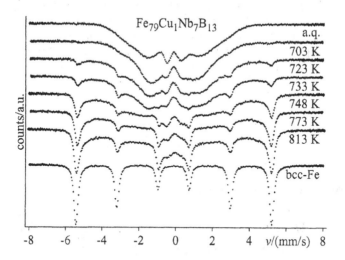

Figure 1. A sequence of room temperature Mössbauer spectra for the alloy $Fe_{79}Cu_1Nb_7B_{13}$ is plotted. Parameter is the annealing temperature. On top the spectrum of the as quenched sample is shown. On bottom a spectrum collected on a foil of pure bcc iron is added for comparison. This Figure demonstrates the increasing content of bcc-Fe in the samples with increasing annealing temperature.

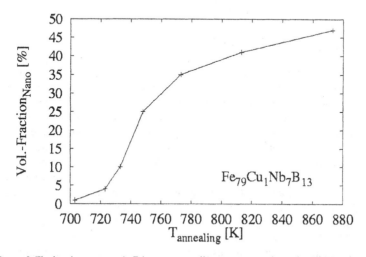

Figure 2. The bcc-iron content (n-Fe) versus annealing temperature determined from areas of the room temperature Mössbauer spectra presented in Figure1.

Formula (1) describes the magnetization $M(T,B_{Ext})$ of a paramagnet vs. temperature T in an external magnetic field B_{Ext} (S= Spin, B_S is the Brillouin function, k_B the Boltzman constant, μ_B=Bohr magneton and μ the maximal magnetic moment $\mu=g\mu_B S$). Following the basic idea of the molecular field approximation a ferromagnetic sample in an external magnetic field is then described by formula (2). Here the Curie temperature T_C of the ferromagnetic sample replaces the usual molecular field constant. In the first step we applied formula (2) to describe the magnetization of the amorphous phase i.e. the as quenched sample. This was working rather well and delivered more or less the same fit parameters for all applied external magnetic fields.

PARAMAGNET

$$\frac{M(T,B_{Ext})}{M(0)} = B_S\left(\left(\frac{\mu_B}{k_B}g_S S B_{Ext}\right)\frac{1}{T}\right) \tag{1}$$

FERROMAGNET in an external FIELD

$$\frac{M(T,B_{Ext})}{M(0)} = B_S\left(\left(\frac{3S}{S+1}\cdot\frac{M(T)}{M(0)}\cdot T_C + \frac{\mu_B}{k_B}g_S S B_{Ext}\right)\frac{1}{T}\right) \tag{2}$$

HANDRICH – KOBE extension (brief notation for $B_{Ext} = 0$

$$B_S(1\pm\Delta) = \frac{1}{2}\left(B_S\left((1+\Delta)\frac{3S}{S+1}\cdot\frac{M(T)}{M(0)}\cdot T_C\right) + B_S\left((1-\Delta)\frac{3S}{S+1}\cdot\frac{M(T)}{M(0)}\cdot T_C\right)\right) \tag{3}$$

4.1. THE MODEL FOR THE SAMPLES' MAGNETIZATION

Handrich and Kobe [7] introduced an extension to the molecular field approximation in which the spacial fluctuations of the exchange interaction i.e. a fluctuation of the molecular field due to the irregular atomic distances in amorphous alloys is introduced. Formula (3) describes the details. Using this description we succeeded to find a very good fit for the amorphous samples for all applied magnetic fields. In the next step the nanograins have to be considered in the model. Without any interaction between the grains and with the amorphous matrix they are described as ideal super-paramagnets in an external magnetic field. In order to describe all samples it is necessary to introduce a coupling of the nanograins to the amorphous matrix and an interaction between the grains. Additional molecular fields realize that as done in formulas (4) and (5). The index 1 describes the amorphous matrix with all relevant paramaters like spin S_1, magnetization M_1, Handrich-Kobe fluctuation Δ_1 and Curie temperature T_{C1}. The index 2 describes the nanograin phase. By this nomenclature T_{K1} describes the coupling between the nanograin and the amorphous phase. T_{C2} describes the Curie temperature of the super-ferromagnetic nanograin phase and T_{K2} the coupling of the amorphous phase to the nanograin one.

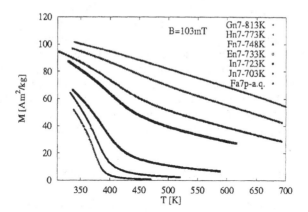

Figure 3. Magnetization measurements performed on the Fe$_{79}$Cu$_1$Nb$_7$B$_{13}$ alloy samples annealed at different temperatures. The measurements were performed in an external magnetic field of 103 mT. A dramatic effect due to the increasing number of bcc-Fe nanograins on the magnetic properties can be observed.

THE MODEL:
PHASE 1 coupled to PHASE 2

$$\frac{M_1(T,B_{Ext})}{M_1(0)} = B_{S1}\left[\left(\frac{3S_1}{(S_1+1)}\cdot\frac{(1\pm\Delta_1)}{M_1(0)}\left(T_{C_1}M_1(T)+T_{K1}M_2(T)\right)+\frac{\mu_B}{k_B}g_{S1}S_1B_{Ext}\right)\frac{1}{T}\right] \quad (4)$$

PHASE 2 coupled to PHASE 1

$$\frac{M_2(T,B_{Ext})}{M_2(0)} = B_{S2}\left[\left(\frac{3S_2}{(S_2+1)}\cdot\frac{(1\pm\Delta_2)}{M_2(0)}\left(T_{C2}M_2(T)+T_{K2}M_1(T)\right)+\frac{\mu_B}{k_B}g_{S2}S_2B_{Ext}\right)\frac{1}{T}\right] \quad (5)$$

TOTAL MAGNETIZATION

$$M(T,B_{Ext}) = M_1(T,B_{Ext}) + M_2(T,B_{Ext}) \quad (6)$$

Equations (4) and (5) were used to fit the magnetization data. The necessary condition was that each set of parameters remains the same independently of the applied external magnetic field. So in fact not only one magnetization curve was fitted but also a complete set belonging to one sample measured in different magnetic fields. The experiments showed that the external magnetic field should not be too low. This seems clear because at very low magnetic fields the sample can exhibit a magnetic domain structure. Our model needs a sample without domain structure, which always can be achieved using high enough external magnetic fields.

From the above given description of the model it can be seen that a lot of simple assumptions have been made. Randrianantoandro et al. [8] stated that the magnetic properties of nanocrystalline materials and their temperature dependences could be described considering an ensemble of single-domain particles distributed within a ferromagnetic or paramagnetic medium. In real particle systems a distribution of particle sizes and shapes as well as different orientations in space (position of easy axis of magnetzation) must be considered. The following energies are present: Magneto-crystalline anisotropy, shape anisotropy, magnetic surface and magnetoelastic anisotropy. For the interactions the magnetic dipole and exchange energy are responsible. To all these energies additionally the thermal energy and the Zeeman energy offered by an applied external magnetic field must be added. In the frame of these meaningful interactions the thermally activated fluctuations of the magnetization in the nanograins must be considered which in the basic case is given by the Néel theory [9].

4.2. RESULTS FROM MAGNETIZATION MEASUREMENTS

From the fits to the magnetization curves for all applied external magnetic fields the parameters described in the model are obtained. For example the magnetic properties of the sample annealed at 748 K (25% n-Fe) are described by $T_{C1}=360$ K, $T_{C2}=629$ K, $T_{K1}=75$ K, $T_{K2}=760$ K, $S_1=14$, $S_2=5186$, $\Delta_1=0.25$, $\Delta_2=0$, $M(0)=M_1(0)+M_2(0)=139$ Am^2/kg, $M_1(0)=0.8*M(0)$. From the fit sequences to the magnetization of the samples it follows that the Curie temperature of the remaining amorphous phase becomes lower with decreasing Fe content (i.e. increasing content of bcc-Fe nanograins). Superparamagnetic behaviour of the nanograins is observed for the $Fe_{79}Cu_1Nb_7B_{13}$ annealed at 703 K (1% n-Fe) and 723 K (5% n-Fe). The first evidence for particle-particle interaction appears for the sample annealed at 733 K (10% n-Fe). It becomes more obvious for the sample annealed at 748 K. Super-ferromagnetic behaviour (forced by the external magnetic field) is also observed for the samples annealed at 813 K (41% n-Fe) and 873 K (47% n-Fe).

5. Mössbauer Effect

We confine ourselves on the Mössbauer effect on ^{57}Fe but the considerations given in this section are of general validity. Usually the transmission spectra collected on ferromagnetic nanostructured alloys are interpreted in terms of so-called hyperfine field distributions. Let's recall shortly what a hyperfine field distribution means:

5.1. HYPERFINE FIELD IN FERROMAGNETS, WHAT ARE DYNAMIC EFFECTS

The nuclear Zeeman effect is responsible for the energy level splitting that in consequence lead to the well-known six-line pattern of "magnetically split lines". The existence of the hyperfine field at the nucleus is the consequence of a rather complex process consisting of spin and orbital contributions of the magnetic moments of atoms and its neighbours in question. Therefore a strong correlation between the thermal fluctuations of the magnetic moment of the iron atom and the hyperfine field exists. The hyperfine field generally is time dependent. Two limiting cases in which well defined and easy to interpret hyperfine field splitting occurs are the cases of fast and slow relaxation. Fast means that the statistical period of fluctuation is short compared to the Larmor precession period of the nuclear magnetic moment in the hyperfine field. In that case the mean value is measured and if there is a preferred orientation of the fluctuating atomic magnetic moment a hyperfine field unequal zero is observed (unfortunately this state is often called "static"). If there is no preferred orientation a thermal collapse of the hyperfine field to zero is observed. The slow relaxation case is just the opposite. The nucleus experiences the momentary value of the hyperfine field. In between the both limiting cases there is the "region" of so-called "relaxation spectra" which in general can't be evaluated in terms of hyperfine field distributions.

The second necessary comment is that a magnetic field is a vector field. But the Mössbauer effect in the described technique is not measuring a vector. It is measuring just a synonymus of the scalar quantity Zeeman energy splitting (please compare with quadrupole energy splitting). The confusion arises because there are six magnetic transitions in case of ^{57}Fe from which this Zeeman energy distribution can be gained. So in a six line spectrum there is redundant information. The evaluation of a Mössbauer spectrum in terms of hyperfine fields is equivalent to a formal transformation of the six lines to a single line transition between two Zeeman levels (like the NMR transition between +1/2 and -1/2). The hyperfine field distribution times a nuclear magnetic momentum delivers the desired Zeeman energy distribution.

5.2. RESULTS FROM MÖSSBAUER SPECTROMETRY

In Figure 4 two sequences of Mössbauer spectra collected versus temperature for the $Fe_{79}Cu_1Nb_7B_{13}$ alloy annealed at 723 K and at 748 K are presented. The first sample has a rather low nanograin content of about 5%. Its nano particles exhibit superparamagnetic behaviour above the Curie temperature of the amorphous matrix. This property immediately can be seen from Figure 4 (left spectra), because the spectral component belonging to the iron nanograins disappears together with the thermal collapse of the hyperfine field splitting of the amorphous phase. This conclusion is supported by the magnetization measurement result.

The second sequence of Mössbauer spectra in Figure 4 belongs to the alloy $Fe_{79}Cu_1Nb_7B_{13}$ annealed at 748 K. The nanograin content is about 25%. This sample exhibits nano particles showing the characteristic line shape indicating particle-particle interaction and relaxation effects as described in the GTLR model [12]. The superparamagnetic behaviour becomes suppressed. Again magnetization measurements confirm this conclusion of a strong coupling between the nanograins, where superferromagnetism forced by the external magnetic field appears.

5.3. NEW INTERPRETATION OF THE SEXTET LINE SHAPES

Relaxation phenomena are considered since the very beginning in nano particle research [3]. The basic model allowing first insight into relaxation phenomena is based on Stoner-Wohlfarth (SW) [10] particles. These ferromagnetic particles consist of one domain. The exchange length determines their maximum size [11].

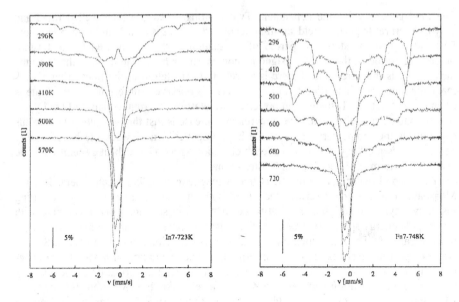

Figure 4: Two sequences of Mössbauer spectra collected at the indicated temperatures. The annealing temperature, which determined the nanograin content, is indicated in the right bottom of each spectra sequence.

The magnetization of a SW particle will change its direction with respect to the particle body in a homogeneous way (coherent reversal) and fluctuate like a single magnetic moment with a value of the order of $(10^4$-$10^5)$ μ_B. These particles exhibit uniaxial anisotropy energy. The particle's magnetic moment is $\mu = M_0 V$ (M_0 = saturation value of the magnetization, K = anisotropy energy density and V = particle's volume, KV = anisotropy energy). In an external magnetic field and a fixed easy direction the energy $E(a,g)$ becomes angle dependent, were $a(alpha)$ is the angle between the magnetization vector M and the applied magnetic field B, $g(gamma)$ represents the angle between the anisotropy energy easy axis and B. If $0<B<B_{Switch} =2K/M_0$ the energy $E(a,g)$ exhibits two minima where the deeper one corresponds to the parallel orientation of the magnetization vector to the external field. These two energy minima lead to the name "two level relaxation".

Afanas'ev and Chuev [12] recently proposed a very new idea for an interpretation of relaxation spectra in a system of interacting SW particles. The interaction causes each particle to experience different energy levels (also in a system of identical particles). If the particle-particle interaction is not too strong, the equilibrium position of the particle's magnetization remains the easy axis. The magnetic moment of the particle is large enough so that even weak interaction with the environment strongly smears out its energy levels. As a consequence the relaxation between the particle's states with opposite directions of magnetic moment never proceed as a transition between the levels with the same energy. On the average the energy levels of all particles will be smeared out over an energy width ΔE. This gives rise to variously shaped relaxation Mössbauer spectra. Very typical is the fact that in a well-resolved six-line structure the lines are highly asymmetric with steep outer and extended inwards tails. Afanas'ev and Chuev treated the magnetic fields of the neighbouring particles as random variables. The corresponding energy shifts ΔE are distributed following a Gaussian form. Details of this new analysis of Mössbauer spectra in nanostructured alloys will be presented in the contribution of O. Hupe et al. (this book).

6. Discussion

The problem of interparticle interactions in nano particle assemblies has been discussed many times. Dormann et al. presented recently a summary [13]. Skorvanek and O'Handley [14] presented a very simmilar study increasing the content of nanograins in Fe-Cu-Nb-Si-B alloys by successive enhancement of annealing temperature. Performing magnetization measurements they showed that increasing interparticle interaction suppresses super-paramagnetic fluctuations and lead to super-ferromagnetic behaviour.

In this contribution the Mössbauer measurements are performed without using an external magnetic field. Spectra collected for each preparation step, which is for a determined number density of nanograins, surprisingly show always that the amorphous phase undergoes a separate magnetic phase transition (identified by the thermal collapse of the spectral component belonging to the amorphous phase above the Curie temperature). The bcc-Fe component, the nanograins, behaves differently. At low nanograin concentration super-paramagnetic behaviour can be observed. In that case the hyperfine field splitting of the spectral component assigned to the nanograins is observed to disappear at the Curie temperature of the amorphous phase.

At higher nanograin concentration a very characteristic change in the line shape of the six-line pattern is observed. This is, following the very recently introduced new interpretation by Afanas'ev and Chuev [12], due to the particle-particle interactions and the dynamic behaviour of the particles' magnetization. The spectral component assigned to the amorphous component exhibits a thermal collapse whereas the nanograin component still exists up to higher temperatures.

In contrast to the Mössbauer spectra the magnetization measurements are always performed in an external magnetic field. The magnetization results lead to the conclusion that a minimal strength of the external magnetic field is necessary to aviod magnetic domain effects. So typically the lowest field applied was 103 mT. This field and the higher external fields seems to have an important influence on the magnetic behaviour of the whole sample which becomes obvious above a critical concentration of nanograins: At low nanograin concentration the magnetization delivers also evidence for super-paramagnetic behaviour of the nanograins above the Curie temperature of the amorphous phase. At higher nanograin concentration a very characteristic change in the magnetization curves is observed: The separate Curie temperature of the amorphous ferromagnetic component vanishes and a kind of collective ferromagnetism of both phases appears. This we will call forced super-ferromagnetism. Its origin is to be seen in the interaction of the nanograins and in the existence of the external magnetic field that polarizes the nanograin magnetic moments. In consequence this polarizes the amorphous matrix, which without the external magnetic field would behave paramagnetic.

7. Conclusions

The combination of Mössbauer effect spectrometry and magnetization measurements delivers a founded and new insight in the properties of modern nanostructured materials. We presented investigations on alloys exhibiting bcc-Fe nanograins embedded in an amorphous matrix gained by suitable annealing of amorphous $Fe_{79}Cu_1Nb_7B_{13}$ alloys. Mössbauer spectrometry and TEM delivered a direct measure about the concentration of the nanograins. The very characteristic shape of the spectral lines of the bcc-Fe spectral component is a mark for particle-particle interaction and was evaluated using a new idea of relaxation spectra [12].

Magnetization measurements and Mössbauer spectrometry delivered evidence for super-paramagnetic behaviour at low nanograin concentrations. At higher nanograin concentrations the particle-particle interaction leads to a collective magnetic behaviour: to the (by the external field) forced super-ferromagnetism.

Acknowledgements
We are grateful to the "Internationales Büro des BMBF" and the Russian Foundation Sponsoring the Domestic Science for supporting our collaboration within the project RUS 97/157. The German authors are thankful to the Deutsche Forschungsgemeinschaft for supporting the experimental work.

References
1. Herzer, G. (1993) Nanocrystalline Soft Magnetic Materials, *Physica Scripta* **T49**, 307-314.
2. Seeger, M and Kronmüller, H. (1996) High-Tech Permanent Magnets, *Z. Metallkde* **87**, 923-933.
3. Mørup, S. (1983) Magnetic Hyperfine Splitting in Mössbauer Spectra of Microcrystalls, *J. Mag. Mag. Mat.* **37**, 39-50.
4. Miglierini, M. and Greneche, J.-M. (1997) Mössbauer spectrometry of Fe(Cu)MB-type nanocrystalline alloys: I. The fitting model for the Mössbauer spectra, *J.Phys.: Condens. Matter* **9**, 2303-2319.
5. Miglierini, M. and Greneche, J.-M. (1997) Mössbauer spectrometry of Fe(Cu)MB-type nanocrystalline alloys: II. The topography of hyperfine interactions in Fe(Cu)ZrB alloys, *J.Phys.: Condens. Matter* **9**, 2321-2347.
6. Kemény, T., Kaptás, D., Balogh, J., Kiss, L. F., Pusztai, T. and Vincze, I. (1999) Microscopic study of the magnetic coupling in a nanocrystalline soft magnet *J. Phys.: Condens. Matter* **11**, 2841-2847.
7. Handrich, K. and Kobe, S. (1970) On the Theory of Amorphous and Liquid Ferromagnets, *Acta Phys. Polonica* **A38**, 819-827; Handrich, K. (1969) Simple model for amorphous and liquid Ferromagnets; *phys. stat. sol* **32**, K55-K58.
8. Randrianantoandro, N. Slawska-Waniewska, A. and Greneche, J.-M. (1997) Magnetic properties in Fe-Cr-based nanocrystalline alloys *J.Phys. Condens. Matter* **9**, 10485-10500.
9. Néel, L. (1949) Theorie du trainage magnetique des ferromagnetiques en grains fins avec applications aux terres cuites, *Ann. Geophys.* **5**, 99-136.
10. Stoner, E. C. and Wohlfarth, E. P. (1948) A mechanism of magnetic hysteresis in heterogeneous alloys *Phil. Trans. Roy. Soc. London*, **240A**, 599-642.
11. Hernando, A. (1999) Magnetic properties and spin disorder in nanocrystalline materials *J. Phys.: Cond. Matter* **11**, 9455-9482.
12. Afanas'ev A. M. and Chuev M. A. (2001) New Relaxation Model for Superparamagnetic Particles in Mössbauer Spectroscopy, *JETP Lett.* **74**, 107-110.
13. Dormann, J. L., Fiorani, D. and Tronc, E. (1999) On the models for interparticle interactions in nanoparticle assemblies: comparison with experimental results, *J. Mag. Mag. Mat.* **202**, 251-267.
14. Skorvanek, I. and O'Handley, R.C. (1995) Fine-particle Magnetism in Nanocrystalline Fe-Cu-Nb-Si-B at Elevated Temperatures, *J. Mag. Mag. Mat.* **140-144,** 467-468.

MOMS – MAGNETIC ORIENTATION MÖSSBAUER SPECTROSCOPY: DETERMINATION OF SPIN ORIENTATIONS IN ION IRRADIATED IRON FILMS

Formation of a Uniaxial Spin Orientation by Ion Implantation into Thin Iron Films

P. SCHAAF[1], A. MÜLLER AND E. CARPENE
Universität Göttingen, Zweites Physikalisches Institut, Bunsenstraße 7/9, 37073 Göttingen, Germany
[1] *Email: pschaaf@uni-goettingen.de;*
URL: www.uni-goettingen.de/~pschaaf

Abstract

Here, a xenon-ion induced magnetic texturing of iron films using depth-sensitive and angle-dependent Conversion Electron Mössbauer Spectroscopy as Magnetic Orientation Mössbauer Spectroscopy (MOMS) in combination with Magneto-optical Kerr Effect (MOKE), Rutherford Backscattering Spectrometry (RBS) and X-Ray Diffraction (XRD) is reported. Three 75 nm thick Fe films containing a 15 nm ^{57}Fe marker layer at different depths were deposited on Si(100) substrates and irradiated under well defined mechanical strain with 200 keV Xe ions. The ion-induced changes of the magnetization were related to the lattice expansion at various depths produced by the implanted xenon. A complete uniaxial orientation of all spins along the mechanical stress axis was found by Mössbauer spectroscopy for the sample with the sensitive marker layer being in the centre of the ferromagnetic film. This agrees with the angular properties of the macroscopic magnetic hysteresis curve measured by MOKE.

1. Introduction

The magnetization and its direction in thin ferromagnetic films is known to depend on a number of preparation parameters, such as thickness, shape, composition, stresses and temperature of the films and its substrates as well as external parameters such as a magnetic field [1,2]. Ion beam techniques are a powerful tool to modify the local stress state of thin films [3,4] and hence may also change the magnetic properties of ferromagnetic films [1-5]. It has been found, indeed, that an in-plane uniaxial magnetic anisotropy can been induced by ion bombardment, either during ion-beam assisted deposition of Fe or Ni films [5,6] or by an ion implantation after deposition [7-10]. Depending on the ion species, energy and fluence, ion irradiations can produce or eliminate magnetic anisotropies [11]. It is known that a change of magnetization, due to

127

M. Mashlan et al. (eds.), Material Research in Atomic Scale by Mössbauer Spectroscopy, 127–136.
© 2003 *Kluwer Academic Publishers. Printed in the Netherlands.*

the spin re-orientation transition, exists as a function of temperature in ferromagnetic materials not only for bulk samples, but also for thin films [5,7]. In the last few years ion beam irradiation of ferromagnetic thin films, tens of nanometers thick, was found to be a very efficient way to induce magnetic anisotropy due to inverse magnetostriction effects [1,2]. In particular, studies on magnetic texturing associated to noble-gas ion irradiation have found in nickel, iron and permalloy films [5,8-11] and its dependence on important parameters like ion energy, ion fluence, external temperature, applied magnetic field and mechanical strain was investigated. Also many different mechanisms have been found as showed in [7,12] for Co/Pt multilayers.

However, in most of the previous investigations the magnetic properties were analyzed using methods probing either the near-surface part of the films, like magneto-optical methods or magnetic force microscopy [7,10,12], or they measure the magnetization of the whole film, like vibrating sample magnetometer or SQUID. This article describes the depth sensitive correlation between magnetization and strain after xenon-ion irradiation in strained ^{57}Fe/Fe layers. The samples were analyzed at room temperature by Rutherford backscattering spectrometry (RBS), θ-2θ X-ray diffraction (XRD) and surface profiling. For the magnetic analyses, we used the longitudinal magneto-optical Kerr effect (MOKE) and Magnetic Orientation Mössbauer Spectroscopy (MOMS) by angle-dependent Conversion Electron Mössbauer spectroscopy (CEMS). The present study focuses on the investigation of the actual spin distribution obtained after the implantation as measured by the MOMS technique, where ^{57}Fe marker-layers, placed in different depths of thin ^{nat}Fe films offer a very elegant way to gather depth-sensitive information on anisotropic magnetic hyperfine fields and spin orientations [13,14]. Up to now this technique has only been applied to characterize structural and electronic properties of single crystals [15,16].

2. Experimental

2.1. SAMPLE PREPARATION

Three special films of 10×5 mm² size have been deposited on ultrasonic-cleaned, 0.5 mm thick Si(100) substrates, each 40×15 mm² in size. The natural iron films were deposited by electron-beam evaporation and the 15 nm thick ^{57}Fe marker-layers by an effusion cell. The marker-layer was placed either directly at the Fe/Si interface (sample I), in the middle of the ^{nat}Fe film (sample M) or at the surface of the film (sample S). The total iron film thickness is 77(3) nm for all three samples. The depositions were performed with the substrate held at room temperature in the same vacuum chamber with a base-pressure of 4×10^{-6} Pa without breaking the vacuum. These as deposited samples have been characterized by RBS, MOMS, XRD and MOKE, exactly in this sequence.

Before irradiation the samples were mounted on a special target holder allowing the defined bending of the Si-substrate to a bending radius $R=1$ m (curvature $1/R=1$ m^{-1}) as controlled by a Dektak surface profiler. According to Stoney [17] this corresponds to a strain of 108(11) GPa in the iron film.

Then, the samples were irradiated at room temperature with 200 keV Xe-ions to a fluence of 1×10^{16} cm^{-2}, provided by the Göttingen ion implanter IONAS [18] at a pressure of about 10^{-5} Pa in the implantation chamber. This implantation energy corresponds to a mean implantation depth of R_p=34 nm with a straggling of ΔR_p=12 nm (SRIM2000 code [19]) and was found to induce the largest change in magnetization [11]. By using an electrostatic X-Y sweeping system (f_x = 120 Hz, f_y = 5 Hz) the ions were homogenously distributed over the irradiated area of 10×10 mm². The ion current was kept at about 0.3 µA, in order to avoid beam heating of the samples. After the irradiation the strain was relaxed by releasing the bending and the samples were again measured by RBS, MOMS, XRD, and MOKE, again in this sequence.

2.2. ANALYTICAL METHODS

All samples were analyzed by X-ray diffraction (XRD) employing Cu-Ka radiation at grazing incidence and also θ-2θ geometry. Phase analyses were also performed at room temperature by Conversion Electron Mössbauer Spectroscopy (CEMS) with a ^{57}Co/Rh source and constant acceleration drive, detecting the conversion electrons emitted from the upper 150 nm of the sample [14] into a He/CH$_4$ gas flow detector. Isomer shifts are given relative to α-Fe, which was also used for velocity calibration. A least-squares fit routine by superimposing Lorentzian lines was used for the analysis of the spectra [14]. Due to the limited thickness of the iron films measured and the high enrichment of the ^{57}Fe marker-layers (96%), almost all of the information is encountered from the marker-layers.

Rutherford backscattering spectroscopy (RBS) was performed with 900 keV He^{2+} ions, using two Si surface barrier detectors positioned at a scattering angle of 165°. The experimental data were analyzed with the RUMP [20] code. The MOKE-set-up is a PCSA-ellipsometer using a photo-elastic modulator and online field measurement [21], with a maximum applied field of about 1500 Oe provided by Helmholtz coil pair.

2.3. MOMS – MAGNETIC ORIENTATION MÖSSBAUER SPECTROSCOPY

Traditionally, conventional Mössbauer spectroscopy is performed with perpendicular (normal) incidence of the γ-beam. For magnetic sextets, the relative line intensities are depending on the angle θ between the γ-direction and direction of the hyperfine field B, the latter being co-linear with the spin direction ($I_{1,6}/I_{3,4}$ is always 3) [22-24]:

$$\frac{I_{2,5}}{I_{3,4}} = \frac{\frac{1}{2}(1-\cos^2\theta)}{\frac{1}{8}(1+\cos^2\theta)} = 4 \cdot \frac{(1-\cos^2\theta)}{(1+\cos^2\theta)} \tag{1}$$

For thin films or foils this results in the line intensity ratios 3:2:1 for a random spin orientation (in space), 3:4:1 for all spin aligned in the film and thus being perpendicular to the γ-beam and 3:0:1 when all spins are pointing out of the film plane, i.e. line 2 and 5 of the sextet disappear. Now, knowing that for thin films and foils the spins are almost completely aligned in the film plane, one might ask if we can resolve the spin distribution within the plane by help of Mössbauer spectroscopy. This becomes possible when we tilt the sample normal against the γ-direction by angle α and then rotate the sample around its normal n by the angle φ. This is sketched in Figure 1.

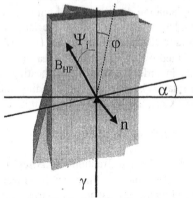

Figure 1. Schematic view of the angles for the MOMS analysis representing the geometry of the angle dependent Mössbauer set-up. Ψ_i indicates the in-plane vectors of the magnetization directions, α is the tilt-angle of the sample in the x-y plane (i.e. the angle between γ-direction and the surface normal n) and φ is the rotation angle of the sample around its normal.

Then the variation of the line intensities with the angle φ gives the information about the hyperfine field orientation in the film plane, i.e. in the ^{57}Fe marker-layers. This variation can be simulated by a linear combination of n distinct magnetic orientations Ψ_i:

$$\frac{I_2}{I_3}(\varphi) = 4\sum_{i=1}^{n} c_i \cdot \frac{1-\sin^2(\alpha)\sin^2(\varphi-\Psi_i)}{1+\sin^2(\alpha)\sin^2(\varphi-\Psi_i)} + 4\left(1-\sum_{i=1}^{n} c_i\right) \cdot \frac{1-\cos^2(\alpha)}{1+\cos^2(\alpha)} \qquad (2)$$

where c_i are the fractions of the spins pointing in direction Ψ_i in the films plane and the remaining spins are pointing out of the plane (along the plane normal). The measurement of the sample for all φ with non-normal incidence is named Magnetic Orientation Mössbauer Spectroscopy (MOMS), resolving the spin distribution in the sample.

Considering an intensity ratio I_2/I_3=4 for the α-iron of the normal incidence spectrum and thus a 100% in-plane alignment of the hyperfine field-vectors within the sample plane, Eq.(2) can be simplified to:

$$\frac{I_2}{I_3}(\varphi) = 4\sum_{i=1}^{n} c_i \cdot \frac{1-\sin^2(\alpha)\sin^2(\varphi-\Psi_i)}{1+\sin^2(\alpha)\sin^2(\varphi-\Psi_i)} \qquad (3)$$

which gives some characteristic variations of the line intensities as shown in Figure 2 for α=45°. There, it becomes obvious that MOMS can really distinguish different spin distributions in an iron film.

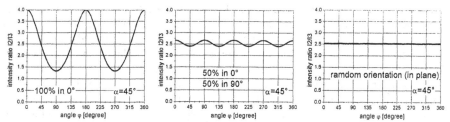

*Figure 2.*Characterisitc patterns for MOMS of thin iron films with an incidence angle of α=45° assuming all spins lying in the plane: left: only one spin direction, middle: two equally populated spin directions with 90° difference, right: random spin orientation in the plane.

This method was employed to investigate the spin distribution in thin iron film on silicon before and after Xe-ion irradiation, which is presented in the following.

3. Results and Discussion

Figure 3 shows the RBS spectra taken of the three samples I, M, and S before and after irradiation. One distinguishes the ^{57}Fe, ^{nat}Fe and Si components of the film and the implanted Xe ions (scaled by factor 5), whose concentration peaks at 30(2) nm, in agreement with the theoretic expectation mentioned before. A small sputtering effect of about 2 nm is visible, e.g. the change of the Si edge. The broadening of the ^{57}Fe marker-layers upon irradiation is not very prominent, thus preserving the depth information of the marker layers. The decrease of the ^{57}Fe signal is within the error limit. In addition to a weak <110> texturing effect of the deposited and irradiated films, the XRD spectra show a small lattice expansion, due to the implanted Xe stored in the iron lattice.

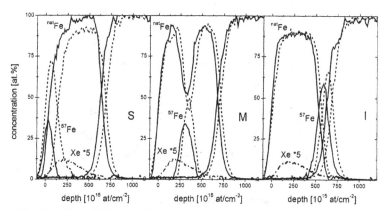

Figure 3. Depth profiles as RBS results for the three samples I, M, and S before (dotted) and after the Xe implantation (solid). The Xe concentration is scaled by a factor of 5.

First, the Mössbauer spectra taken with normal incidence were fitted using mainly the α-iron sextet, before implantation only sample I showed some interface contributions (17% of iron silicide) in addition to the pure iron. At the interface to the substrate (sample I), we added a second sextet and a doublet of together about 31%

fraction, indicating the increasing silicon fraction in the iron after implantation. After the irradiation we observed also the formation of a paramagnetic iron silicide (FeSi) with parameters in good agreement with [25,26]. At the surface (sample S) we observed the formation of some iron oxide after implantation, while for this sample S and sample M the implanted Xe is visible as a small additional sextet with 8(2)% relative area and B_{HF}=26(2) T representing iron atoms having a Xe atom as nearest neighbor. All these analyses of the Mössbauer spectra taken with normal incidence of the γ-beam indicate a complete in-plane orientation of the magnetization direction in the samples before and after irradiation, the values are given in Table I. Only sample M shows a smaller percentage of in plane alignment, but still within in the error limit. Maybe, in this case some spins also point out of the plane.

Now, for the CEMS measurements a proportional-counter was used, that surrounded the sample and could be tilted with respect to the γ-beam (α=0°-80°) and also could be rotated around the sample-normal (φ=0°-360°) as indicated already in Figure 1. With that, the samples have been also measured with α=45° and for angles φ ranging from 0° to 180° (due to the symmetry, the angles 180°-360° show the same and were omitted). Some of these spectra for sample M are shown in Figure 4.

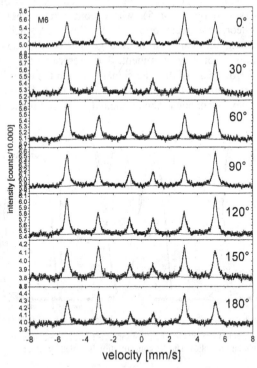

Figure 4. MOMS series of sample M for α=45° and at various sample angles φ (indicated).

It is clearly seen, that there is a nice variation of the line intensities with the angle j. After analyzing the spectra and taking the line intensity for the α-Fe subspectrum, the variations of all samples were achieved and these are displayed in Figure 5.

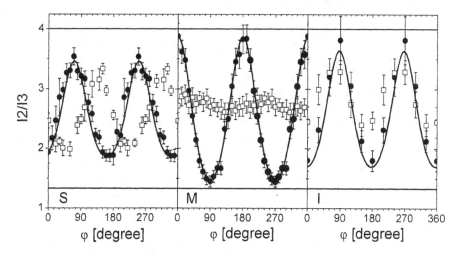

Figure 5. Intensity variations for the three samples as obtained by MOMS (□ before implantation, ● after implantation). The lines are fits with Eq.(3) and parameters are given in Table I. The theoretical minimum (1.3333) and maximum (4) is indicated.

Figure 5 presents the results of the angle dependent Mössbauer analysis (MOMS) on the α-Fe fraction of the three samples. There, the 0°-direction is the direction of the external strain during implantation. The solid lines are the results of fits with Eq. (3) for the irradiated samples and considering two in-plane terms of the hyperfine field direction. The parameters are summarized in Table I.

Before implantation, indicated by the hollow squares, the samples show a different behavior. At the Fe/Si-interface (sample I) one can clearly observe an oscillation of the I_2/I_3 ratio, indicating a preferred spin direction. A possible interpretation is the alignment of the field caused by lattice mismatches of Si and Fe which leads to permanent inverse magnetostriction effects [1]. The substrate orientation was observed by Laue-images and found to be (011) in 0°-direction. The best fit of the data was achieved when using a linear combination of two vectors, including an angle of 66°.

In the middle of the film (sample M) with the ^{57}Fe embedded in the ^{nat}Fe, no preferred hyperfine field direction is significant before implantation. The only disturbance of the film structure resulting in strains could occur due to the different deposition methods of the neighboring chemically and magnetic identical layers, but the spin orientation is isotropic. The behavior at the surface (sample S) is similar to the interface (sample I). It shows a preferred hyperfine field direction with a rather strong intensity. The preferred direction is at 71(3)° with a fraction of 80(2)%.

The situation after irradiation is indicated by the filled symbols in Figure 5. Sample I shows the smallest effect. The intensity and the direction of the oscillations change only slightly, what might have different reasons. First of all, the penetration depth of the Xenon is only 34(12) nm. Therefore, less than 8% Xe deposits their energy in this part of the film. The most obvious change of the hyperfine field parameters takes place in the middle of the film (M). The φ-dependence changes from a nearly isotropic distribution to a nearly perfect (95(1)%) alignment in 0° direction of the sample. At this depth, we also have the highest Xe-content after irradiation, indicating the strong dependence of the texturing effect on this parameter. Combined with the relaxation of the internal strains of the as deposited film during irradiation, the Xe induces a small widening of the lattice, as already mentioned as a result of the XRD-analysis, resulting in a strong inverse magnetostriction effect. The widening is much larger in the middle of the film than at the Fe/Si interface and induces a net compressive strain, resulting in the 0° alignment of the B_{HF}-direction [10]. At the surface (S), a similar behavior can be observed as at the interface: The amplitude and the direction of the φ-dependence changed, but here the change of the angle is much larger than in the case of sample I. All ions pass through this surface layer, resulting in the highest energy deposition. But as can be seen in Figure 2, there is almost no Xenon remaining in this part, resulting again in lower lattice deformations. All results of the MOMS experiments are summarized in Table I.

TABLE 1. MOMS and MOKE results for the three samples before and after ion implantation.

Sample	before implantation MOMS			after implantation MOMS			MOKE
	$I_2/I_3(\alpha=0°)$	Ψ_i	c_i	$I_2/I_3(\alpha=0°)$	Ψ_i	c_i	easy axis
I	3.93(28)	70(8)°	68(5)%	4.19(19)	87(4)°	86(3)%	63(3)°
		160(8)°	32(5)%		153(3)°	14(3)%	
M	3.78(21)	33(10)°	5(3)%	4.00(18)	0(3)°	95(1)%	1(3)°
		75(40)°	3(2)% *		90(3)°	5(1)%	
S	4.16(10)	136(4)°	74(2)%	4.23(33)	71(3)°	80(2)%	101(4)°
		226((4)°	26(2)%		161(3)°	20(2)%	

* remaining part is randomly distributed in the plane.

To compare the results with the macroscopic magnetic properties of the samples we made an angle dependent longitudinal MOKE-analysis. The measurements are shown in Figure 6 and the results also given in Table I. Before implantation, the magnetic remanence, as obtained from the hysteresis curves, are nearly isotropic. After implantation we obtain a rather strong magnetic anisotropy with a fourfold and a uniaxial contribution as can be seen in the polar plots. Since MOKE probes the first 20-30 nm of the film, at least for sample S the information range of CEMS and MOKE are the same. One would expect a similar preferential direction of the magnetization, since we would expect the hyperfine field and the easy axis to coincide. It is not yet clear, why the methods give different results, but MOMS gives an instantaneous ('non-destructive') picture of the spin distribution, the MOKE analysis demands the application of a strong external field and thus immanently changes actively the spin distribution. Further investigations are under way in this direction. Nevertheless, for the middle of the film (sample M), MOMS and MOKE results are in perfect agreement, although the MOKE signal from this film-depth is very small. At the interface I, the measurements are not directly comparable, since they probe non-overlapping parts of the film. As can be seen in Table I, the uniaxial part of the anisotropy is not equal in the three samples.

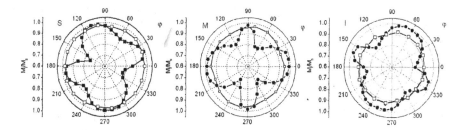

Figure 6. Polar plots of the film remanence as measured by MOKE (sample S, M, and I, from left).

This behavior is not understood so far and further experiments have to be carried out in that direction.

4. Conclusions

In conclusion, MOMS seems to be a very interesting method to study magnetic anisotropies - with the help of ^{57}Fe marker-layers also depth dependent - and resolving the spin distribution in a non-destructive way. Directly after the irradiation, the magnetization of the film is obviously quite well aligned in certain axes for all depths. Up to now we were not able to find a correlation between the MOMS and MOKE-analysis for the surface and the interface layers, but we found a very good agreement for the region in the middle of the film coinciding exactly with the external strain axis.

Acknowledgements

This work is supported by the Deutsche Forschungsgemeinschaft (DFG) within the SFB 602 in Göttingen under grant B4.

References

1. Sander, D. (1999) Rep. Progr. Phys. 62, 809.
2. Chikazumi, S. (1997) *Physics of Ferromagnetism, second edition*, Clarendon press, Oxford.
3. Heinrich, B, and Bland, J. A. C. (1994) in: Ultrathin Magnetic Structures, Springer, Berlin, p. 21.
4. Huber, A., and Schäfer, R. (1998) *Magnetic Domains – the Analysis of Magnetic Microstructures*, Springer, Berlin.
5. Lewis, W. A., Farle, M., Clemens, B. M., and White, R. L. (1994) Magnetic and structural modifications in Fe and Ni films prepared by ion-assisted deposition, *J. Appl. Phys.* **75**, 5644-5646.
6. Farle, M., Saffari, H., Lewis, W. A., Kay E., and Hagstrom, S. B. (1992) In-plane magnetic anisotropies in polycrystalline Ni films induced by Xe bombardment during growth, *IEEE Trans. Magn.* MAG-28, 2940-2942.
7. Bernas, H., Devolder, T., Chappert, C., Ferré, J., Kottler, V., Chen, Y., Vieu, C., Jamet, J.P., Mathet, V., Cambril, E., Kaitasov, O., Lemerle, S., Rousseaux, F. , and Launois, H. (1999) Ion beam induced magnetic nanostructure patterning, *Nucl. Instr. Meth* **B 148**, 872.
8. Neubauer, M., Reinecke, N., Uhrmacher, M., Lieb, K.-P., Münzenberg, M., and Felsch, W. (1998) Ion-beam induced magnetic anisotropies in iron films, *Nucl. Instr. Meth.* **B 139**, 332-337.
9. Gupta, R., Müller, G. A., Zhang, K., Uhrmacher, M., Schaaf, P., and Lieb, K.-P., to be published.

136

10. Zhang, K., Lieb, K.-P., Schaaf, P., Uhrmacher, M., Felsch, W., and Münzenberg, M. (2000) Ion-beam-induced magnetic texturing of thin nickel films, *Nucl. Instr. Meth.* **B 161-163**, 1016-1021.

11. Lieb, K.-P., Zhang, K., Müller, G.A., Schaaf, P., Uhrmacher, M., Felsch, W., and Münzenberg, M. (2001) Magnetic textures in thin ion-irradiated Ni and Fe films, *Act. Phys. Pol.* **A 100**, 751-760.

12. Devolder, T., Chappert, C., Chen, Y., Cambril, E., Bernas, H., Jamet, J.P., and Ferré, J. (1999) Sub-50 nm planar magnetic nanostructures fabricated by ion irradiation, *Appl. Phys. Lett.* **74**, 3383-3385.

13. Zemcik, T., Kraus, L., and Zaveta, K. (1989) Mössbauer-spectroscopy of creep-annealed soft magnetic amorphous-alloys, *Hyp. Int.* **51**, 1051-1059.

14. Schaaf, P. (2002) Laser Nitriding of Metals, *Progress in Materials Science* **47**, 1-161.

15. Woike, Th., Imlau, M., Angelov, V., Schefer, J., and Delley, B. (2000) Angle-dependent Mössbauer spectroscopy in the ground and metastable electronic states in Na-2[Fe(CN)(5)NO] center dot 2H(2)O single crystals, *Phys. Rev.* **B 61**, 12249-12260.

16. Reuther, H., Behr, G., and Teresiak, A. (2001) Determination of the hyperfine parameters of α-FeSi$_2$ by angle dependent Mössbauer spectroscopy on single crystals, *J. Phys.: Condens. Matter* **13**, L225-L229.

17. Stoney, G.G. (1909) *Proc. Roy. Soc.* (London) **A 82**, 172.

18. Uhrmacher, M., Pampus, K., Bergmeister, F.J., Purschke, D., and Lieb, K.-P. (1985) Energy calibration of the 500 kV heavy-ion implanter IONAS, *Nucl. Instr. Meth.* **B 9**, 234-242.

19. Ziegler, J.F., Biersack, J.P., and Littmark, U. (1985) *The stopping and range of ions in solids*, Pergamon Press, New York.

20. Doolittle, L.R. (1985) Algorithms for the rapid simulation of Rutherford backscattering spectra, *Nucl. Instr. Meth.* **B9**, 344-351.

21. Müller, A. (2002) PhD thesis, University of Göttingen.

22. Gonser, U. (1975) *Mössbauer Spectroscopy*, Springer, New York..

23. Wertheim, G.K., Wernick, J.H., and Buchanan, D.N.E. (1966) *J. Appl. Phys.* **37**, 333.

24. Wegener, H. (1966) *Der Mößbauereffekt und seine Anwendungen*, Bibliographisches Institut, Mannheim; Wertheim, G.K. (1964) *The Mössbauer Effect: Principles and Applications*, Academic Press, New York.

25. Ogale, S.B., Joshee, R., Godbole, V.P., Kanetkar, S.M., and Bhide, V.G. (1985) Ion-beam mixing at Fe-Si interface - an interface-sensitive conversion electron Mössbauer spectroscopic study, *J. Appl. Phys* **57**, 2915-2920.

26. Fanciulli, M., Rosenblad, C., Weyer, G., Svane, A., Christensen, N.E., von Kanel, H. and Rodriguez, C.O. (1997) The electronic configuration of Fe in β-FeSi$_2$, *J. Phys.: Condens. Matter* **9**, 1619-1630.

MÖSSBAUER EFFECT IN IRON-BASED NANOCRYSTALLINE ALLOYS

An Attempt to Evaluate the Spectra in a Generalized Two Level Relaxation Model

O. HUPE[1], M. A. CHUEV[2], H. BREMERS[1], J. HESSE[1], A. M.
AFANAS'EV[2], K. G. EFTHIMIADIS[3] AND E. K. POLYCHRONIADIS[3]
[1] *Institut für Metallphysik und Nukleare Festkörperphysik, Technische Universität Braunschweig, Mendelssohnstrasse 3, D-38106 Braunschweig, Germany*
[2] *Institute of Physics and Technology, Russian Academy of Sciences, Nakhimovskii pr. 34, 117218 Moscow, Russia,*
[3] *Department of Physics, Aristotle University, GR54006 Thessaloniki, Greece.*

1. Introduction

Nanocrystalline ferromagnetic alloys of the Fe-Cu-Nb-B type consist of a ferromagnetic amorphous matrix in which nanograins are embedded. These nanograins normally are so small (typical diameter in the range of 5-30 nm) that, if they were free, their magnetic relaxation time would fall into the time window of the ^{57}Fe Mössbauer spectrometry and relaxation phenomena would become visible. It is well known that Mössbauer spetroscopy easily perceives relaxation processes on the time scale (10^{-11} s to 10^{-6} s). Normally the nanograins are coupled to the ferromagnetic matrix and an interaction between these two components must be regarded, too. This coupled situation influences the thermally driven fluctuation of the nanograins' magnetization and often led to the very common interpretation of Mössbauer spectra in terms of hyperfine field distributions (also distributions of quadrupole splitting and isomer shift). Due to the high degree of complexity such spectra consist of many spectral components and an analysis becomes a rather complex task. Recently Miglierini and Greneche [1,2] proposed a new idea to interpret such hyperfine field distributions. From their paper follows that in the spectra information is available about the so-called interface between the nanograins and the residual matrix, too. Nevertheless contrary arguments (Balogh et al. [3]) and also the idea of relaxation influence on the linewidth are present in literature (Kemeny et al. [4]).

The situation may change dramatically when the temperature of the sample becomes enhanced and the Curie temperature of the residual amorphous matrix is exceeded. So collecting Mössbauer spectra versus temperature a change in the magnetic behaviour of the nanograins is expected. Here it seems to be true that the physics of the alloy will depend on the mean distance of the nanograins. Fortunately the alloy system Fe-Cu-Nb-

137

M. Mashlan et al. (eds.), Material Research in Atomic Scale by Mössbauer Spectroscopy, 137–146.
© 2003 *Kluwer Academic Publishers. Printed in the Netherlands.*

B allows investigations of samples with different contents of bcc-iron nanograins by changing the annealing temperature.

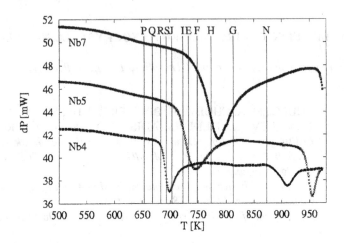

Figure 1. The heat production dP during nano-crystallization of the $Fe_{(86-x)}Cu_1Nb_xB_{13}$ (x=4,5,7) alloys versus temperature. The orthogonal lines indicated with P, Q, R,... mark the annealing temperature applied to each sample for one hour in He-gas atmosphere. Nb7, Nb5 and Nb4 indicate the Nb content.

2. The Samples in Question

The experiments are performed on ferromagnetic $Fe_{(86-x)}Cu_1Nb_xB_{13}$ (x=4,5,7) alloys. In Figure 1 the heat production dP during nano-crystallization of these three alloys versus temperature is plotted. Nb7, Nb5 and Nb4 indicate the Nb content in at%. The orthogonal lines with letters E, F, G, ... indicate the time sequence of sample's preparation and the annealing temperature applied for one hour in He-gas atmosphere in order to produce a defined amount of bcc-Fe nanograins. Due to the different amounts of crystallites (while the grain size nearly remains the same) their mean distance changes and therefore the strength of their interaction is influenced.

Mössbauer spectroscopy results presented here are chosen only from the alloys with 7 at% Nb. The amorphous ribbons (thickness about 24 μm, width of 14.8 mm) were provided by the Vacuumschmelze GmbH, Hanau, Germany (Dr. G. Herzer). One favourable feature of this alloy system is that the appearing nanocrystallites consist of pure bcc-Fe. So their spectral component in the Mössbauer spectrum is expected to be "simple". We found by transmission electron microscopy (TEM) that the nanograins in the samples considered are nearly of the same size with a rather narrow distribution (Figure 2). Therefore we used the relative amount of the Mössbauer spectral component for bcc-Fe as an objective measure for the number of nanograins in the samples after different annealing (which is true also in the case when the Mössbauer-Lamb factor is different for the bcc and amorphous phases).

In Figure 2 TEM photos of the En7 (annealed 1h at 733 K, left) and of the Fn7 (annealed 1h at 748 K, right) samples are shown. The rather narrow size distribution of their nanograins can directly be observed. Due to the etching process needed for thin-

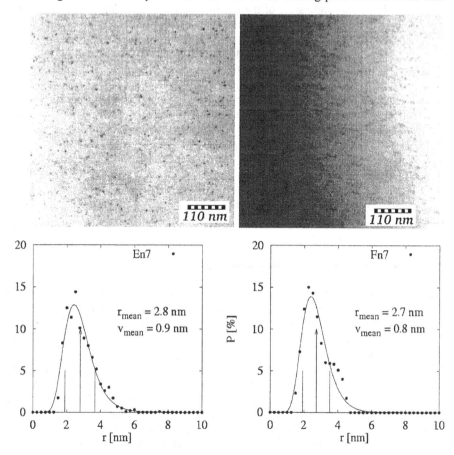

Figure 2. TEM photos of the En7 (annealed 1h at 733 K, left) and of the Fn7 (annealed 1h at 748 K, right) samples are presented. In the lower part of the Figure the corresponding size distributions of the nanograins are shown.

ning the foils for TEM measurements each sample exhibits different thickness, which causes the impression that the particle density is increasing looking from the bright to the darker parts. In the bottom of Figure 2 the results of a computer analysis of the size distribution is presented, too. These results show that there is no markedly change in the size distribution enhancing the annealing temperature. The particles radius distribution was determined by scanning the digitalized TEM photo with self developed software (K. Efthimiadis) and fitted afterwards by a lognormal distribution.

3. The Mössbauer Spectra

The spectra are collected in a wide temperature range, covering both the regions below and above the Curie temperature of the amorphous matrix. The highest temperature of measurement is still below the annealing temperature to prevent further crystallisation. Room temperature spectra are taken before and after the measurements to check this. In Figure 3 a temperature sequence of Mössbauer spectra collected on both samples $Fe_{79}Cu_1Nb_7B_{13}$ annealed at 733 K (sample En7) and 748 K (sample Fn7) are shown. We have chosen these examples to demonstrate a very typical spectral line shape for the bcc-Fe component often observed in similar alloy systems. The lines are asymmetric with very sharp outer and smoothly smeared inward tails. Lines of this unusual form have been observed in many studies of systems with superparamagnetic particles reported earlier [5-9] and could not fit the conventional two-level relaxation model without taking into account a rather wide distribution over the particles' sizes. An alternative to interpret such Mössbauer spectra was recently suggested by Afanas'ev and Chuev [10]. Their contribution considers the thermally driven fluctuation of the nanoparticle's magnetic moment regarding the particle-particle interaction in the frame of Stoner-Wohlfarth single domain particles

4. The Generalized Two-level Relaxation Model (GTLR)

The magnetic moment of a Stoner-Wohlfarth particle (SW) [11], with particles fixed in space, changes its direction in a homogeneous way (coherent reversal). Normally there is no interaction considered between SW particles. The magnetization vector fluctuates like a single magnetic moment with a very large value being typically (10^4-10^5) μ_B (Bohr magnetons). SW particles exhibit uniaxial magnetic anisotropy energy (one easy axis). The equilibrium position of the particle's magnetization is given by a competition of the anisotropy energy and the energy due to the applied external magnetic field. Both the easy axis and the external magnetic field define a plane in which the magnetization vector of the particle is located. Plotting the energy of the magnetization vector versus its angle with respect to the external magnetic field (or easy axis) two minima are observed (if the field strength does not exceed the so-called switching field strength).

The deeper minimum corresponds to the more or less parallel orientation of the magnetization vector to the external field. Because of these two energy minima thermal activation can lead to "jumps" from one orientation to the other. Therefore a description of this process is called as the two level relaxation model.

Very recently Afanas'ev and Chuev [10] proposed a new idea for interpretation of relaxation spectra on systems of interacting SW particles. As a consequence the particles exhibit two different energy levels (also in a system of identical particles) because the particle-particle interaction is causing magnetic fields acting at each particle position. If the interaction is not too strong, the equilibrium position of the particle's magnetization without external magnetic field remains the easy axis. Because of the large

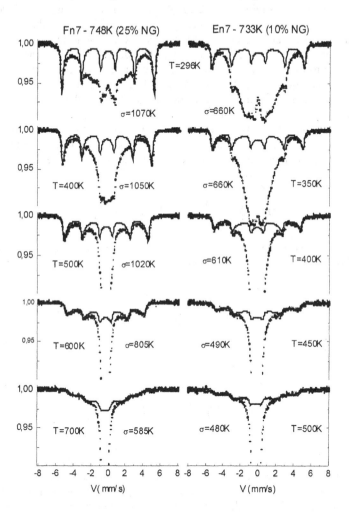

Figure 3. Mössbauer spectra of nanocrystalline ferromagnetic Fe$_{79}$Cu$_1$Nb$_7$B$_{13}$ samples with different content of nanograins: 25% NG (left) and 10% NG (right). Solid lines are calculated within the generalized two-level relaxation model for the outermost hyperfine magnetic component corresponding to NG of the same size but with different energy distribution widths σ.

values of the magnetic moments of the particles in nanostructured systems even particle-particle distances on the order of some particle diameters lead to interactions, which strongly smears out the particles energy levels. As a consequence the relaxation between the particle's states with opposite directions of the magnetization vector never proceed as a transition between the levels with the same energy. On the average the energy levels of all particles will be smeared out over an energy width ΔE, which may be rather large and comparable to the thermal energy. In [10] the magnetic fields caused by the neighbouring particles are treated as random variables, and the corresponding energy shifts are described as being distributed like a Gaussian form.

In their paper Afanas'ev and Chuev showed that depending on the energy width ΔE, in comparison to the thermal energy $k_B T$, this gives rise to variously shaped relaxation Mössbauer spectra of very characteristic spectral line forms in the six-line ^{57}Fe patterns. The most representative is the fast relaxation limit when the fluctuation of magnetization in the nanograins is very fast in comparison to typical Mössbauer nuclei Larmor precession time and the hyperfine field H_{hf} becomes time average (1), were w_1 and w_2 are the normally unequal quasi-equilibrium populations for the magnetization vector of the particles to point in the preferred orientation or in the opposite one.

$$\overline{H}_{hf}(\Delta E) = \left[w_1(\Delta E) - w_2(\Delta E) \right] \cdot H_{hf} \tag{1}$$

In the simplest case of weak interaction $\Delta E << U_0$, (U_0=anisotropy energy barrier for the particle) they can be written in the form [10]:

$$w_{1,2}(\Delta E) = \frac{\exp(\mp \Delta E / k_B T))}{\exp(\Delta E / k_B T)) + \exp(-\Delta E / k_B T))} \tag{2}$$

This simplification leads to (3)

$$\overline{H}_{hf}(\Delta E) = \tanh\left(\frac{\Delta E}{k_B T} \right) \cdot H_{hf} \tag{3}$$

which results in a well-resolved six-line hyperfine structure at non-zero ΔE in the fast relaxation limit and, as a result, to appearence of highly asymmetrical lines with steep outer and extended inwards tails, just as observed in the experiments. For this case a completely new fit spectral function $S(v)$ was proposed in [10] were v stands for the velocity, $v_{Mm} = (\mu_e M - \mu_g m) H_{hf} c / E_\gamma$ indicates 'static' positions of spectral lines of the 'static' hyperfine structure. I_{Mm} are the relative transition intensities, Γ_0 is the natural line width. A spread over the interaction parameter ΔE is represented by a distribution function $P(\Delta E, \sigma)$ of the width σ.

$$S(v) = \sum_{M,m} I_{Mm} \int_{-\infty}^{\infty} \frac{1}{\left[v - v_{Mm} \tanh(\Delta E / k_B T)\right]^2 + \Gamma_0^2 / 4} P(\Delta E, \sigma) d(\Delta E) \qquad (4)$$

The Mössbauer measurements shown in Figure 3 were performed in the temperature range from 300 K up to 700 K. The samples consist of embedded nanograins with an average grain size of about 5.6 nm but due to different annealing with different nanograin content and therefore with different particle distances. The numerical evaluation was performed using the DISCVER [12] program.

5. The Fit Results

For the fit procedure incorporating Eq. (4) within an assumption of the Gaussian shape of the distribution $P(\Delta E, \sigma)$ it was used that the line-sextet corresponding to the bcc-Fe nanograins is well resolved which allows one to carry out a separate fit without fitting simultaneously the spectrum corresponding to the remaining amorphous phase and the interface. In the same time all the spectra taken at different temperatures for each of two $Fe_{79}Cu_1Nb_7B_{13}$ samples were fitted simultaneously, which exhibit highly asymmetric lines with steep outer and extended inwards tails as shown in Figure 3. Solid lines are calculated within the generalized two-level relaxation model for the outermost hyperfine magnetic component corresponding to nanograins of the same size with different energy distribution widths σ.

Figure 3 demonstrates a very good description of the outermost lines of all the spectra collected. Also the resolved magnetic hyperfine structure with lines of strongly asymmetrical shape mentioned above at temperatures higher than the Curie temperature of amorphous phase are evidenced for both samples.

Remembering that solid lines in Figure 3 are calculated within the GTLR model for the only hyperfine magnetic component corresponding to nanograins of the same size, one can understand that in this case there is no need to introduce a broad distribution of particle's size at least for this nanograin phase even at higher temperatures. This is because the relaxation behaviour of the interacting particles results in the characteristic broadening of the spectral lines. In real samples of course a particle size distribution will appear. The GTLR model does not deny principally the particle's distribution over sizes but taking into account the interparticle interaction should strongly modify the shape of the distributions obtained by conventional interpretation of Mössbauer spectra.

The temperature dependences of the hyperfine fields (top) and energy distribution widths $\sigma = k_B T_{Width}$ (bottom) presented as T_{Width} in Figure 4 were obtained from the fits to the spectra presented in Figure 3. Rectangular symbols (top) correspond to the bulk values of hyperfine field in pure iron plotted for comparison.

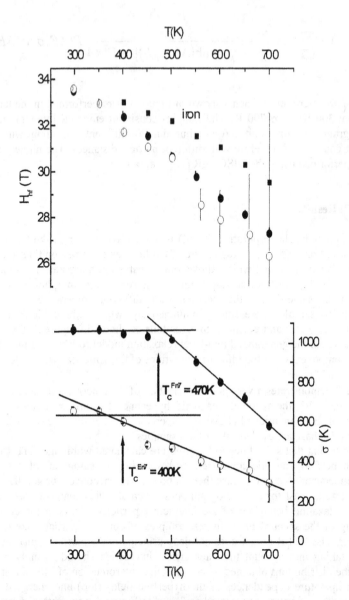

Figure 4. Temperature dependences of the hyperfine fields (top) and energy distribution widths σ (bottom) obtained from fitting the outermost lines corresponding to bcc-iron nanograins of the spectra shown in Figure 3: Sample Fn7-748K with 25% nanograins (closed circles) and sample En7-733K with 10% nanograins (open circles). The rectangular symbols (top) correspond to the bulk values of hyperfine field in pure iron. Arrows indicate the Curie temperature of the corresponding amorphous phases estimated from Mössbauer thermal scan measurements.

As Figure 4 shows, despite of considerable difference between the temperature evolutions of Mössbauer spectra for the two samples presented, the hyperfine field values for nanograins in these samples are practically the same up to higher temperatures and slightly deviate from the temperature dependence of bulk values of hyperfine field in pure iron plotted for comparison, too. On the other hand, the differences in the temperature evolution of the spectra for the two samples as well as the forms of this evolution mentioned above seem to be governed by the temperature dependence of energy distribution width σ characterizing interparticle interactions. At least, it is obviously seen in Figure 4 that the $\sigma(T)$ dependences exhibit remarkable changes in the interactions just above the Curie temperature of the amorphous phase for both the samples, which again evidences for an essential interrelation between the magnetic behaviour of nanograins and amorphous matrix, as previously suggested by [1,2].

The description of the measured spectra in the limit of fast relaxation is a very good approximation. Fitting the spectra by the general expression (formula (2) in [10]) with assuming the Gaussian energy distribution leads to an estimation for the statistical fluctuation frequency $p_0 = (1.0 \pm 0.8) \cdot 10^{11} s^{-1}$ and $U_0 = (1100 \pm 300)$ K, the anisotropy energy of the particle's magnetic moment.

In conclusion the generalized two-level relaxation model (GLTR) proved to be rather efficient in describing specific shapes of Mössbauer spectra of nanocrystalline ferromagnetic alloys and actually represents a new approach to analysis of the spectra taking into consideration the interparticle interaction in not so complicated form. Of course, there still remain a number of problems. First of all, the GTLR model must be extended to the case of time-dependent energy shifts ΔE that should be correlated over mutually interacting superparamagnetic particles. Then, such an extended model should be realized as a computer program for fitting experimental spectra taking into account different magnetic phases in a self-consistent way.

Even a detailed quantitative analysis of Mössbauer spectra collected on nanocrystalline ferromagnetic alloys obviously requires the development of efficient computational procedure taking into account both the relaxation of interacting particles and their distribution over sizes. Nevertheless the importance of the new relaxation model was proved by this real experiment.

Acknowledgements

We are grateful to the "Internationales Büro des BMBF", Bonn, and the Russian Ministry of Science and Technology, Moscow, for supporting our collaboration within the project RUS-157-97. The German authors are thankful to the Deutsche Forschungsgemeinschaft for supporting the experimental work.

References

1. Miglierini, M. and Greneche, J.-M. (1997) Mössbauer spectrometry of Fe(Cu)MB-type nanocrystalline alloys: I. The fitting model for the Mössbauer spectra, *J.Phys.: Condens. Matter* **9**, 2303-2319.

146

2. Miglierini, M. and Greneche, J.-M. (1997) Mössbauer spectrometry of Fe(Cu)MB-type nanocrystalline alloys: II. The topography of hyperfine interactions in Fe(Cu)ZrB alloys, *J.Phys.: Condens. Matter* **9**, 2321- 2347.
3. Balogh, J., Bujdoso, L., Kaptás, D., Kemény, T., Vincze, I., Szabo, S. and Beke, D. (2000) Mössbauer-study of the interface of iron nanocrystallites, *Phys. Rev. B* **61**, 4109-4116.
4. Kemény, T., Kaptás, D., Balogh, J., Kiss, L. F., Pusztai, T. and Vincze, I. (1999) Microscopic study of the magnetic coupling in a nanocrystalline soft magnet *J. Phys.: Condens. Matter* **11**, 2841-2847
5. Mørup, S. (1994) Superferromagnetic nanostructures, *Hyperfine Interact.* **90**, 171-185.
6. Mørup, S. and Tronc, E. (1994) Superparamagnetic relaxation of weakly interacting particles, *Phys. Rev. Lett.* **72**, 3278-3281.
7. Tronc, E., Prené, P., Jolivet, J.P., d'Orazio, F., Lucari, F., Fiorani, D., Godinho, M., Cherkaoui, R., No-guès, M. and Dormann, J.L. (1995) Magnetic behaviour of γ-Fe₂O₃ nanoparticles by Mössbauer spectros-copy and magnetic measurements, *Hyperfine Interact.* **95**, 129-148.
8. Dormann, J.L., D'Orazio, F., Lucari, F., Tronc, E., Prené, P., Jolivet, J.P., Fiorani, D., Cherkaoui, R. and Noguès, M. (1996) Thermal variation of the relaxation time of the magnetic moment of γ-Fe₂O₃ nanopar-ticles with interparticle interactions of various strengths, *Phys. Rev. B* **53**, 14291-14297.
9. Tronc, E., Ezzir, A., Cherkaoui, R., Chanéac, C., Noguès, M., Kachkachi, H., Fiorani, D., Testa, A.M., Grenèche, J.-M. and Jolivet, J.P. (2000) Surface-related properties of γ-Fe₂O₃ nanoparticles, *J. Magn. Magn. Mater.* **221**, 63-79.
10. Afanas'ev A. M. and M.A.Chuev M. A. (2001) New Relaxation Model for Superparamagnetic Particles in Mössbauer Spectroscopy, *JETP Lett.* **74**, 107-110.
11. Stoner, E. C. and Wohlfarth, E. P. (1948) A mechanism of magnetic hysteresis in heterogeneous alloys *Phil. Trans. Roy. Soc. London*, **240A**, 599-642.
12. Afanas'ev, A.M. and Chuev, M.A. (1995) Discrete forms of Mössbauer spectra, *JETP* **80**, 560-567.

NANOCRYSTALLINE $Fe_{81-x}Ni_xZr_7B_{12}$ (x = 10 - 40) ALLOYS INVESTIGATED BY MÖSSBAUER SPECTROSCOPY

M. KOPCEWICZ[1] AND B. IDZIKOWSKI[2]
[1]Institute of Electronic Materials Technology, Wólczyńska 133, Warsaw, Poland
[2]Institute of Molecular Physics, Polish Academy of Sciences, Smoluchowskiego 17, Poznań, Poland

1. Introduction

Nanocrystalline alloys, produced by partial devitrification of melt-spun amorphous precursor, exhibit a two-phase structure with nanocrystalline bcc grains dispersed in the residual amorphous matrix. This was first established for FINEMET [1] followed by NANOPERM [2,3] and HITPERM [4]. The soft magnetic properties are due to the reduction of effective magnetic anisotropy randomly averaged by exchange interaction and is related to the refinement of the grain size [5].

Crystallization behaviour of the amorphous precursors has attracted a lot of attention in recent years. The search for new nanocrystalline alloys with improved magnetic and mechanical properties, favourable for technical applications, continues. So far, almost all soft magnetic nanocrystalline alloys consisted of a homogeneous nanostructure of bcc-Fe(Si), bcc-Fe or bcc-(FeCo) phases embedded in an amorphous matrix. We prepared novel amorphous $Fe_{81-x}Ni_xZr_7B_{12}$ (x = 10 - 40) alloys in which the nanostructure was formed, as expected, by controlled annealing. The significant presence of Ni in these alloys may result in the formation of fcc-type phases in the course of annealing. It is of particular interest to study such alloys and to attempt an understanding of the details of the complex crystallization process leading to the formation of fcc-based soft magnetic nanostructured alloys. Structural changes induced by thermal treatment of amorphous precursors were characterized by Mössbauer spectroscopy (in transmission geometry and also by using the conversion electron Mössbauer spectroscopy – CEMS) supplemented by differential scanning calorimetry (DSC) and X-ray diffraction (XRD) measurements. The crystallization of Ni-containing amorphous alloys was studied earlier (e.g., [6,7]) but these alloys were unsuitable for nanocrystalline alloys. However, the nanocrystalline grains were found in the rapidly heated amorphous $Fe_{15}Ni_{60}Si_{10}B_{15}$ and $Fe_{20}Ni_{55}Si_{10}B_{15}$ systems [7]. The only Ni-containing amorphous system, $Fe_{40}Ni_{38}Mo_4B_{18}$, in which the nanostructured phase was clearly formed was studied by the XRD technique [8].

147

M. Mashlan et al. (eds.), Material Research in Atomic Scale by Mössbauer Spectroscopy, 147–158.
© 2003 Kluwer Academic Publishers. Printed in the Netherlands.

2. Details of Experiments

Amorphous $Fe_{81-x}Ni_xZr_7B_{12}$ (x = 10 - 40) alloys were prepared by rapid quenching from the melt in an Ar protective atmosphere. The ribbons were 3-4 mm wide and 20-25μm thick. All quenched alloys were fully amorphous, as verified by X-ray diffraction (XRD) and Mössbauer measurements. The nanocrystalline alloys were produced by annealing the amorphous precursors at temperatures ranging from 440°C to 670°C for 1 h in vacuum. Annealing at higher temperatures (up to 800°C) leads to the formation of phases with poor soft magnetic properties. The differential scanning calorimetry (DSC) measurements were performed to establish the crystallization temperatures of the first and second crystallization steps for which the DSC 404 Netzsch calorimeter was used. Conventional Mössbauer measurements were done for all quenched and annealed alloys. For samples with x=30 annealed at 670°C and 700°C and with x=40 annealed at 595°C - 670°C the Mössbauer measurements were performed at a low temperature range (80K - 300K). The isomer shifts relate to the α-Fe standard.

All quenched and annealed samples were also investigated by conversion electron Mössbauer spectroscopy (CEMS), which provided information on the near surface properties of the samples. The CEMS measurements were carried out at room temperature by using a gas-flow electron counter with He-6%CH$_4$ gas in which the investigated sample was placed.

The Mössbauer measurements were supplemented with X-ray diffraction measurements, which were performed at room temperature by using the Seifert diffractometer with Co K$_\alpha$ or Cu K$_\alpha$ radiation.

3. Results and Discussion

3.1. DSC MEASUREMENTS

Crystallization of amorphous $Fe_{81-x}Ni_xZr_7B_{12}$ (x = 10 - 40) alloys, as investigated by DSC linear-heating curves, clearly shows two crystallization stages with two characteristic crystallization temperatures. Such behaviour is typical for amorphous alloys serving as precursors for the formation of nanocrystalline alloys [9,10]. The crystallization temperatures for the first and second stages depend on the alloy composition. They decrease from about 510°C and 730°C for low Ni-content (x=10) to about 480°C and 650°C for high Ni-content (x=40), Fig. 1. All DSC measurements were carried out at a heating rate of 20 K/min.

3.2. CONVENTIONAL MÖSSBAUER AND CEMS STUDIES OF $Fe_{81-x}Ni_xZr_7B_{12}$ ALLOYS

The conventional Mössbauer measurements performed in transmission geometry at room temperature clearly reveal the changes in the microstructure of amorphous $Fe_{81-x}Ni_xZr_7B_{12}$ (x=10 - 40) alloys induced by annealing. The crystallization behaviour depends on the alloy composition, therefore the results obtained for low Ni content (x=10 and 20) and high Ni content (x=30 and 40) will be discussed separately.

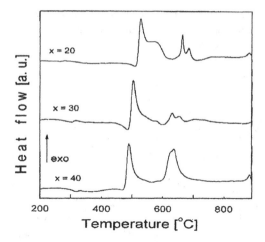

Figure 1. DSC measurements for x=20, 30 and 40 samples.

3.2.1. *Alloys with low Ni content (x=10 and 20)*

As a typical example the Mössbauer measurements in transmission geometry and the corresponding CEMS results are shown for $Fe_{61}Ni_{20}Zr_7B_{12}$ alloy in Fig. 2. The quenched alloy is fully amorphous (Fig. 2a, 2g). The crystallization behaviour of the alloys with x=10 and 20 resembles closely to that observed earlier for other nanocrystalline alloys, e.g., $Fe_{93-x-y}Zr_7B_xCu_y$ [11] and Fe-M-B-Cu (M:Ti, Ta Nb, Mo) [12, 13]. Up to an annealing temperature of T_A= 470°C, bulk crystallization is not observed in the spectra measured at transmission geometry (Fig. 2b). However, the surface crystallization is already evident for T_A= 470°C in both the x=10 and x=20 samples (Fig. 2h). This strongly suggests that crystallization starts at the surface of the amorphous ribbons, most probably because of different sample composition at the surface region as compared to the nominal bulk composition. Enhanced surface crystallization of nanocrystalline alloys has been attributed in the past [9] to boron depletion at the surface and the corresponding decrease in the crystallization temperature. Fairly strong bulk crystallization is observed after annealing at T_A= 520°C in both the x=10 and x=20 samples (Fig. 2c). The resulting crystalline phase is almost pure bcc-Fe (hyperfine field $H_{hf} \approx 33$ T). The surface crystallization is enhanced at all annealing temperatures in comparison to the bulk one, as seen in Fig. 2. Higher annealing temperatures (T_A= 570°C and 620°C) result in an increase of the spectral contribution of the sextet corresponding to the Ni-containing bcc-Fe phase. The hyperfine field slightly increases to about 33.5 - 34 T, which suggests that some Ni atoms are incorporated into the bcc-Fe phase. However, a new paramagnetic spectral component with a quadrupole doublet (QS≈0.60 mm/s, IS≈+0.04 mm/s) clearly appears in the spectra recorded for the samples annealed at T_A= 620°C for x=10 and at T_A= 570°C and 620°C for x=20 (Fig. 2d). A possible explanation can be the residual amorphous matrix whose Curie temperature is greatly reduced because of reduced Fe content in the amorphous phase due to the formation of the nanocrystalline Fe-rich phase. Additionally the spectral contribution of

150

the second sextet (with $H_{hf} \approx 30$ T), most probably related to the interfacial regions between the nanocrystalline grains and the residual amorphous matrix, enhances with increasing annealing temperature, up to $T_A = 620°C$.

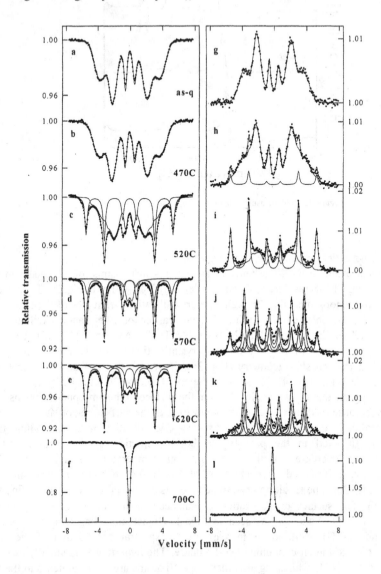

Figure 2. The transmission and CEMS spectra for the x=20 sample annealed at indicated temperatures.

At peak annealing temperatures ($T_A = 800°C$ for x=10 and $T_A = 700°C$ and $800°C$ for x=20) a single line spectral component appears in the transmission spectra. This component, whose spectral contribution is only marginal for the x=10 sample, becomes the only spectral component for x=20 sample (Fig. 2f). The single line component is not dependent on temperature. The spectra measured at 300K and 80K are virtually identical, except for a shift of the spectrum due to the second order Doppler shift which changes from about -0.044 mm/s to +0.060 mm/s at 300K and 80K, respectively. The linewidth is fairly small ($\Gamma \approx 0.35$ mm/s) and temperature independent. This spectral component appears at annealing temperatures exceeding the second crystallization stage (see DSC results, Fig. 1) and most probably originates from the fcc-FeNi phase paramagnetic at room temperature.

The CEMS spectra recorded for the x=10 and 20 samples annealed at T_A= 570°C and 620°C in addition to the bcc-Fe(Ni) phase show a large volume fraction of tetragonal iron borate (Fe_2B and Fe_3B) [10,14] which dominate in the spectra (Fig. 2j, 2k). The Fe_2B and Fe_3B phases were only formed at surfaces of the annealed samples. They were not found in the transmission measurements. The CEMS spectrum of x=20 sample annealed at peak temperatures shows only a single line spectral component (Fig. 2l) resembling that observed in Fig. 2f.

3.2.2. Alloys with high Ni content (x=30 and 40)

The crystallization process proceeds in a distinctly different manner in the samples with higher Ni content (x=30 and 40, Figs. 3 and 4, respectively) as compared with the samples containing less Ni (x=10 and 20) discussed above. Both the x=30 and 40 samples remain fully amorphous after annealing at T_A= 440°C. Annealing at T_A= 470°C causes the formation of the bcc-Fe phase accompanied by the residual amorphous matrix in both samples (sextet with narrow lines and $H_{hf} \approx 33$ T accompanied by the broad sextet due to hyperfine field distribution with the average $H_{hf} \approx 18$ T, Fig. 3b), but the crystalline bcc-Fe phase is more abundant in the x=30 sample than in the x=40 (about 13% of the total spectral area as compared with only 5% for x=40). The relative content of this phase in the x=30 sample dramatically increases with increasing annealing temperature, up to about 62% (Figs. 3c) and 75% for T_A= 570°C and 595°C, respectively. The hyperfine field resulting from this sextet slightly increases from 33 T for the pure bcc-Fe phase for T_A= 470°C (Fig. 3b) to 33.4 T for T_A= 595°C which suggests that some Ni atoms reside in the bcc structure.

For sample with x=30 annealed at 520°C-595°C another spectral component appears (Fig. 3c) - a quadrupole doublet with QS≈0.57 mm/s and IS≈+0.06 mm/s similar to that observed for x=10 and x=20. The spectral contribution due to retained amorphous matrix strongly decreases. In addition to the sextet related with the bcc phase additional broad sextet with $H_{hf} \approx 29$ T appears in Fig. 3c. This spectral contribution probably originates from the fcc-(Fe,Ni) phase secondary to annealing.

For the x=40 sample the relative spectral contribution of the bcc-Fe sextet slightly increases with increasing annealing temperature (to 15% for T_A= 520°C) and then decreases to 4% for T_A= 570°C and finally disappears for T_A= 582°C (Fig. 4b, 4c). A broad sextet due to hyperfine field distribution which prevails in the spectrum for the sample annealed at T_A= 582°C may correspond to the bcc-FeNi-type phase and to the $(FeNi)_{23}B_6$ phase, as suggested by the XRD measurements.

152

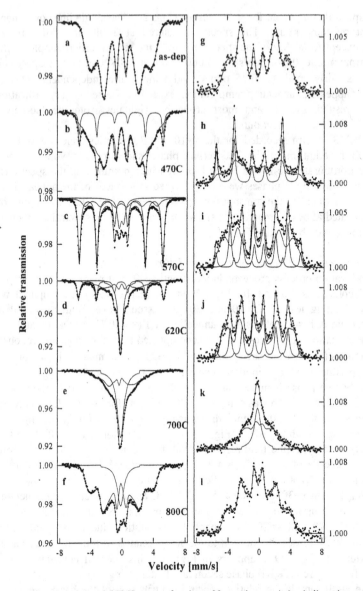

Figure 3. The transmission and CEMS spectra for the x=30 sample annealed at indicated temperatures.

The hyperfine field distribution, however, contains a significant peak at about 20T due to the retained amorphous phase. The corresponding CEMS spectra show quite similar behaviour of the surface crystallization. However, the formation of the crystalline component at the surface is stronger for the x=30 sample at $T_A \leq 582°C$ (Fig. 3) than for the x=40 sample (Fig. 4). The CEMS spectra of the x=30 sample clearly show the formation of the bcc-Fe phase at the surface of the sample annealed at $T_A \leq$ 520°C.

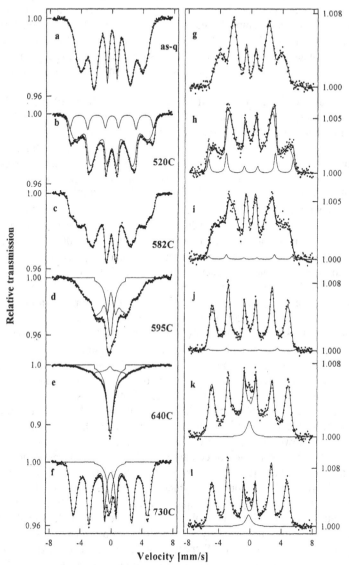

Figure 4. The transmission and CEMS spectra for the x=40 sample annealed at indicated temperatures.

However, at higher annealing temperatures (T_A= 570°C and 595°C the Fe$_2$B phase appears in addition to the bcc-Fe phase.

The XRD diffraction patterns of the x=40 sample in the quenched state and annealed at temperatures ranging from 582°C to 800°C show that the quenched sample is completely amorphous as revealed by the broad diffused peak in the diffractogram. At $T_A= 582°C$ the reflections from the cubic $(FeNi)_{23}B_6$ phase appear together with the amorphous halo. The sample annealed at $T_A= 640°C$ consists mainly of the $(FeNi)_{23}B_6$ phase. Annealing at higher temperatures ($T_A= 700°C$ and $800°C$) causes the formation of the fcc-(Fe,Ni) phase which dominates in the XRD patterns for $T_A= 800°C$. The lattice constants determined for the x=40 sample correspond well to the $(FeNi)_{23}B_6$ (1.051 nm and 1.068 nm for $T_A= 582°C$ and $640°C$, respectively) and to the fcc-FeNi phase (0.350 nm for $T_A= 800°C$).

Annealing at a temperature range of $620°C$ - $730°C$ for the x=30 and $595°C$ - $670°C$ for the x=40 samples, i.e., at temperatures close to the second peak in DSC curve (Fig. 1) causes a complete change in the Mössbauer spectra - a broad single-line component accompanied by the "magnetic wings" (Fig. 3e, 4d, 4e). A single-line component appears for the x=30 sample after annealing at $T_A= 595°C$. A single line is very broad (linewidth $\Gamma \approx 1.0$ mm/s) with isomer shift IS= -0.063 mm/s. Its spectral contribution increases from about 7% to over 50% for $T_A= 595°C$ and $700°C$, respectively. Higher annealing temperature causes a decrease in this spectral component to 40% at $T_A= 730°C$ and finally to 15% at $T_A= 800°C$ (Fig. 3f). The decrease in the spectral contribution of this nonmagnetic component at $T_A= 730°C$ and $800°C$ is associated with the simultaneous increase in the contribution of the magnetically ordered component, first in the form of the stronger magnetic wings (at $T_A= 730°C$) and then as a fairly well resolved broadened hyperfine field component (at $T_A= 800°C$, Fig. 3f). Similar features are observed for the x=40 sample. A single-line component appears in the spectrum of the sample annealed at $T_A= 595°C$ (spectral fraction about 25%, Fig. 4d). This fraction strongly increases to 45% and 60% for $T_A= 620°C$ and $640°C$ (Fig. 4e), respectively. For $T_A= 670°C$ the magnetically split component is partly restored and at $T_A \geq 700°C$ it prevails in the spectra (Fig. 4f). The corresponding CEMS spectra recorded for the x=30 and 40 samples do not reveal such a dominance of the single-line component. Only for the x=30 sample annealed at $T_A= 700°C$ the CEMS spectrum shows a shape similar to the transmission spectrum consisting of the dominating single-line component (Fig. 3k). Also the shape of the CEMS and transmission spectra for the $T_A= 800°C$ are similar (Figs. 3l and 3f). However, the spectra recorded for the $T_A= 620°C$ - $670°C$ are clearly different. The CEMS spectra do not contain a single-line component (which dominates in the transmission spectra) and consist of the broadened sextet with the typical hyperfine field reduced to about 28T which may suggest that the fcc-FeNi and/or $(FeNi)_{23}B_6$ phase is formed on the sample surfaces. Similar differences between the surface and bulk crystallization can be observed for the x=40 sample (Fig. 4). While the transmission spectra are dominated by a single-line component at $T_A= 595°C$ - $640°C$ (Fig. 4d, 4e) the corresponding CEMS spectra contain a broadened sextet with a typical hyperfine field of about 28T. Only at peak annealing temperatures ($T_A \geq 730°C$) the transmission and CEMS spectra become quite similar (Fig. 4f, 4l).

The origin of the single-line spectral component is difficult to explain. As suggested by the XRD measurements the dominant phase present in the x=40 sample annealed at $582°C$- $640°C$ is the $(FeNi)_{23}B_6$. Identification of this phase from Mössbauer data is difficult because the parameters given by various authors differ significantly. In some cases the $Fe_{23}B_6$ phase is claimed to be paramagnetic (spectrum consists of a quadrupole doublet, [15]), what seems rather unlikely, whereas in other cases this phase is magnetically ordered with hyperfine fields of about 28.8 T and 23.8 T [14]. Since in our study the phase formed is most probably $(Fe_{1-y}Ni_y)_{23}B_6$ the spectral component relating to this phase consists of a broad sextet which may show up in the hyperfine distribution as a peak at about 28 T. We observed such a case for the spectrum in Fig. 4c for T_A= 582°C. However, the XRD results suggest that at T_A= 582°C and 640°C the phase prevailing in the sample is the same, but the Mössbauer spectra recorded for these T_A are completely different (Fig. 4c, 4e) thereby making the phase identification more complicated. The single line observed for the x=30 and 40 samples (Figs. 3d, 3e, 4d 4e) is very broad and certainly does not correspond to the nonmagnetic fcc-FeNi phase observed for the x=20 sample (Fig. 2). Such a crystallization behaviour, in which a phase exhibits a broad single line spectral component, has not been observed for any nanocrystalline alloy known to the authors. The shape of the spectra consisting of such a single line accompanied by magnetic wings may suggest a relaxation origin. In order to verify the possibility of a relaxation origin of this spectral component the Mössbauer measurements were performed at a low temperature range (300K - 80K). Such Mössbauer measurements were performed for the x=30 sample annealed at T_A= 670°C and 700°C and for the x=40 sample annealed at T_A= 595°C, 620°C, 640°C and 670°C.

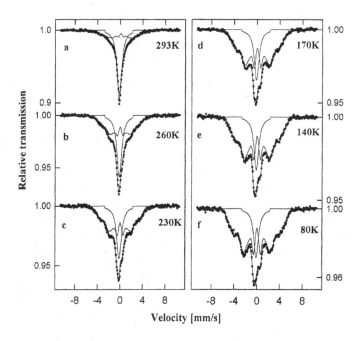

Figure 5. Low temperature Mössbauer spectra of x=40 sample (T_A=640°C).

The typical behaviour of the transformation of a single line to magnetic pattern at decreasing temperature is shown in Fig. 5. The spectral fraction of a single line strongly decreases with decreasing temperature and the magnetic component increases (Fig. 5).

The single line transforms gradually to a magnetic pattern (Fig. 6), which rules-out a simple para- to ferromagnetic phase transition. The observed gradual transformation of the single line to the magnetic pattern vs. temperature suggests a superparamagnetic relaxation origin of this line. A similar behaviour was observed in all other samples. The spectral contribution of the single-line component is shown in Table I.

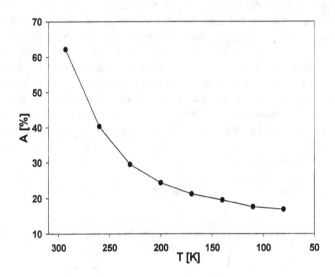

Figure 6. Temperature dependence of a single line component in the spectra shown in Fig. 5.

Most probably at favourable temperatures (T_A= 620°C - 640°C) the crystalline phase of the FeNi-type (($Fe_{1-y}Ni_y)_{23}B_6$ or fcc-(Fe,Ni)) is formed as very small nanograins separated in a residual matrix, probably of iron-poor NiZrB-type, that may allow a superparamagnetic relaxation. However, the presence of such an intergranular phase would be very difficult to detect by any technique. Superparamagnetic relaxation in nanocrystalline alloys, when a residual nonmagnetic amorphous matrix separates the nanograins, has been recently suggested [16].

The CEMS measurements, which in most cases do not reveal a single-line component, seem to support the superparamagnetic origin of this spectral component. In the surface regions of the samples, where the fcc-FeNi phase is formed, the relaxation effects do not occur (e.g., because of too large fcc nanograins) or are suppressed by the additional anisotropy related to the stress induced by enhanced surface crystallization.

Another possible interpretation of the single-line spectral component may be related to the formation of ($Fe_{1-y}Ni_y)_{23}B_6$ or fcc-$Fe_{1-y}Ni_y$ phases with a continuous variation in the Ni content which may result in a corresponding change of the Curie temperatures for these phases from above to below room temperature. The transformation of the single-line component vs. temperature in such a case could be related to the gradual transition through the Curie points for various Ni-content in the crystalline phases.

Given that the interpretation of the origin of the single-line spectral component in terms of the superparamagnetic relaxation seems to be the most plausible, other possibilities for elucidation of this novel phenomenon should not be left unexplored.

Table I. Spectral contribution of the single-line component.

Ni-content x	Annealing temperature T_A (°C)	Temperature of the measurement (K)	Spectral contribution of the single-line component (%)
30	670	300	50
		80	27
	700	300	55
		80	30
40	595	300	21
		80	10
	620	300	45
		80	15
	640	300	60
		80	17
	670	300	40
		80	20

4. Conclusions

A detailed study of the crystallization process of the amorphous $Fe_{81-x}Ni_xZr_7B_{12}$ (x = 10 - 40) alloys revealed a two-step crytstallization behaviour for all compositions. The crystallization process, studied by the transmission Mössbauer measurements and the conversion electron Mössbauer spectroscopy (CEMS), revealed clear differences between bulk and surface crystallization. At low Ni-content the CEMS spectra show the formation of the Fe_2B and Fe_3B phases, which were not observed for bulk crystallization, leading to the formation of the bcc-Fe(Ni) phase (at annealing temperatures T_A < 700°C) and finally to the paramagnetic fcc-FeNi phase, both in the bulk and at the surface. For higher Ni-content (x=30, 40) the broad single-line spectral component appears at medium annealing temperatures, most probably secondary to superparamagnetic relaxation of nanograins in the $(FeNi)_{23}B_6$ and fcc-FeNi phases. Gradual transformation of the single-line component to the magnetic pattern at decreasing temperature was observed in the low temperature transmission spectra. However, the CEMS spectra for the x=40 sample do not reveal the single-line spectral component. A possible explanation could be that the superparamagnetic relaxation is suppressed by the stress-induced anisotropy relating to enhance surface crystallization.

158

Formation of the $(FeNi)_{23}B_6$ and fcc-FeNi phases was confirmed by XRD measurements.

Acknowledgement

Financial support from NEDO (Japan) "Nanopatterned magnets" grant is gratefully acknowledged.

References

1. Yoshizawa, Y., Oguma, S., and Yamauchi, K. (1988) New Fe-based soft magnetic alloys composed of ultrafine grain structure, *J. Appl. Phys.* **64**, 6044-6046.
2. Suzuki, K., Ktaoka, N., Inoue, A., Makino, A., and Masumoto, T. (1990) High saturation magnetization and soft magnetic properties of FeZrB alloys with ultrafine structure, *Mat. Trans. JIM* **31**, 743-746.
3. Makino,. A., Inoue, A., and Masumoto, T. (1995) Nanocrystalline soft magnetic Fe-M-B (M=Zr, Hf, Nb) alloys produced by crystallization of amorphous phase, *Mat. Trans. JIM* **36**, 924-938.
4. Willard, M.A., Laughlin, D.E, McHenry, M.E., Thoma, D., Sickafus, K., Cross, J.O., and Harris, V.G. (1998) Structure and magnetic properties of $(Fe_{0.5}Co_{0.5})_{88}Zr_7B_4Cu_1$ nanocrystalline alloys, *J. Appl. Phys.* **84**, 6773-6777.
5. Herzer, G. (1993) Nanocrystalline soft magnetic materials, *Phys. Scr.* **T49**, 307-314.
6. Gonser, U., Bauer, H.-J., and Wagner, H.G. (1987) Amorphous and microcrystalline structures obtained from fast quenching of FeNiB melts, *J. Magn. Magn. Mater.* **70**, 419-420.
7. Kopcewicz, M., Jackiewicz, E., Załuski, L., and Załuska, A. (1992) Crystallization and structural relaxation of amorphous FeNiSiB alloys due to rapid heating, *J. Appl. Phys.* **71**, 3997-4008.
8. Li, J., Su, Z., Wang, T.M., Ge, S.H., Hahn, H., and Shirai, Y. (1999) Microstructure of nanostructured $Fe_{40}Ni_{38}Mo_4B_{18}$ alloy, *J. Mater. Res.* **34**, 111-114.
9. Kopcewicz, M., and Grabias, A. (1996) Mössbauer study of the surface crystallization of amorphous and nanocrystalline $Fe_{81}Zr_7B_{12}$ alloy, *J. Appl. Phys.* **80**, 3422-3425.
10. Kondoro, J.W., and Campbell, S.J. (1990) Crystallization of Fe-Ni-B amorphous alloy, *Hyperfine Inter.* **55**, 993-1000.
11. Kopcewicz, M., Grabias, A., and Williamson, D.L. (1997) Magnetism and nanostructure of $Fe_{93-x-y}Zr_7B_xCu_y$ alloys, *J. Appl. Phys.* **82**, 1747-1758.
12. Idzikowski, B., Baszyński, J., Skorvanek, I., Mueller, K.-H., and Eckert, D. (1998) Microstructure and magnetic properties of amorphous and nanocrystalline $Fe_{80}M_7B_{12}Cu_1$ (M: Nb, Ti or Mo) alloys, *J. Magn. Magn. Mater.* **177-181**, 941-942.
13. Miglierini, M., Kopcewicz, M., Idzikowski, B., Horvath, Z.E., Grabias, A., Skorvanek, I., Dłużewski, P., and Daroczi, Cs.S. (1999) Structure, hyperfine interactions and magnetic behavior of amorphous and nanocrystalline $Fe_{80}M_7B_{12}Cu_1$ (M: Mo, Nb, Ti) alloys, *J. Appl. Phys.* **85**, 1014-1025.
14. Fernandez van Raap, M.B., Sanchez, F., and Zhang, Y.D. (1995) On the microstructure and thermal stability of rapidly quenched Fe-B alloys in the intermediate composition range between the crystalline and amorphous state, *J. Mater. Res.* **10**, 1917-1926.
15. Arshed, M., Siddiquem M., Anwar-ul-Islam, M., Ashfaq, A., Shamim, A., and Butt, N.M. (1996) Isochronal crystallization of Metglass $Fe_{83}B_{17}$ using Mössbauer effect and resistivity measurements, *Solid State Commun.* **98**, 427-430.
16. Kemeny, T., Kaptas, D., Balogh, J., Kiss, L.F., Pustai, T., and Vincze, I. (1999) Microscopic study of the magnetic coupling in a nanocrystalline soft magnet *J. Phys.: Condens. Matter* **11**, 2841-2847.

CORROSION OF Fe-BASED NANOCRYSTALLINE ALLOYS

J. SITEK, K. SEDLAČKOVÁ AND M. SEBERÍNI
Department of Nuclear Physics and Technology
Slovak University of Technology, Ilkovičova 3, 81219 Bratislava, Slovakia

Abstract

Amorphous and nanocrystalline FINEMET and NANOPERM were investigated after corrosion treatments. An accelerated laboratory corrosion test was performed in sulfuric acid solution. The samples were additionally exposed to the outdoor atmosphere in industrial and rural environments. To evaluate the changes in phases and magnetic microstructure, the technique of ^{57}Fe Transmission Mössbauer Spectrometry (TMS) was used. Furthermore, Conversion Electron Mössbauer Spectrometry (CEMS) was employed to obtain information from the corroded surface layers.

The results point to the different behavior of the amorphous and nanocrystalline samples. From the relative decrease of the absorption area after acid treatment indicating the mass loss of the Fe-containing phases, it is obvious that in the case of FINEMET, nanocrystalline sample is more corrosion-resistant than its amorphous precursor. This tendency was found to be in contrast to NANOPERM.

The analysis of the outdoor exposed samples confirmed this behavior and identified the iron compound preferably formed on the corroded sample surfaces as ferric oxyhydroxide FeOOH, most probably in the form of γ-FeOOH (lepidocrocite).

1. Introduction

Nanocrystalline materials attract attention because they show excellent soft magnetic properties. Recently, attention has been given to the iron-based nanocrystalline alloys of FINEMET (FeCuNbSiB), NANOPERM (FeZrB), and HITPERM (FeCoZrB) [1]. Nanocrystalline alloys show a good combination of low coercive force and high permeability, which present as very interesting characteristics for several applications. Nevertheless, small attention has been paid to the corrosion behavior of Fe-based nanocrystalline materials. Corrosion resistance of amorphous and partially crystallized alloys has been reported [2-5] after treatment in sulfuric acid water solution. Methods like mass loss measurement, potentiometry, SEM, EDS, XRD, XPS, and magnetic measurements have been employed to perform the analysis of the corrosion mechanism.

In this work, we report a Mössbauer spectroscopy study of the Fe-based nanocrystalline alloys and their amorphous counterpart attacked by corrosion. We have

M. Mashlan et al. (eds.), Material Research in Atomic Scale by Mössbauer Spectroscopy, 159–166.
© 2003 *Kluwer Academic Publishers. Printed in the Netherlands.*

studied the influence of the corrosion treatment on the structural and magnetic properties in the bulk of the materials as well as on the surface.

2. Experimental

The samples of $Fe_{73.5}Nb_3Cu_1Si_{13.5}B_9$ (FINEMET) and $Fe_{87.5}Zr_{6.5}B_6$ (NANOPERM) were studied in the amorphous ('as cast') and nanocrystalline state. The amorphous precursors were prepared by a planar flow casting method at the Institute of Physics of the Slovak Academy of Sciences in Bratislava. Ribbon-shaped precursors of about 20-30 μm thick were annealed in vacuum at the temperature of 540 °C for 1 hour to obtain a nanocrystalline-structured system.

In order to quantify the corrosion resistance of a material, it is a common practice to submit the material to a harsher environment than it is normally encountered in the service. It is expected that in this way, the corrosion process will be accelerated. In this respect, the samples were treated in 0.1 M and in 0.5 M solution of sulfuric acid for 24 hours. Concurrently, direct atmospheric exposure of the samples in rural and industrial environments was carried out.

The transmission Mössbauer spectra as well as conversion electron Mössbauer spectra were taken using a conventional Mössbauer spectrometer with a $^{57}Co(Rh)$ source at room temperature. All recorded spectra were evaluated using the NORMOS DIST and NORMOS SITE program, respectively.

3. Results and Discussion

3.1. ACCELERATED LABORATORY TEST OF FINEMET

Results from the mass loss measurements [2-4] indicate that the corrosion resistance of both, 'as cast' and nanocrystalline samples accelerate with increasing content of the silicon (Figure 1.).

For the samples with higher Si content, the presence of nanocrystalline state leads to a decrease in mass loss when compared to the 'as cast' samples. This indicates that partial crystallization improves corrosion resistance. The higher corrosion resistance of the nanocrystalline alloys can be attributed to the higher diffusion rate in this state, which leads to a higher deposition rate of Si on the surface and thus to the growth of a thicker SiO_2 film.

Transmission Mössbauer spectra of the untreated 'as-cast' and nanocrystalline FINEMET alloys, compared with those etched in the 0.5 M water solution of the sulfuric acid are depicted in Figure 2.

From the spectra of the 'as cast' samples, one can directly observe a rapid decrease of the absorption area after sulfuric acid treatment. This fact confirms a significant mass loss of iron from the structure. On the other hand, nearly indistinguishable change of the spectrum area of nanocrystalline sample confirms its better corrosion resistance.

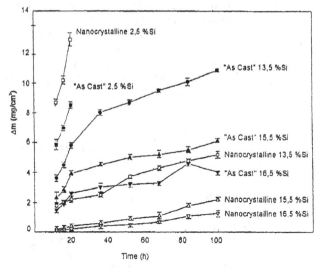

Figure 1. Values of mass loss in 0.1 M water solution of sulfuric acid as a function of immersion time [2].

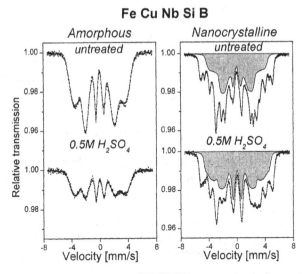

Figure 2. Transmission Mössbauer spectra of FINEMET, untreated and after treatment in 0.5 M sulfuric acid solution.

From further Mössbauer parameters, it is valuable to follow the orientation of the net magnetic moment, average hyperfine magnetic field of amorphous remainder, the distribution function of the internal magnetic field and the relative area of crystalline (amorphous) part, respectively. The orientation of the net magnetic moment is described by means of the A_{23} parameter, which corresponds to the relative area of the 2nd and 5th lines with respect to the 3rd and 4th lines. A rapid decrease in the A_{23} parameter from

2.27 to 1.38 observed for 'as cast' samples implies the tendency of the net magnetic moment to turn out of the ribbon plane after sulfuric acid treatment. The A_{23} parameter estimated for nanocrystalline samples displayed the same tendency, i.e. it has fallen from 3.27 up to 1.83.

Concerning the amorphous/crystalline component abundance, we observed a decrease in the crystalline part from 50 to 40 %. Other Mössbauer parameters remained almost unchanged.

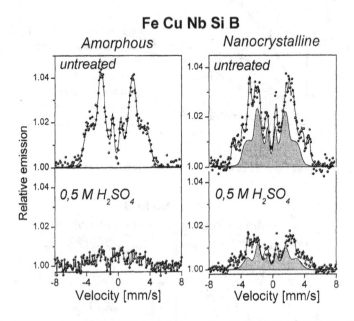

Figure 3. CEMS spectra of FINEMET, untreated and after treatment in 0.5 M H_2SO_4 solution.

CEMS spectra of FINEMET treated in sulfuric acid solution are depicted in *Figure 3*. The spectrum of the 'as cast' sample exhibited, in spite of the long measurement time, very poor statistic. This indicates that the great deal of the iron atoms were etched out from the sample surface.

The nanocrystalline sample showed after treatment a typical spectrum shape, however with decreased area of the spectrum. This confirms that the surface of the nanocrystalline sample contains protective SiO_2 film, which probably inhibits further etching of iron.

The results of mass loss measurements of different NANOPERM-type alloys ($Fe_{84}Nb_7B_9$, $Fe_{83}Nb_{3.5}Zr_{3.5}B_9Cu_1$, $Fe_{84}Zr_7B_9$) are shown in *Figure 4*. The results indicate a decrease in corrosion resistance with partial crystallization of the 'as cast' alloy. This is an opposite tendency as observed in case of FINEMET. Obtained results also show that Nb containing alloys have a higher corrosion resistance than the Zr containing samples.

For the NANOPERM alloy of $Fe_{87.5}Zr_{6.5}B_6$ we observed a complete dissolution after 24-hours immersion in the 0.5 M and even in the 0.1 M sulphuric acid.

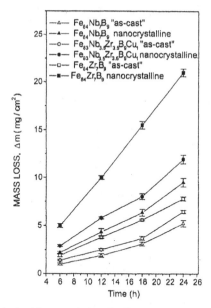

Figure 4. Mass losses in 0.1 M water solution of sulfuric acid as a function of immersion time [5].

3.2. ATMOSPHERIC CORROSION

Mössbauer spectra of amorphous and nanocrystalline FINEMET taken in transmission geometry revealed only delicate changes in the Mössbauer parameters. The most evident is the decrease in the A_{23} parameter observed for amorphous structure (from 2.27 to 1.82 in the industrial environment), which indicates that the net magnetic moment turns out of the ribbon plane likewise observed for accelerated test in H_2SO_4. In the case of the nanocrystalline specimens, the changes in the Mössbauer parameter varied within the statistical error.

A pronounced compositional phase transformation can be observed from the surface analysis carried out by CEMS measurement. Conversion electron Mössbauer spectra of the amorphous FINEMET before and after exposure to the rural and industrial environments for 2 months are shown in *Figure 5*. The doublet presented in the spectra was ascribed to the Fe(III) in the form of ferric oxyhydroxide, most probably in the form of γ-FeOOH (lepidocrocite), which is a typical starting corrosion compound in the iron containing materials. As one could expect, the sample influenced by the SO_2-polluted industrial atmosphere comprised considerably higher content of γ-FeOOH (49 %) than the sample exposed to the rural atmosphere (6 %). Concerning the surface of the nanocrystalline samples, no conspicuous changes followed as far as the sensitivity of the used method reaches. The presence of γ-FeOOH was not observed either, what might be again due to the formation of thicker layer of protective SiO_2 film.

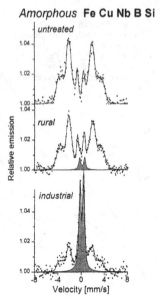

Figure 5. CEMS spectra of amorphous FINEMET after 2-months weathering in industrial and rural areas.

Transmission Mössbauer spectra of the NANOPERM, untreated and after treatment in the industrial and rural areas, are shown in *Figure 6*. The *'as cast'* specimens do not exhibit relevant signs of weathering, however the FeOOH contribution within 2 % is discernible.

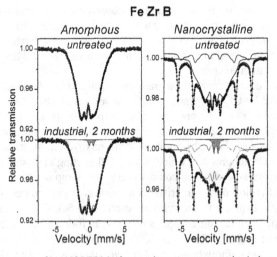

Figure 6. TMS spectra of NANOPERM after outdoor treatment under industrial conditions.

The spectra of the nanocrystalline samples were fitted using three components, i.e. the crystalline, interfacial and the amorphous rest. We observed a pronounced influence of the weathering on the bulk structural arrangement. The results show a decrease in the amorphous component (from 69 % to 48 % in rural and to 39 % in industrial area) at the detriment of the crystalline part (increased from 23 % to 34 % in rural and 41 % in industrial environment) and the interface as well (from 8 % to 14 % and 15 %, respectively). The contribution of the lepidocrocite was found to be 4 % in the rural and 5 % in the industrial areas.

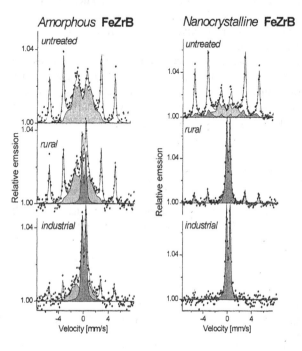

Figure 7. CEMS spectra of amorphous and nanocrystalline NANOPERM after exposure to the industrial and rural atmosphere.

Significant changes were revealed by CEMS measurements either (*Figure 7.*). 19 % of the γ-FeOOH was found on the surface of the *'as cast'* samples after 2-months corrosion in rural area and even 45 % under industrial conditions. The nanocrystalline samples contained 80 % of lepidocrocite after exposure in rural environment. The aggressive industrial environment induced complete covering of the nanocrystalline sample surface exclusively by lepidocrocite.

4. Conclusion

The mass loss of iron indicated after sulfuric acid treatment, as well as the changes in the orientation of the net magnetic moment and surface properties followed by CEMS, have certified the corrosion behavior of FINEMET after accelerated corrosion test. The higher corrosion resistance of nanocrystalline FINEMET as compared with its amorphous counterpart might be attributed to the quicker passive SiO_2 film formation.

The results obtained after atmospheric corrosion have shown that beginning of the corrosion process is followed by the formation of lepidocrocite in the rural as well as in the industrial areas for both FINEMET and NANOPERM-type alloys. Higher amount of γ-FeOOH created under industrial condition point at the corrosion-accelerating properties of SO_2-comprising atmosphere. The influence of the formed corrosion products and passive layers, respectively, on magnetic microstructure was proposed. Therefore, during forthcoming corrosion cycles, the expected conversion of γ-FeOOH into maghemite and goethite, respectively, and its further impact on structural and magnetic behavior will be followed.

The observed corrosion-induced changes affecting the macroscopic properties, harboring considerable interest from practical applications perspective, suggest the onward investigation of the corrosion mechanism and its prevention. In order to preserve the excellent soft magnetic properties of the studied alloys, an optimization of the material composition, and eventually an engagement of some surface modification techniques are desirable.

Acknowledgement

The authors are indebted to P. Duhaj for supplying the master ribbons. The work was supported by the grant VEGA 1/7631/20 and partly by VEGA 1/8305/01.

References

1. Barandiarán, J.M. (1994) Magnetic and Transport Properties of Fe-based Nanocrystalline Materials, *Physical Scripta* T55 - 194
2. Souza, C.A.C., Kiminami, C.S. (1997) Crystallization and Corrosion Resistance of Amorphous FeCuNbSiB, *J.Non-Cryst. Sol.* **219**, 155 - 159
3. Szewieczek, D., Tyrlik-Held, J., Paszenda, Z. (1998) Corrosion Investigations of Nanocrystalline Iron Based Alloy, *J.Mat. Proc. Tech.* **78**, 171 - 176
4. Souza, C.A.C., Kuri, S.E., Politti, F.S., May, J.E., Kiminami, C.S. (1999) Corrosion Resistance of Amorphous and Polycrystalline FeCuNbSiB Alloys in Sulphuric Acid Solution, *J.Non-Cryst. Sol.* **247**, 69 - 73
5. Souza, C.A.C., de Oliveira, M.F., May, J.E., Botta, F.W.J., Mariano, N.A., Kuri, S.E., Kiminami, C.S. (2000) Corrosion Resistance of Amorphous and Nanocrystalline Fe-M-B (M=Zr, Nb) Alloys, *J.Non-Cryst. Sol.* **273**, 282 - 288

MÖSSBAUER AND ATOMIC FORCE MICROSCOPY OBSERVATIONS OF MODIFIED SURFACES OF Fe-Si STEEL

Y. JIRÁSKOVÁ[1], O. SCHNEEWEISS[1], C. BLAWERT[2], AND P. SCHAAF[3]

[1] Institute of Physics of Materials, Academy of Sciences of the Czech Republic, Žižkova 22, CZ-616 62 Brno, Czech Republic

[2] ZFW gGmbH Clausthal, Sachsenweg 8, D-38678 Clausthal-Zellerfeld, Germany, prezent address: GKSS Forschungszentrum Geesthacht GmbH, Institut für Werkstoffforschung, Max-Planck-Strasse 1, D-21502 Geesthacht, Germany

[3] Universität Göttingen, II. Physikalisches Institut, Bunsenstrasse 7/9, D-37073 Göttingen, Germany

Abstract

Detailed surface structure investigations by microscopic methods and microhardness measurements of modified Fe-6at.%Si steel samples are presented here. It is found that a preceding mechanical and heat treatment of the samples influence drastically the surface morphology, phase composition and properties after ion implantation at 300°C in pure nitrogen plasma and in a mixed plasma of nitrogen and silicon.

1. Introduction

Compositionally modulated surfaces are of great importance from the point of view of practical applications. Many different methods are used to synthesize surfaces without influencing the bulk properties of materials. Ion implantation is one of them. After ion implantation, the surface composition can be found to be altered, leading to the formation of stable or metastable alloys or the formation of new phases in the metal surface. As a result, the surface properties, such as hardness, wear and/or corrosion resistance, chemical stability, optical or magnetic properties, biocompatibility etc., can be improved. The present investigations were focused on two topics:

i) to correlate the surface properties of the Fe-Si samples prior to the nitrogen implantation with the phase composition and the surface properties after implantation,

ii) to resolve the influence of silicon on the plasma assisted thermochemical process.

A considerable effect of the structural properties of a pure α-iron surface prior to the implantation on the resulting surface properties after incorporation of nitrogen by plasma immersion ion implantation was shown in Ref. [1]. This led to a question,

167

M. Mashlan et al. (eds.), Material Research in Atomic Scale by Mössbauer Spectroscopy, 167–176.
© 2003 *Kluwer Academic Publishers. Printed in the Netherlands.*

whether bcc iron based alloys, modified by nitrogen implantation, are also similarly sensitive to mechanical and/or heat treatment preceding the surface modification. The Fe-6at.% Si steel was chosen as a well-known transformer material. Plasma immersion ion implantation has been carried out in order to investigate the effect of introducing N into the surface of Fe-Si samples and to investigate a reaction of crystalline silicon with nitrogen plasma with the aim to form a non-ferromagnetic isolating surface layer.

2. Experimental Details

Discs of 25 mm in diameter and 3 mm thick were prepared from Fe-6at.% Si steel sheet. The samples were grinded, polished to a 1 μm diamond mirror finish and cleaned in ethylalcohol using an ultrasonic bath. Half of samples denoted "A" was kept in this mechanically treated state, the second half, denoted "B", was heat treated at 800°C for 10 h in hydrogen followed by slow cooling down to room temperature, in order to remove strains due to sample machining.

Plasma immersion ion implantation (PIII) was performed in a cold wall chamber for 3 h at 300 °C. HV pulses of 100 μs length and -30 kV amplitude were used. The plasma was generated at a nitrogen pressure of 3×10^{-3} mbar and 300 W rf-power. The temperature of 300 °C was controlled by the repetition rate of the HV pulses, resulting in an equilibrium frequency of 150 Hz, thus the average dose was 2.4×10^{18} at/cm^2. A part of samples of group "A" (AI) as of group "B" (BI) was implanted in pure nitrogen plasma, the other parts (AII, BII) were implanted in nitrogen plasma containing silicon sputtered from solid silicon placed together with the steel samples on the working table.

The phase composition of the surface was determined by Mössbauer spectroscopy using conversion electrons (CEMS) for an analysis of surface layers, approximately 200 nm thick. The measurements were performed at room temperature using ^{57}Co in a Rh-matrix as a source. The calibration was done against the standard pure α-iron. Computer processing of measured spectra was carried out using a least squares fitting procedure with the CONFIT program package [2]. The discrete single- and double-line components represent the paramagnetic (pm) phases, the six-line components the ferromagnetic ones (fm). The phases were identified on the basis of spectral hyperfine parameters, hyperfine magnetic induction B and/or distribution of hyperfine inductions ΔB (fm), isomer shift IS (fm, pm) and quadrupole splitting EQ (fm, pm) by comparing with existing literature data. The phase contents were determined from the corresponding sub-spectra intensities I (fm, pm), i.e. from the iron atom fractions supposing identical Lamb-Mössbauer factor for all phases present. X-ray diffraction measurements were performed using Cu-K$_\alpha$ radiation at room temperature and an incident angle of 2°.

The nitrogen depth profiles were measured by Rutherford backscattering spectroscopy [3] using 2.2 MeV protons and a scattering angle of 170°. This method is suitable for surface analysis in depth range up to 10 μm. The spectra were evaluated by GISA 3.991 interactive PC program package [4].

Atomic force microscopy (AFM) was done on TMmicroscopes (US) in contact mode at room temperature in air.

The hardness measurements by nanoindentation were performed employing a Fischerscope HV100 ultra-microhardness tester at loads 300 mN and 1000 mN.

3. Results and Discussion

3. 1. UN-NITRIDED STATE

3.1.1. *Mechanical and heat treatment*

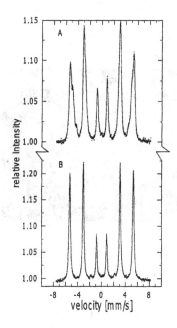

Figure 1. CEMS spectra of Fe-6at.%Si samples in mechanically (A) and heat treated (B) state.

The CEMS spectra taken from the mechanically, A, and heat treated, B, surfaces of Fe-6at.%Si samples are depicted in Fig. 1. A comparison of both spectra reflects a lower Si content in the surface layers of heat-treated sample B at first sight. The spectrum consists of 95 % of sub-spectra representing the Fe atoms with no Si atoms in the 1st nearest neighborhood (nn) and no and/or 1 Si atom in the 2nd nn. The sub-spectrum with B = 30 T for Fe atoms with 1 Si in the 1st nn has only ~ 3 % intensity. This result corresponds to approximately 1 at.% Si content in the surface of heat treated samples B [5]. A decrease in Si content could be ascribed to formation of SiO_2 surface layer not detectable by Mössbauer spectroscopy. A small amount of a component with the hyperfine induction of 21 T is cementite [6] due to surface contamination by carbon probably from annealing furnace. This contamination was confirmed by XPS analysis after Ar sputtering.

The relative abundance of sub-spectra in the surface layers of mechanically treated samples A does not correspond only to randomly distributed Si atoms in Fe-Si solid solution. A good agreement is obtained by supposing approximately 41% of short range ordered micro-regions (A_7B) and 59% of randomly ordered Si atoms. Low content of oxide FeO was detected at the surface of sample A only. The hyperfine parameters obtained by analysis of CEMS spectra are summarized in Table 1.

An AFM image of topography of mechanically treated Fe-Si surface is seen in Fig. 2a. It shows big scratches over the studied zone arising from grinding and polishing. Few light protuberances represent probably oxides and/or other surface contamination. No grains are observed at this surface in contrast to the AFM image of heat-treated surface in Fig. 2b. Also here, the similar scratches after the mechanical treatment and light protuberances are seen next to small grains.

TABLE 1. Hyperfine parameters of CEMS sub-spectra and their assignment to phases in the mechanically and heat treated surfaces of Fe-6at.% Si samples.

		Fe(Si)			Fe(1)	Fe(2)	Fe(3)	FeO	Fe$_3$C
			Fe(0,0) + Fe(0,1)						
Mechanically	B [T]	34.9	33.4	32.3	30.7	28.4	23.0		
treated A	ΣI [%]		56.9		22.4	16.6	2.0	2.1	-
Heat	B [T]	34.4	33.3	32.7	30.7		-	-	21.1
treated B	ΣI [%]		95		2.8	-	-	-	2.2

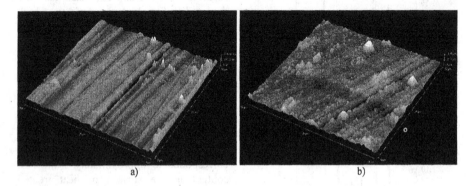

Figure 2. Surface of Fe-6at.% Si samples after a) mechanical (A) and b) heat (B) treatment as seen by AFM.

A higher microhardness measured at the mechanically treated surface A at load of 300 mN (2739±32) Nmm^{-2} in comparison to (2337±27) Nmm^{-2} at the heat treated surface B is conditioned partly by preserved higher Si content in the A sample and partially influenced by dislocation substructure formed due to the mechanical treatment and causing a surface hardening. An increased grain size after heat treatment at given conditions, annealed dislocation substructure and the decrease of Si content can lower the microhardness measured at sample B. The same difference in microhardness was observed at load of 1000 mN. As seen in Fig. 3 the maxima of microhardness in the heat treated sample B at loads 300 and 1000 mN, respectively, are shifted into a depth of ~500 nm reflecting also the softening of the surface layers due to the lower Si content as detected by CEMS.

3.2. NITRIDED STATE

Nitrogen depth profiles are influenced by implantation up to a depth of approximately 50 nm whereas deeper in the bulk a diffusion process prevails. Structurally different surfaces of Fe-Si samples preceding nitriding are reflected also in a different uptake of nitrogen, chemical and phase composition, and properties of surface layers after implantation as documented hereafter.

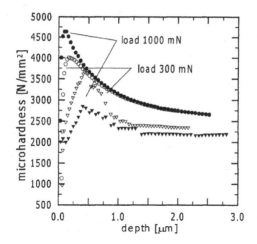

Figure 3. Microhardness depth profiles in the mechanically (circles) and heat (triangles) treated Fe-6 at.%Si samples.

3.2.1. *Implantation in pure nitrogen plasma.*

Nitrogen depth profiles if no additional silicon (AI, BI) is used during implantation are seen in Fig. 4a. The high nitrogen uptake (> 30 at.%) was obtained in the close surface layers for both samples. The content of nitrogen in the mechanically treated surface (AI) decreases within the first 170 nm steeply to 19 at.%, and then shows the constant value up to 295 nm followed by approximately exponential decrease into the bulk. The whole curve of depth profile for heat-treated sample (BI) is shifted to higher values of nitrogen content. It means that the structure of an annealed surface offers better conditions for higher nitrogen uptake. This is in contradiction to results obtained for pure Fe samples where the nitrogen uptake was higher in the mechanically treated surface [1]. The reason can be found in different surface structures of heat-treated Fe and Fe-Si samples. While the annealing of pure Fe samples in vacuum contributed to annealing out of vacancy type defects confirmed by lifetime positron annihilation measurement [1], the heat treatment of Fe-Si has contributed to annealing out of stresses and defects and caused formation of SiO_2 surface oxide and depletion of Fe-Si surface by Si atoms. On the other hand, the formation of SiO_2 has reduced formation of iron oxides. The absent Si atoms were replaced by either vacancies and/or hydrogen atoms and they acted as traps for N atoms during an implantation process. This could be an explanation for higher nitrogen uptake in the heat-treated surface. The higher N content in the B sample is reflected in formation of high nitrogen nitride phases as documented below.

Glancing angle X-ray diffraction results for mechanically and heat-treated Fe-Si samples are depicted in Fig. 5a. They reflect the phase composition from depth comparable with CEMS. The broad peak reflections are visible at the mechanically treated surface (AI). This can suggest a number of different fine-grained phases with similar lattice constant, which cannot be resolved properly, or highly deformed structure acting as an amorphous structure, which does not allow for a better phase decomposition. Some peaks are consistent with those ones assigned to α-Fe-Si and ε-

Fe-N in the surface of heat-treated sample BI. The γ'-Fe₄N phase was detected only in the surface of BI sample.

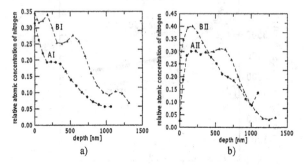

Figure 4. Nitrogen depth profiles in the mechanically (A) and heat (B) treated Fe-6 at.%Si samples nitrided by PIII: a) without Si (I) and b)with Si (II) in implantation chamber.

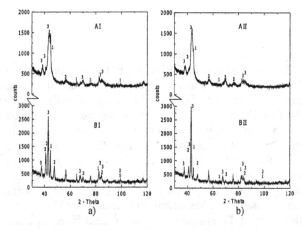

Figure 5. X-ray diffraction patterns of the mechanically (A) and heat (B) treated Fe-6 at.%Si samples nitrided by PIII: a) without Si (I) and b) with Si (II) in implantation chamber.

The mean values of hyperfine induction and total intensities for individual phases detected in CEMS spectra, Fig. 6a, are summarized in Table 2. A higher nitrogen uptake in the surface of heat treated sample (BI) gives rise to the ε-nitride phases, mainly paramagnetic ε-Fe₂N phase with the hyperfine parameters IS = 0.43 mm/s, EQ = 0.28 mm/s which are in good agreement with literature data [7, 8]. Similarly, the other nitrides were determined according to their hyperfine parameters published by other authors [8 and Refs. therein]. Contrary to nitriding of pure Fe samples [1] another paramagnetic components were detected in the surfaces of both Fe-Si samples. Two paramagnetic sub-spectra with the same isomer shifts of 0.36 mm/s and quadrupole splittings of 0.58 and 1.03 mm/s were detected in the CEMS spectrum of mechanically treated sample and three sub-spectra with IS = 0.31; 0.51; 0.48 mm/s and EQ = 0.52; 1.43; 0.85 mm/s, respectively, were found in the surface of heat treated sample. It can be supposed that the detected doublets could represent the Fe^{3+} in Fe-Si oxides. An increase in mean value of hyperfine induction of α-Fe-Si phase is due to a decrease in Si content in the close surface layers and it represents an indirect proof of SiO_2 formation at the surface as seen by AFM.

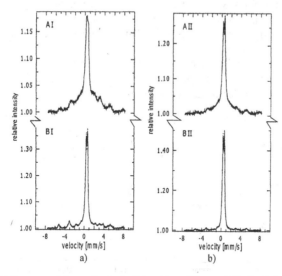

Figure 6. CEMS spectra of the mechanically (A) and heat (B) treated Fe-6 at.%Si samples nitrided by PIII: a) without Si (I) and b) with Si (II) in implantation chamber.

TABLE 2. Surface phase composition of nitrided Fe-Si samples obtained by Mössbauer phase analysis of CEMS spectra; $B_{mean}[T]/\Sigma I[\%]$, fm-ferromagnetic phase, pm-paramagnetic phase.

Phase State	α-Fe-Si (fm)	γ'-Fe$_4$N (fm)	ε-Fe$_{3-x}$N (fm)	ε-Fe$_2$N (pm)	Fe^{3+} in Fe-Si oxides (pm)	α-Fe$_2$O$_3$ (fm)
Mechanically treated (AI)	29.6/12.0	-	15.6/38.7	-/27.6	-/18.7	48.8/3.0
Heat treated (BI)	31.1/8.6	24.0/10.7	13.1/15.0	-/49.9	-/15.7	-

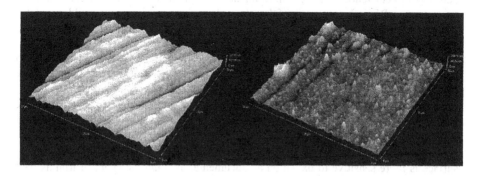

Figure 7. AFM of nitrided surface of Fe-6at.% Si samples; mechanically (AI-left) and heat (BI-right) treated.

An AFM image of surfaces after nitrogen implantation in Fig. 7 documents a conservation of morphology prior implantation. Temperature of PIII treatment has contributed to formation of probably SiO_2 and/or Si_3N_4 films at AI surface (left), which could not be detected by CEMS. The BI surface shows the formation of small nitrides.

The microhardness has increased at both loads (300 mN and 1000 mN) after nitriding of mechanically treated surface only. This is connected probably with a more pronounced oxidation (SiO_2) of AI sample as observed by AFM in comparison to BI and /or with additional stresses induced by high N content in the mechanically treated surface and producing an extra hardening. The results of microhardness measurements for the load of 300 mN at different surface states are depicted in Fig. 8.

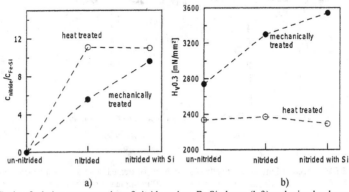

a)   b)

Figure 8. Ratio of relative representation of nitride and α–Fe-Si phases (left) and microhardness at 300 mN (right) in dependence on surface treatment.

3.2.2. *Implantation in nitrogen plasma with Si addition.*

Fig. 4b represents nitrogen depth profiles if silicon is present during implantation. The curves display a slightly higher N uptake and a greater nitrogen penetration depth than the curves for conventional PIII treatment without Si-addition, see Fig. 4a, at both samples. It documents a reaction of silicon with nitrogen plasma resulting in an increased efficiency of nitriding process. Similar effect was observed by rare earth (RE) addition into conventional plasma nitriding and PIII processes [9]. The mechanism of Si analogous to RE influence is not fully clear. An enhanced N content could suggest a catalytic effect of silicon or deposition and decomposition of metastable silicon nitrides on the surface.

X-ray diffraction patterns, Fig. 5b, show only very small differences as compared to patterns in Fig. 5a. Slightly lower intensity of peaks for α-Fe-Si (peaks "1") and γ'-Fe₄N (peaks "2") phases on behalf of ε-Fe₃₋ₓN phase (peaks "3") can be observed.

CEMS spectra of samples nitrided by presence of Si are depicted in Fig. 6b, the results of phase analysis are summarized in Table 3. An enhanced N content determined by RBS is reflected by increased amount of paramagnetic ε-Fe₂N phase with high content of nitrogen (~33at.%) at both samples (AII, BII). The mechanically treated surface is more sensitive to oxidation as documented by higher content of iron as well as iron-silicon oxides.

TABLE 3. Mössbauer phase analysis of CEMS spectra, $B_{mean}[T]/\Sigma I[\%]$, of Fe-Si samples implanted in nitrogen plasma with Si addition. fm-ferromagnetic phase, pm-paramagnetic phase.

Phase State	α-Fe-Si (fm)	ε- $Fe_{3-x}N$ (fm)	ε- Fe_2N (pm)	Fe^{3+} in Fe-Si oxide (pm)	α-Fe_2O_3 (fm)
Mechanically Treated (AII)	29.3/6.7	14.1/31.5	-/35.1	-/21.7	50.3/5
Heat treated (BII)	30.2/6.0	14.9/7.2	-/70.2	-/16.6	-

Detailed observation of AFM images in Fig. 9 shows a slight increase in grain size at the surface of BII sample and vanishing scratches at the AII surface. The analyses of both surfaces by EDX have shown round light objects with high content of Si, which, according to our experiences, represent SiO_2. The other dark objects also with high amount of Si could be only speculatively ascribed to Si_3N_4 because no analysis of light elements (nitrogen, oxygen) was possible by our equipment. No similar objects were observed at the surfaces implanted in pure nitrogen plasma.

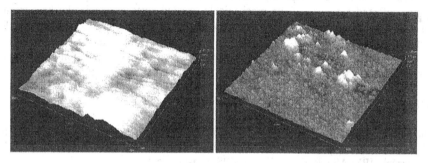

Figure 9. AFM images of Fe-6at.% Si surfaces treated by PIII in N+Si; mechanically treated (AII- left) and heat (BII-right) treated.

The microhardness measured at the AII surface, (3542 ± 56) N/mm² for load 300 mN, was higher in comparison to previous states, while it has slightly decreased to (2294 ± 34) N/mm² at the BII surface. The same was obtained also for the load of 1000 mN. No important differences in composition of α-Fe-Si matrix and nitrides were detected by XRD and CEMS. Therefore, this opposite tendency could be ascribed to different surface morphologies due to formation of SiO_2, Si_3N_4 and/or other types of Fe^{3+}-Fe-Si oxides. More detailed investigations are desirable.

4. Conclusions

The present investigations of α-Fe-6at.%Si sample with differently prepared surface structures have shown:

♦ an influence of the original structural state of a surface on the phase composition and surface properties after its modification by nitrogen plasma immersion ion implantation at 300 °C,

♦ in opposite to pure α-Fe [1] the higher nitrogen uptake accompanying by formation of high nitrogen nitrides were observed in the heat treated surface. This can be related with different surface states of samples. While the heat-treating of pure α-Fe has led to annealing out of defects a depletion of the surface layers of heat-treated α-Fe-Si sample by Si atoms was observed by CEMS and confirmed by microhardness. A formation of Si vacancies as attractive traps for implanted nitrogen atoms could be an explanation for higher nitrogen uptake,

♦ no continuous surface film of Si_3N_4 was proved after nitrogen implantation with addition of silicon. Nevertheless, the reaction of silicon with nitrogen plasma contributed positively to enhanced N content and an increase in thickness of nitrided surface layer. These investigations go on.

♦ the mechanically treated surface is more sensitive to oxidation,

♦ the microhardness has increased at the mechanically treated samples after nitriding in pure N plasma as well as by addition of silicon. The opposite tendency was found at heat-treated sample. It can be supposed: ♦ ♦ that an influence of additional stresses due to nitrogen supersaturation of surface layers is more pronounced in the highly defected structure of the mechanically treated surface than in the heat treated surface. This can contribute to an extra hardening, ♦ ♦ the microhardness of nitrided surfaces is more sensitive to the surface morphology given by different formation of SiO_2, Si_3N_4 and/or other types of Fe^{3+}-Fe-Si oxides than to a composition of Fe-Si and nitride phases.

Acknowledgements

The authors express their thanks to Dr. V. Peřina, INP, AS CR Řež near Prague, CZ, for RBS, to Dr. M. Vůjtek, Palacký University Olomouc, CZ, for AFM, and Mgr. M. Vondráček, IPM AS CR Brno, CZ, for XPS measurements.

This work was supported by the Czech Ministry of Education, Youth and Sports under the Project No. ME 373, by the Project No. S2041105 of AS CR, and by DFG Project Scha632.

References

1. Jirásková, Y., Blawert, C., Schneeweiss, O., Peřina, V., and Macková, A. (2002) Defect and magnetic phases at nitrided iron surfaces, *phys. stat. sol.* **189**, 971-977.
2. Žák, T. (1999) Confit for Windows 95, in M. Miglierini, D. Petridis (eds.), *Mössbauer Spectroscopy in Material Science*, Kluwer Academic Publishers, Dordrecht, pp. 385-390.
3. Chu, W.K., Mayer, J.W., and Nicole, A. (1978) *Backscattering Spektrometry*, Acad. Press New York.
4. Saarilahti, J. and Rauhala, E. (1992) Interactive personal-computer data analysis of ion backscattering spectra, *Nucl. Instrum, Methods Phys. Res.* **B64**, 734 -738.
5. Schneeweiss, O. (1979) *Atomic ordering in FeSi alloys, PhD thesis, Institute of Physics of Materials*, Brno (in Czech).
6. Schaaf, P., Wiesen, S., and Gonser, U. (1992) Mössbauer study of iron-carbides: Cementite $(Fe,M)_3C$ with various manganese and chromium contents, *Acta Metall.* **40**, 373-379.
7. Kopcewitz, M., Jagielski, J., Turos, A., and Williamson, D.L. (1992) Phase transformations in nitrogen implanted α-iron, *J. Appl. Phys.* **71**, 4217 – 4226.
8. Schaaf, P. (2002) *Laser nitriding of metals* **47**, 1-161.
9. Cleugh, D., Blawert, C., Steinbach, J., Ferkel, H., Mordike, B.L., and Bell, T. (2001) Effects of rare earth additions on nitriding of EN40B by plasma immersion ion implantation, *Surf. Coat. Technol.* **142-144**, 392-396.

MÖSSBAUER INVESTIGATION OF SURFACE PROCESSING BY PULSED LASER IRRADIATION IN REACTIVE ATMOSPHERES

E.CARPENE[1], M. KAHLE, M. HAN AND P. SCHAAF[2]
*Universität Göttingen, Zweites Physikalisches Institut, Bunsenstrasse 7/9,
37073 Göttingen, Germany*
[1]*Email: ecarpen@gwdg.de*
[2]*URL: www.uni-goettingen.de/~pschaaf*

Abstract

Irradiations of iron (Armco, Fe > 99.85%) substrates in controlled gaseous atmospheres (nitrogen and methane) have been performed with nanosecond pulses of an excimer laser, leading to the formation of nitride and carbide surface layers, respectively, with different stoichiometry and crystallographic structures. The evolution and abundance of each phase is correlated to the experimental parameters such as the number of laser pulses, the laser fluence and the ambient gas pressure.

The capability of CEMS to distinguish the atomic surrounding in combination with its sensitivity to the surface makes it an indispensable tool for a proper investigation and optimisation of such surface treatments. Additional analytical techniques have been used in order to have complementary information. Ion beam analysis (Rutherford backscattering spectrometry and resonant nuclear reaction analysis) was performed to measure the depth profiles of nitrogen and carbon, while X-ray diffraction and Rietveld refinement were employed to characterize the crystallographic morphology of the irradiated samples.

The results revealed that under proper experimental conditions, the laser treatment can produce almost homogeneous layers of nitride/carbide, where the composition and the structure are determined only by the parameters of irradiation.

1. Introduction

Nitriding and carburisation of metals is a well-known and widely used process to improve the mechanical and tribological properties of a metal surface. It consists on the incorporation of nitrogen or carbon into the metallic matrix leading to the formation of solid solutions and/or nitride/carbide phases. Many methods have been developed in order to introduce atomic nitrogen and carbon into the metal, such as plasma or gas treatments and ion beam assisted techniques. Among all, laser nitriding [1] and carburisation [2] has the advantage of being a very fast and precise process. Moreover, due to the strong light absorption in metals the heat-affected zone is typically of the

177

M. Mashlan et al. (eds.), Material Research in Atomic Scale by Mössbauer Spectroscopy, 177–186.
© 2003 *Kluwer Academic Publishers. Printed in the Netherlands.*

order of few microns, making this technique suitable for surface treatments, without affecting the bulk properties of the matrix. In the present contribution, the results of UV nanosecond laser irradiation of iron substrates in controlled nitrogen and methane atmospheres will be presented, with special emphasis on the phase analysis where Mössbauer spectroscopy is extremely sensitive.

2. Experiments

Samples of pure iron (ARMCO, purity > 99.85 %) have been cut into slices of 1.5 mm thickness and mechanically polished (SiC grinding papers with mesh 1200, 2000, 4000 and 1 μm diamond paste) to a mirror-like finishing. The laser irradiations were carried out with a Siemens XP2020 XeCl pulsed excimer laser ($\tau = 55$ ns FWHM, $\lambda = 308$ nm) focusing the beam through a fly-eye homogenizer lens. The laser fluence was set to 4 J/cm^2 on a spot area of 5×5 mm^2. The samples, placed inside a chamber equipped with a quartz window and mounted on a X-Y table, were irradiated with a meandering scan (the notation n×m indicates the overall spot overlap). Prior to the irradiation, the chamber was evacuated (residual pressure ~10^{-5} mbar) to reduce oxygen contamination and then filled with nitrogen (purity > 99.999 %) or methane (purity > 99.5 %) gas at the desired pressure between 0.1 and 10 bar. The Rutherford backscattering spectroscopy (RBS) and the resonant nuclear reaction analysis (RNRA) were carried out at the IONAS acceleration facility in Göttingen [3]. RBS was performed with a 900 keV He^{++} beam at a scattering angle of 165°, while the reaction $^{15}N(p,\alpha\gamma)^{12}C$ at a resonance energy of 429.6 keV was used for the nitrogen profiling with RNRA. X-ray diffraction was measured with a Bruker AXS diffractometer equipped with Cu anode and a grazing incidence attachment. The incident angle was set to 1° or 2°. Mössbauer spectroscopy was measured at room temperature in conversion electron and X-ray mode (CEMS and CXMS) using a $^{57}Co(Rh)$ source with constant acceleration drive. The electrons were detected in He/CH_4 gas flow proportional counter, while the X-rays in an Ar/CH_4 gas toroidal detector. The spectra, stored in a 1024 multichannel analyser, were fitted with a least square routine using superimposed lorentzian lines. The calibration was performed with a 25 μm iron film to which all isomer shifts are referred.

3. The Fe-C and Fe-N Systems

The systems Fe-C and Fe-N have been extensively studied in the past century and their equilibrium binary phase diagram are reported in Figure 1. The maximum solubility of C in α-Fe (bcc) is ~0.1 at.% at 740°C, while it reaches the value of ~9 at.% in γ-Fe (fcc) at 1150°C. The only stable carbide is θ-Fe_3C (also known as cementite) that crystallizes in the orthorhombic structure Pnma. Under equilibrium conditions, a carbon content higher than 25 at.% leads to the precipitation of graphite in the iron matrix. Other metastable Fe-C phases have been reported in the literature [4,5] such as Fe_7C_3, χ-Fe_5C_2 (Hagg carbide, monoclinic structure) and ε-FeC_x (typically with $1/3 < x < 1/2$, hexagonal structure).

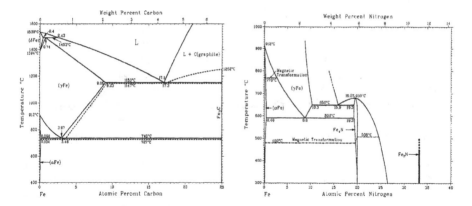

Figure 1. Equilibrium phase diagram of the Fe-C system (left) and the Fe-N system (right).

The maximum solubility of N in α-Fe (bcc) is ~0.4 at.% at 592°C, while it reaches the value of ~10.3 at.% in γ-Fe (fcc) at 650°C. The stoichiometric γ'-Fe$_4$N phase (also known as roaldite) has a narrow region that extends up to 680°C, crystallizing in the cubic P$\bar{4}$3m structure, while the ξ-Fe$_2$N compound has orthorhombic form. The ε-FeN$_x$ phase has structure similar to the ε-FeC$_x$ carbide and a wide compositional range. The existence of the metastable α"-Fe$_{16}$N$_2$ phase has been reported as well.

The Mössbauer characterization of the α- and γ- phases for both the Fe-C and the Fe-N systems as well as the stoichiometric phases θ-Fe$_3$C, γ'-Fe$_4$N and ξ-Fe$_2$N are well established [5,6]. Recently, an exhaustive Mössbauer investigation on the ε-FeN$_x$ system has been reported, showing the detailed dependence of all Mössbauer parameters on the nitrogen concentration [7]. A similar analysis for the ε-FeC$_x$ system was done on carbides extracted from the metal matrix [5]. As it will be shown in the following section, laser irradiation of iron in N$_2$ gas induces the formation of α, γ and ε phases, while irradiation in low pressure (0.1 bar) CH$_4$ atmosphere leads to the formation of a ε-FeC$_x$ surface layer and higher pressures induce the formation of cementite, demonstrating the capability of this technique in the synthesis of peculiar phases.

4. Results and Discussion

4.1. IRRADIATION IN N$_2$ ATMOSPHERE

The Mössbauer spectra of the sample irradiated with an 8×8 meandering scan in nitrogen atmosphere at various pressures are reported in Figure 2 with the corresponding fits. The central component has been attributed to γ-Fe(N) austenite: it consists of a singlet ($\langle$IS$\rangle$ = -0.09 mm/s) representing an iron site with no nitrogen neighbours and a doublet ($\langle$IS$\rangle$ = 0.15 mm/s, $\langle$QS$\rangle$ = 0.31 mm/s) due to an iron atom with one nitrogen neighbour. At low nitrogen pressure also the magnetic components of the α'-Fe(N) martensite could be distinguished.

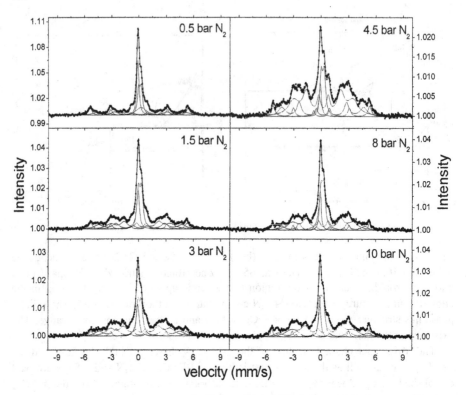

Figure 2. Mössbauer spectra (CEMS) and fits of the iron samples irradiated at different nitrogen pressures (8×8 meandered scan).

Besides, a broad central doublet with isomer shift of about 0.3 mm/s was attributed to paramagnetic ε-FeN$_x$ nitride. For pressures exceeding 1 bar, the ε-FeN$_x$ nitride becomes more abundant, reaching the maximum amount at 4.5 bar. It has been fitted with two broad magnetic components representing the hexagonal iron sites with 1 and 2 nitrogen nearest neighbours ($\langle IS_1 \rangle = 0.17$ mm/s, $\langle HF_1 \rangle = 28.5$ T, $\langle IS_2 \rangle = 0.25$ mm/s, $\langle HF_2 \rangle = 20.1$ T). The relative Mössbauer area of each phase as function of the ambient pressure is reported in Figure 3. The fraction of martensite and non-reacted iron rapidly decreases with increasing pressure, stabilizing to the average value of 10 % for pressures higher than 1.5 bar, while the austenitic phase and the ε-FeN$_x$ nitride have a complementary behaviour (the amount of one changes at expenses of the other).

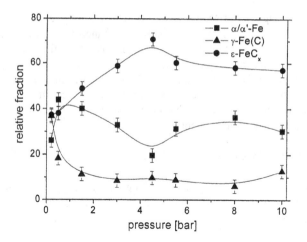

Figure 3. Phase fraction vs. nitrogen pressure as obtained from the Mössbauer analysis.

Figure 4 reports the average nitrogen content estimated from the RNRA depth profiles averaged over the first 150 nm from the surface.

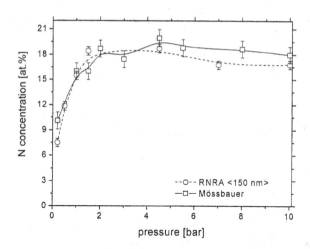

Figure 4. Average nitrogen content vs. nitrogen gas pressure obtained from the RNRA depth profiles (dash line) and from the Mössbauer analysis (continuous line).

The result is compared with the nitrogen content evaluated from the Mössbauer analysis according to the following procedure: the amount of nitrogen in the austenite was obtained from the relative intensities of the doublet and the singlet according to the binomial distribution [8]. The nitrogen concentration in the ε phase was estimated assigning a weight of 1/7 (Fe_6N, one nitrogen nearest neighbour) to the paramagnetic component and to sextet with higher hyperfine field, while a weight 1/4 (Fe_3N, two nitrogen nearest neighbours) to the sextet with lower hyperfine field, obtaining the average stoichiometry of ε-FeN$_{0.27}$ (or ε-Fe$_{3.7}$N). Since the nitrogen content in the martensite is negligible it was not considered. The recoilless fractions of 0.36 [9], 0.6 and 0.8 were used for the ε phase, the austenite and the ferrite, respectively. The excellent agreement in the nitrogen contents as estimated from Mössbauer analysis and RNRA indicates that the stoichiometric assignments are correct. It should be mentioned that after annealing at about 250°C in vacuum or in air, the nitride phases obtained with the laser treatments convert almost completely to γ'-Fe$_4$N [10,11].

4.2. IRRADIATION IN CH$_4$ ATMOSPHERE

The Mössbauer spectra of the sample irradiated with an 11×12 meandering scan in methane atmosphere at various pressures are reported in Figure 5 with the corresponding fits.

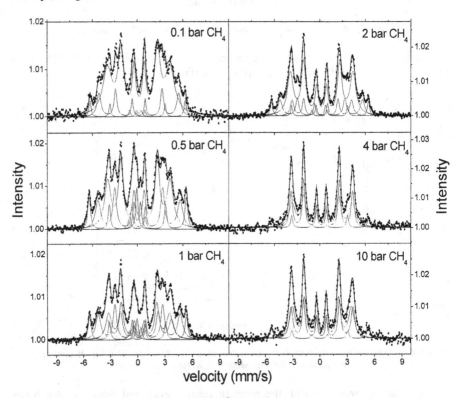

Figure 5. Mössbauer spectra (CEMS) and fits of the iron samples irradiated at different methane pressures (11×12 meandered scan).

The central component, attributed to γ-Fe(C) austenite, is detectable only for pressures lower than 2 bar. Similar to the γ-Fe(N) austenite, it consists of a singlet (⟨IS⟩ = −0.06 mm/s) and a doublet (⟨IS⟩ = 0.01 mm/s, ⟨QS⟩ = 0.70 mm/s). Besides the typical sextet of non-reacted iron, broad magnetic components are observed over the entire pressure range. The ε-FeC$_x$ carbide was identified by two sextets with the same interpretation as given for the ε-FeN$_x$ nitride and very similar hyperfine parameters. Nevertheless, a gradual increase in the isomer shift of the lower hyperfine field component was observed with increasing pressures. Simultaneously, the relative area of the higher hyperfine field component gradually dropped, disappearing from the samples treated in CH$_4$ pressures higher than 4 bar.

TABLE 1. Mössbauer parameters used for the ε-FeC$_x$ and θ-Fe$_3$C phase mixture.

	ε-FeC$_x$ (x=1/6)	ε-FeC$_x$ (x=1/3)	θ-Fe$_3$C
IS (mm/s)	0.15	0.17*	0.205*
HF (T)	28.0	21.1	21.1

* constrained values

As confirmed by GIXRD measurements, this behaviour is due the gradual increase of the relative fraction of θ-Fe$_3$C at expenses of the ε-FeC$_x$ phase with increasing methane pressure. The Mössbauer analysis has been performed employing two distinct subspectra for the low hyperfine field component, constraining their isomer shift to 0.17 mm/s for ε-Fe$_3$C (two carbon neighbours) and to 0.205 mm/s for θ-Fe$_3$C. The average values of the Mössbauer parameters used to fit the ε-FeC$_x$ and θ-Fe$_3$C mixture are reported in Table I. In high methane pressure, only the cementite phase is produced, and the two characteristic sextets (corresponding to the two different crystallographic sites) can be clearly distinguished.

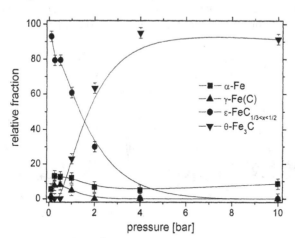

Figure 6. Phase fraction vs. methane pressure as obtained from the Mössbauer analysis.

The relative areas of all phases as function of the CH$_4$ pressure, according to the Mössbauer fitting, are reported in Figure 6. The fraction of non-reacted iron is about 10 % regardless of the ambient pressure, while the amount of cementite increases at expenses of the ε-FeC$_x$ carbide.

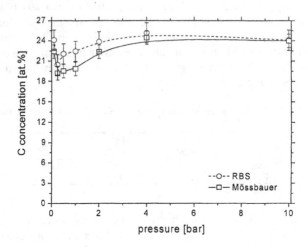

Figure 7. Average carbon content vs. methane gas pressure obtained from the RBS depth profiles (dash line) and from the Mössbauer analysis (continuous line).

The austenitic phase drops simultaneously to the ε phase and both are not observed for irradiations performed above the 2 bar of CH$_4$. As for the laser treatment in nitrogen atmosphere, the carbon content measured by RBS and extrapolated from the Mössbauer analysis (considering the proper recoilless fractions of each phase) are in good agreement, confirming the stoichiometric assignments of the different phases and obtaining the mean value x = 0.28 for the ε-FeC$_x$ phase (i.e. ε-Fe$_{3.6}$C).

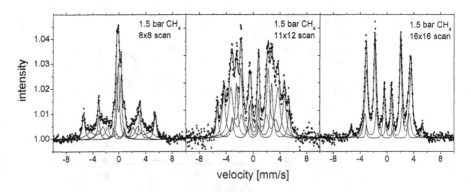

Figure 8. Mössbauer spectra (CEMS) and fits of the iron samples irradiated in 1.5 bar CH$_4$ with different spot overlap.

The dependence of the carburisation efficiency on the laser spot overlap has been investigated in the samples irradiated in 1.5 bar CH$_4$. The Mössbauer spectra of the samples meandered with 8×8, 11×12 and 16×16 scans are shown in Figure 8 and the corresponding phase fractions are shown in Figure 9.

The 8×8 scan in 1.5 bar methane gas produces results comparable to the irradiation performed in nitrogen atmosphere under identical experimental conditions (same pressure, same spot overlap, same laser fluence). A similar amount of austenite, ε-FeC$_x$ and non-reacted Fe is obtained. With the 11×12 scan the γ-Fe(C) is not observed and the sample is a mixture of non-reacted iron, ε-FeC$_x$ and θ-Fe$_3$C phases. In the sample irradiated with 16×16 meander scan only the cementite is observed [2], as obtained with higher methane pressure (4 bar), but lower spot overlap (11×12).

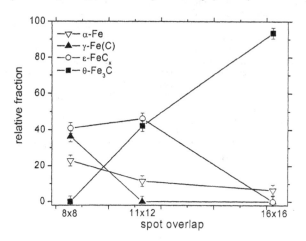

Figure 9. Phase fraction vs. spot overlap as obtained from the Mössbauer analysis.

5. Conclusions

Laser irradiations of iron samples in nitrogen and methane atmospheres leads to a successful incorporation of nitrogen and carbon, respectively, in the metallic matrix. In nitrogen atmosphere with a spot overlap of 8×8, a mixture of ferritic, austenitic and ε-FeN$_x$ ($\langle x \rangle = 0.27$) phases is produced, with average nitrogen concentration roughly independent of the gas pressure above 2 bar. The relative amount of ε-FeN$_x$ phase changes at expenses of the austenitic phase and it reaches a maximum for the treatment performed in 4.5 bar. Laser irradiation with a spot overlap of 11×12 in low methane gas pressure (0.1 bar) leads to the formation of a ε-FeC$_x$ ($\langle x \rangle = 0.28$) surface layer, that gradually transforms into cementite with increasing pressure. Above 4 bar the layer is purely θ-Fe$_3$C. A similar result is obtained with lower gas pressure (1.5 bar), but higher spot overlap (16×16) and the thickness of the carbide layer is estimated to be of the order of 1 μm [2].

186

References

1. Schaaf, P. (2002) Laser nitriding of metals, *Progress in Materials Science* **47**, 1-161

2. Carpene, E. and Schaaf, P. (2002) Formation of Fe_3C surface layers by laser plasma cementation, *Applied Physics Letters* **80**, 891-893.

3. Uhrmacher, M., Pampus, K., Bergmeister, F., J., Purschke, D, and Lieb, K.-P., (1985) Energy calibration of the 500 kV heavy ion implanter IONAS, *Nuclear Instruments and Methods* **B9**, 234-242.

4. Kunze, J. (1990) *Nitrogen and carbon in iron and steel*, Academic-Verlag, Berlin.

5. Ron, M. (1980) Iron-carbon and iron-nitrogen systems, in R. L. Cohen (ed.), *Applications of Mössbauer spectroscopy II*, Academic Press, New York.

6. Schaaf, P., Wiesen, S. and Gonser., U. (1992) Mössbauer study of iron-carbides: cementite $(Fe,M)_3C$ with various manganese and chromium contents, *Acta Metall.* **40**, 373-379.

7. Schaaf, P., Lieb, K.-P., Carpene, E., Han, M. and Landry, F. (2001) Laser-produced iron nitrides seen by Mössbauer spectroscopy, *Czechoslovak Journal of Physics* **51**, 625-650.

8. Schaaf, P., Illgner, C., Niederdrenk, M. and Lieb, K.-P. (1995) Characterization of laser-nitrided iron and sputtered iron nitride films, *Hyperfine Interactions* **95**, 199-225.

9. Mathalone, Z., Ron, M., Shechter, H. (1970) Effective Debye temperature of precipitated carbides, *Applied Physics Letters* **17**, 32-34.

10. Han, M., Carpene, E., Landry, F., Lieb, K.-P. and Schaaf, P. (2001) The thermal stability of laser produced iron nitrides, *Journal of Applied Physics* **89**, 4619-4624.

11. Carpene, E., Landry, F., Han, M., Lieb, K-P. and Schaaf, P. (2002) Investigation of the thermal stability of laser nitrided iron and stainless steel by annealing treatments, *Hyperfine Interactions*, in print.

APPLICATION OF MÖSSBAUER SPECTROSCOPY AND POSITRON ANNIHILATION SPECTROSCOPY FOR TESTING OF NEUTRON-IRRADIATED REACTOR STEELS

V. SLUGEŇ, J. LIPKA, J. HAŠČÍK, R. GRÖNE, I. TÓTH, P. UVÁČIK, A. ZEMAN AND K. VITÁZEK

Department of Nuclear Physics and Technology, Slovak University of Technology, Ilkovičova 3, 81219 Bratislava, Slovakia

Abstract

The study is focused on the application of Mössbauer spectroscopy (MS) and positron annihilation spectroscopy (PAS) in the evaluation of the microstructure parameters of materials used in nuclear industry. The practical applications of these methods are documented on the evaluation of degradation processes currently in practise in nuclear power plant (NPP) reactor pressure vessel (RPV) steels.

The samples originating from the Russian 15Kh2MFA and Sv10KhMFT steels, commercially used at WWER-440 reactors, were irradiated near the core at NPP Bohunice (Slovakia) with neutron fluences ranging from 7.8×10^{23} m^{-2} to 2.5×10^{24} m^{-2}. The systematic changes in the MS and PAS spectra were observed mainly during the first period (1-year stay in irradiation containers under operating conditions by "speed factor" of about 10). A possible explanation could be due to the changes caused by precipitation of elements like Cu or Cr mainly in carbides to the surface. These MS results confirm that the close environment of Fe atoms in b.c.c. lattice of RPV steels remains almost stable after initial changes and perhaps could be correlated with the ductile-brittle transition temperature curve from mechanical tests.

1. Introduction

The main goal from a nuclear-industry perspective is to enhance safety and reliability of nuclear power plants (NPP). The degradation of reactor pressure vessel (RPV) steel is a very complicated process dependent on many factors (thermal and radiation treatments, chemical compositions, preparing conditions, ageing, etc.). This topic has been the subject of many comprehensive works [1-7], however, the related neutron embrittlement and microstructural changes remain incompletely understood. The effect of intensive fluxes of neutrons results in considerable changes of material structures and properties. In particular, the development of fine scale radiation-induced defects which impede the dislocation motion under applied stress, known as irradiation embrittlement, leads to mechanical properties degradation which can result in partial loss of plasticity and in an increase incidence of brittle fracture [2,5,6]. Defects are formed from vacancies

187

M. Mashlan et al. (eds.), Material Research in Atomic Scale by Mössbauer Spectroscopy, 187–198.

and interstitials created in collision cascaded processes. Those point defects surviving the cascades migrate freely through the crystal lattice, interacting with each other and with solute atoms in the matrix and also with dislocation substructure and precipitates. These irradiation-induced diffusion processes result in the formation of point defect clusters, dislocation loops and precipitates [8].

Properties of the reactor pressure vessel (RPV) steels and influence of thermal and neutron treatment on these properties are routinely investigated by macroscopic methods such as Charpy V-notch and tensile tests [1, 2, 4, 5, 9]. A number of semi-empirical laws, based on macroscopic data, have been established, but, unfortunately, these laws are not completely consistent with all data and do not provide the desired accuracy. Therefore, many additional test methods, summarised in [10] have been developed to unravel the complex microscopic mechanisms responsible for RPV steel embrittlement. According to our experimental possibilities and practical experiences, the Mössbauer spectroscopy (MS) results were compared mostly with positron annihilation spectroscopy (PAS) [11-26] and with transmission electron microscopy [18, 20, 24, 27, 28].

2. Reactor Pressure Vessel Steels and Mössabauer Spectroscopy

Mössbauer spectroscopy (MS) measures hyperfine interactions and these provide valuable and often unique information about the magnetic and electronic state of the iron species, their chemical bonding to co-coordinating ligands, the local crystal symmetry at the iron sites, structural defects, lattice-dynamical properties, elastic stresses, etc. [29, 30, 31]. In general, a Mössbauer spectrum shows different components if the probe atoms are located at lattice positions, which are chemically unequivalent. From the parameters that characterize a particular Mössbauer sub-spectrum it can, for instance, be established whether the corresponding probe atoms reside in sites which are not affected by structural lattice defects, or whether they are located at defect-correlated positions. In this respect, however, it is almost imperative to combine Mössbauer measurements with other research methods, which preferably are sensitive to the nature of the defect properties. Combining the results of MS with various other techniques [32] on the same samples seems to be a promising approach for such study. Differences between Mössbauer spectra obtained from different Eastern as well as Western types of RPV steels were already discussed in detail [33].

All RPV-steel samples investigated show the typical Mössbauer spectra of a steel with low alloy-element concentration, the main features being the presence of 2-3 magnetically split subspectra with isomer and quadrupole shifts close to 0 mm/s and, in certain cases at Western RPV-steels, a weak superimposed doublet component. The first sextet (with the largest absorption area and hyperfine field H_{hf} around 33.2 T) is assigned to Fe-atoms in the iron matrix that are not surrounded by foreign atoms in their close-neighbour shells (the so-called 'unperturbed' component), while the other magnetically split subspectrum (with H_{hf} in the range 29.0-30.7 T) are associated with iron atoms surrounded in their first or second neighbour shells by alloying elements (the 'perturbed' component). The isomer shift $\delta^{(Fe)}$ and quadrupole splitting $2\varepsilon_Q$ for both magnetically split subspectra was found to be close to 0 mm/s. Therefore, it was decided to fix $2\varepsilon_Q$ to 0 mm/s during the analyses.

In contrast to the MS results of the Western-type steels, the Eastern RPV steels had to be described by three or four sextets (four, if the pure α-Fe component was fixed at 33.0T). The most prominent gaussian is associated with the 'unperturbed' component, while the remaining gaussians are associated with the 'perturbed' component and are suggested to be due to iron atoms surrounded by one, two or three alloying elements in their first neighbour shell, respectively. The quadrupole shift obtained for both subspectra was again close to 0 mm/s, so that it was fixed at 0 mm/s during the analysis. The doublet is completely absent in the Eastern base material. The comparison of Mössbauer spectra of different non-irradiated commercially used RPV steels is described in details in [23]. In general, the thermal treatment, the irradiation and the post-irradiation heat treatment of Western RPV-steels do not affect significantly the MS parameters (see Fig.1).

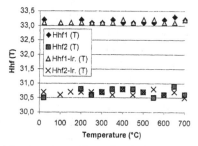

Figure 1. Influence of isochronal annealing and equivalent post-irradiation heat treatment on the change of MS parameters of IAEA reference quality RPV-steel (JRQ). Hhf1, Hhf2 and Hhf1-Ir, Hhf2-Ir indicate values of hyperfine field of MS sub-spectra before and after irradiation, respectively.

Results relevant to the comparison between irradiated and non-irradiated RPV-steel specimens are reported in details in [34].

Only few studies indicating MS to be a potentially interesting tool to investigate the microstructural aspects of irradiation embrittlement of RPV steels have been performed so far [29, 31, 34]. For this reason it was decided to investigate three RPV-steel samples with slightly different chemical compositions to determine the applicability of MS to the problem of RPV steel embrittlement. The H_{hf} distribution of the spectra enables to distinguish between the different steel types. Small differences in carbon concentrations between Western base and weld metal are reflected in the small area fraction of the doublet for the weld metal. Differences between Eastern and Western type RPV steels are reflected in the overall shape of the derived H_{hf}-distribution profile. The larger fraction of the 'perturbed' area for the Eastern steel, the differences in H_{hf} and δ values and the absence of a carbide doublet subspectrum are all due to the fact that the overall alloy-element concentration (especially for Cr and V) for the Eastern steels is larger than for Western-type steels. This interpretation was supported also by results from transmission electron microscopy and positron-annihilation measurements [24, 27]. Vanadium carbides (V_xC_y) formation was confirmed also using the small angle neutron scattering (SANS) [35].

3. Irradiation Experiments

In the framework of Extended Surveillance Specimen Program 24 specimens, designed especially for MS measurement, were selected and measured in "as received" state, before their placement into the core of the operated nuclear reactor [25]. For this program and measurement of high-irradiated RPV-specimens the one-dimensional angular correlation positron-annihilation spectrometer was developed in our department [36]. Results from this measurement are reported in [37]. Additionally, the positron annihilation lifetime spectra were measured using pulsed low energy positron system at the University of Bundeswehr in Munich (Germany). Both techniques were upgraded for measurement of irradiated specimens, where disturbing ^{60}Co contribution was a limiting factor for measurements in the past. The positron annihilation results were initially reported in [26].

Room temperature Mössbauer spectroscopy measurements were carried out in transmission geometry on a standard constant accelerator spectrometer with a ^{57}Co source in Rh matrix [29]. The absorbers consisted of 25-40 μm thick foils. Due to higher neutron embrittlement and ageing sensitivity of WWER 440 (V-230) nuclear reactors, our study has been focused on the Russian 15Kh2MFA steel used in this older WWER-440 reactor type.

Specimens were measured before their placement in the irradiation chambers, near the core of the operated nuclear reactor, and after 1, 2 and 3 years of irradiation (neutron fluence ranging from 7.8 10^{23} m^{-2} up to 2.5 10^{24} m^{-2}). Taking into account the enhancement of the irradiation due to the close proximity of the irradiation chambers to the reactor core ("accelerating" factor of about 10 at the most loaded position), the radiation treatment of the specimens after 3 years is equivalent to about 30 years of real RPV-steel (projected lifetime of RPV). The chemical composition and the irradiation conditions of the studied RPV-steel specimens are shown in Table 1 and Table 2.

TABLE 1. Chemical composition of used RPV steel 15Kh2MFA specimens in wt% [14].

Element	C	Si	Mn	Cr	Ni	Mo	V	S	P	Co	Cu
Base material	0.140	0.31	0.37	2.64	0.20	0.58	0.27	0.017	0.014	0.019	0.091
Weld	0.048	0.37	1.11	1.00	0.12	0.39	0.13	0.013	0.043	0.020	0.103

TABLE 2. Irradiation conditions of specimens at the 3rd unit of NPP Bohunice (Slovakia).

Sample	Material	Time of irradiation [eff. days]	Neutron Fluence (E_n>0.5 MeV) [m^{-2}]	Total activity [kBq]	Thickness of sample [μm]
ZMNF	Base metal, non-irradiated	0	0	0	60
ZM1Y	Base metal, 1 year irradiated	280	7.81E23	62	50
ZM2Y	Base metal, 2 year irradiated	578	1.64E24	109	40
ZM3Y	Base metal, 3 year irradiated	894	2.54E24	89	30
ZMNA	Sample ZMNF annealed 1 hour in vacuum at 385°C	0	0	0	60
ZKNF	Weld, non-irradiated	0	0	0	55
ZK1Y	Weld, 1 year irradiated	280	7.81E23	30	45
ZK2Y	Weld, 2 years irradiated	578	1.64E24	48	25
ZK3Y	Weld, 3 years irradiated	894	2.54E24	110	47
ZKNA	Sample ZKNF, annealed 1 hour in vacuum at 385°C	0	0	0	60

The irradiation temperature measured using melting monitors (special materials with well-defined melting temperature) placed inside special containers, reached values in the region of 285-298 °C. Neutron monitors measured the level of neutron fluence.

Mössbauer spectra, which correspond to the basic and weld material samples, show typical behaviour of dilute iron alloys and can be described with three [23] or four sextets (this work). In comparison to other RPV steels, the 15Kh2MFA steel contains more Cr and V. According to the latest knowledge, V is known as a predominant element to enhance the tensile properties of material. However, the toughness and weld crack sensitivity of the material are also affected by the addition of this element. Thus no V should be added, even though a maximum content of 0.03 or 0.05% is allowed in the material specifications. It has been said that Cu and P affect irradiation damage, which is evaluated by the shift in Charpy impact curve [27]. In response to this requirement, the target values of Cu and P content of 0,08% max. and 0,008% max., respectively are maintained in the last types of RPV-steels [38-40].

The reference MS spectra were compared with the irradiated state of identical specimens. These specimens were placed in nuclear operated reactor in the NPP Bohunice unit-3 and unit-4 with the aim to perform identical thermal and radiation treatment of RPV surveillance specimens as is true in reality.

The total specific activity of the first batch of specimens (sample k716 BM-I with the weight of 25.6 mg) was 3.2×10^7 Bq/g. Contributing elements are presented in Table 3.

TABLE 3. Activities of the most detected nuclides in RPV-steel specimen k716 BM-I (25.6 mg).

Nuclide	Sb 124	Co 58	Co 60	Cr 51	Fe 59	J 131	Mn 54	Na 24
Activity [Bq]	6438	1673	160750	22622	47952	37660	544790	143
Error	$\pm 1 \times 10^2$	$\pm 1 \times 10^2$	$\pm 2 \times 10^2$	$\pm 3 \times 10^2$	$\pm 2 \times 10^2$	$\pm 2 \times 10^3$	$\pm 3 \times 10^2$	$\pm 1 \times 10^2$

For the most suitable fitting model, we used the four components with fixed sextet No2, which corresponds to the pure α-iron with hyperfine field $H_{hf2} = 33.0$ T. MS parameters as areas under sextets (A_x) and hyperfine fields ($H_{hf,x}$) of RPV-steel specimens are selected in Tables 4 and 5. Typical Mössbauer spectra of base metal of RPV specimens after 1-year stay in operated nuclear reactor, treated by the neutron fluence $\Phi(E_n>0.5$ MeV)$=7.8 \times 10^{23}$ m^{-2} are shown in Fig. 2.

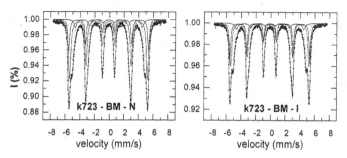

Figure 2. Comparison of Mössbauer spectra of RPV- base material in non-irradiated and irradiated state.

TABLE 4. Comparison of RPV-steel specimens in non-irradiated and irradiated state at 3rd unit NPP Bohunice (abbreviations BM-N stand for: base material - not irradiated, BM-I base material – irradiated – numbers of years, WM-N weld material - not irradiated, WM-I weld material – irradiated – numbers of years). A_{1-4} is a contribution for each of the component in MS spectrum. ΔA_{1+} and ΔA_{2-} represent the increase or decrease of this contribution after irradiation. Parameter $H_{hf\,2}$ was fixed at the value 33.0 T.

Specimen	A_1 [%]	A_2 [%]	ΔA_{1+} [%]	ΔA_{2-} [%]	A_3 [%]	A_4 [%]	$H_{hf\,1}$ [T]	$H_{hf\,3}$ [T]	$H_{hf\,4}$ [T]
k716 BM-N (ZMNF)	24.3	35.0			33.3	7.4	33.8	30.6	28.5
k716 BM-I-1y (ZM1Y)	31.6	25.1	7.3	9.9	37.3	6.0	33.7	30.7	28.6
k723 WM-N (ZKNF)	17.2	41.9			34.1	6.8	33.8	30.6	28.5
k723 WM-I-1y (ZK1Y)	23.8	34.0	6.6	7.9	35.3	6.9	33.7	30.6	28.4
2 years									
k721 BM-N	27.1	30.7			35.2	6.9	33.8	30.6	28.3
k721 BM-I-2y (ZM2Y)	31.1	25.4	4.0	5.3	36.9	6.6	33.7	30.6	28.1
k725 WM-N	20.5	41.4			30.8	7.3	33.8	30.7	28.6
k725 WM-I-2y (ZK2Y)	21.6	40.7	1.1	0.7	29.6	8.1	33.6	30.6	28.5
3 years									
k720 BM-N	26.2	32.5			33.7	7.6	33.8	30.7	28.6
K720 BM-I-3y (ZM3Y)	33.4	25.4	7.2	7.1	35.2	6.0	33.7	30.6	28.1
k728 WM-N	16.5	42.4			35.4	5.7	33.8	30.6	28.3
k728 WM-I-3y (ZM3Y)	19.0	40.1	2.5	2.3	34.6	6.3	33.7	30.6	28.1
Accuracy	±0.8	±0.8	±0.8	±0.8	±0.8	±0.8	±0.1	±0.1	±0.2

TABLE 5. Comparison of RPV-steel specimens in irradiated and non-irradiated state at 4th unit NPP Bohunice.

Specimen	A_1 [%]	A_2 [%]	ΔA_{1+} [%]	ΔA_{2-} [%]	A_3 [%]	A_4 [%]	$H_{hf\,1}$ [T]	$H_{hf\,3}$ [T]	$H_{hf\,4}$ [T]
k731 BM-N	24.6	33.3			35.4	6.7	33.8	30.6	28.3
k731 BM-I-1y	28.4	28.0	3.8	5.3	36.0	7.6	33.7	30.6	28.2
k735 WM-N	18.1	44.0			31.1	6.8	33.7	30.6	28.7
k735 WM-I-1y	23.0	38.2	4.9	5.8	32.7	6.1	33.7	30.6	28.4
2 years									
k733 BM-N	27.1	33.0			34.0	5.9	33.8	30.6	28.3
k733 BM-I-2y	31.7	27.6	4.6	5.4	34.3	6.4	33.7	30.6	28.2
k739 WM-N	17.2	46.1			29.1	7.7	33.8	30.7	28.9
k739 WM-I-2y	20.4	43.0	3.2	3.1	30.2	6.3	33.6	30.6	28.4
3 years									
k734 BM-N	26.1	36.8			31.5	5.7	33.8	30.6	28.4
k734 BM-I-3y	31.1	29.7	5.0	7.1	32.2	7.0	33.7	30.6	28.4
k740 WM-N	17.8	44.9			32.1	5.2	33.8	30.6	28.6
k740 WM-N-3y	20.5	42.5	2.7	2.4	31.7	5.3	33.7	30.6	28.1
Accuracy	±0.8	±0.8	±0.8	±0.8	±0.8	±0.8	±0.1	±0.1	±0.2

The most significant change after neutron irradiation is observable in the areas of the first two components (see Tables 4 and 5). The deterioration of RPV-steel specimens due to fast neutron bombardment is shown in the decrease of the pure α-iron component (fixed in all analysis at $H_{hf,2}$ = 33.0 T). The significant decrease (up to 10%) was observed in all specimens. This decrease was balanced by increased H_{hf1} values for the first component, which can be assigned to the contribution of complex of other atoms where Cr is dominant. A comparison of results from both units and between the base and weld material is shown in Fig.3a – Fig.3d.

Significant changes were observed between base and weld materials, but the behaviour of MS parameters due to irradiation are similar. After relatively intensive jumps in the values after the first year of irradiation, these changes were not consistent due to increased irradiation treatment and remained almost stable.

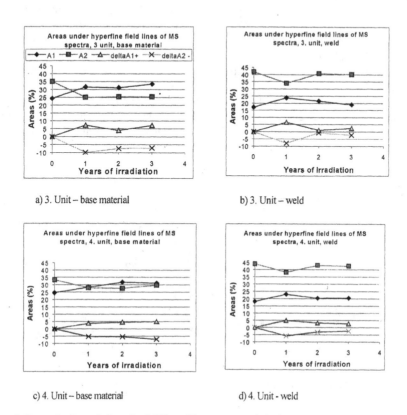

a) 3. Unit – base material   b) 3. Unit – weld

c) 4. Unit – base material   d) 4. Unit - weld

Figure 3. Changes in areas under hyperfine field lines of first two sextets of Mössbauer spectra measured on original irradiated RPV-steels from NPP Bohunice (Slovakia) in frame of Extended surveillance specimen program. ΔA_{1+} and delta ΔA_2. present increase of first and decrease of second component, respectively.

4. Application of the PAS Techniques

Since the year 1985, positron annihilation spectroscopy (PAS) has been repeatedly used in the study of RPV steels [5, 11-26]. The positron lifetime (PL) technique is a well-established method for studying open-volume type atomic defects and defect-impurity interactions in metals and alloys. The lifetime of positrons trapped at radiation-induced vacancies, vacancy-impurity pairs, dislocations, microvoids, etc. is longer than the one of free positrons in the perfect region of the same material. As a result of the presence of open-volume defects, the average positron lifetime observed in structural materials is found to increase with radiation damage [15, 19].

Specimens measured by MS and PAS techniques were identical. For the first time the pulsed low-energy positron system (PLEPS) [41,42] was used for the investigation of neutron-irradiated RPV-steels. This system enables the study of the micro structural changes in the region from 20 to 550 nm (depth profiling) with small and very thin (<50 μm) specimens, therefore reducing the disturbing ^{60}Co radiation contribution to the lifetime spectra to a minimum [26]. Such a disturbance is the limiting factor for the investigation of highly irradiated RPV specimens with conventional positron lifetime systems. Several approaches to tackle the problem of the ^{60}Co prompt-peak interference with the physical part of the positron lifetime spectra have been considered so far [11,21,43].

The time resolution of PLEPS was about 240 ps FWHM. All lifetime spectra of irradiated RPV specimens contained about 3×10^7 events at a peak to background ratio in the range between 30:1 and 100:1 [26]. Experimental results for the mean positron lifetime τ_m after various irradiation treatments are shown in Fig.4.

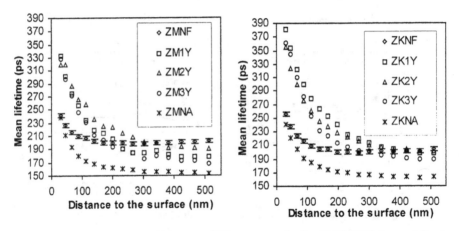

Figure 4. Comparison of mean lifetimes τ_m of different neutron-irradiated 15Kh2MFA (base metal) and Sv10KhMFT (weld) steel specimens.

τ_m is plotted versus the mean positron implantation depth. The characteristic decrease of τ_m with increasing positron implantation depth is typical for measurements with a low energy pulsed positron beam. It is due to the back diffusion of positrons to the surface and subsequent trapping at a surface state or in the oxide layer.

Each lifetime spectrum can be well analysed by two lifetime components, which below a depth about 300 nm remain fairly constant over the full depth range (see Fig.5). The shorter lifetime τ_1 of about 170 ps is the dominant steel component (most likely iron monovacancies) with an intensity of about 97% in the bulk. The longer lifetime τ_2 with intensity of about 3% or less and a value of about 400-500 ps can be assigned to the contribution of large vacancy clusters. The intensity of this component is surprisingly much higher (up to 10-12 %) for those specimens, which were not thermally treated (reference specimens ZMNF and ZKMF), indicating that large vacancy clusters are already present in the material before irradiation.

The interesting results for the irradiated specimens, in respect to the defect structure in the bulk, presented in Fig. 5, are the almost constant value of τ_1 for the weld alloy as well as the oscillating behaviour of τ_1 for the base alloy, where the resulting lifetime indicate that after one year and three years irradiation obviously only dislocations are present, whereas after two years of neutron irradiation a mixture of iron mono- and di-vacancies is produced.

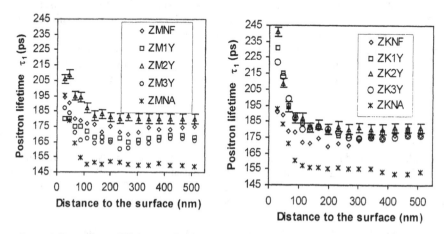

Figure 5. Comparison of lifetimes τ_1 of RPV steel specimens (base metal and weld) after different neutron irradiation.

5. Discussion

Generally, interactions between the fast neutron and the lattice atom can transfer energies ranging from a few eV to tens of keV. Indeed, a high fraction of the primary knock-on atoms (PKA) have energies above a few eV. These PKA lose energy though interaction with both the electrons and atoms of the solid. If the energy transferred to a lattice atom is greater than some threshold value, E_d (typically >40eV), then the atom will be displaced from its lattice site, creating a Frenkel defect, i.e. a vacancy and an interstitial. If the PKA energy is much greater than a few keV the PKA is able to displace many atoms, rapidly entering a regime where the collisions occur in every lattice spacing (displacement cascade). It is important to realise, that in the evolution of a cascade not only a heavily damaged region containing a large number of displacement atoms can be established, but that during the subsequent evolution considerable point defect motion, recombination and clustering may occur as well [7].

Both vacancy and interstitial point defects are expected to be mobile in the temperature range of most operating pressure vessels. However, they are also expected to interact with solute atoms. The key interstitial impurity in RPV steel is Carbon [10,28]. The partial or complete trapping of self-interstitial by C solutes will cause heterogeneous cluster nucleation and a fine cluster distribution.

Mössbauer spectroscopy is a powerful analysis technique, which precisely identifies the close environment of Mössbauer nucleus (Fe). Further, in steels containing residual levels of elements such as copper, which are in supper-saturated solution, radiation-enhanced diffusion will occur at these temperatures, which leads to the formation of small clusters, which can again harden the matrix. Generally, the thermal treatment together with neutron irradiation lead to microstructure consisting of small clusters (<5nm in diameter) which create obstacles to the free movement of dislocations thereby producing an increase in the yield stress, hardness and the ductile-brittle transition temperature of the material [8, 39].

Their relative areas are close to the theoretical values calculated from a random distribution model of impurities in a b.c.c. structure (5% of 12 elements in total). Results confirmed MS sensitivity to detect also small differences in chemical composition or preparing technology of RPV steel samples. In comparison with Western types of RPV steels such as A533 Cl.1 and A508 Cl.3, the doublet fraction ascribed as Mn and/or Cr- substituted cementite is completely absent in 15Kh2MFA. Here are formed probably mainly $Cr_{23}C_6$, Cr_7C_3 and VC carbides. [34,40].

6. Conclusion

This work was focused mostly on the influence of irradiation on the changes in hyperfine fields of Mössbauer spectra obtained from original RPV surveillance specimens. We aimed to obtain scaled MS results that could be correlated to another destructive (Charpy-V tests, hardness measurements) or non-destructive methods (positron annihilation techniques, transmission electron microscopy) [26].

Significant differences between the base and weld materials can be explained by different chemical compositions and/or different preparing technology and were observed also in the past [29,31,40]. The trendline in the changes due to irradiation is almost the same in both materials. It seems that the expected changes in material microstructure (precipitation of elements like Cu or Cr mainly in carbides to the surface) occurred mainly during the first period (1-year stay in irradiation containers in operating conditions by "speed factor" of about 10). These results confirm that the close environment of Fe atoms in b.c.c. lattice of RPV steels after initial changes remain almost stable and perhaps could be correlated to the trendline of ductile-brittle transition temperature (DBTT) curve obtained from mechanical tests or with the defects density curve obtained from the transmission electron microscopy studies. In dependence on the increased neutron fluence (upto $\Phi(E_n>0.5 \text{ MeV})=1\times10^{25}$ m^{-2}) the dislocation density as well as the average defect diameter remain stable after a first increase of about 20-30% [28].

On the other hand, the isochronal annealing of 2 selected irradiated specimens performed on the 400, 475 and 520 °C did not cause return of the MS parameters back to the starting positions. It means that the radiation-induced changes observable in MS spectra were not re-annealed in such complete way as the re-annealing of the point defects observed by TEM [18, 28].

The results from the present extensive study of RPV surveillance specimens indicate that MS could be a suitable technique for the evaluation of some microstructural changes in RPV-steels and, in combination with other spectroscopic methods, can contribute to an increase of

NPPs operational safety and lifetime prediction. However, MS is not a magic wand solving all substantial questions in neutron embrittlement and material ageing. This investigation will continue also in the next phase and specimens treated in reactor up to 10 years will be studied. Such loaded specimens will be equivalent to real RPV steel after more than 50 years of operation and will be surely interesting for the RPV lifetime management.

Acknowledgement

We would like to thank IAEA (Research contracts SLR10994, SLR11120), VEGA 1/7631/20, SE-EBO Bohunice, plc. and NPPRI Trnava for invaluable support. Special thanks to EU for support in the frame of project FU 00082-2000.

References

1. Davies, L.M. (1999) A comparison of western and eastern nuclear reactor pressure vessel steels, *Int. J. Press. Vess, & Piping* **76**, 163-208.
2. Nikolaev, Y.A., Nikolaeva, A.V., Kryukov, A.M., Shtrombakh, Y.I. and Platonov, P.A. (2000) In: *Proc. of the TACIS Workshop on the RPV Life Predictions*, PCP3-ENUCRA-D4.
3. Ghoneim, M.M. and Hammad, F.H. (1997) Pressure vessel steels: influence of chemical composition on irradiation sensitivity, *Int. J. Press. Vess. & Piping* **74**, 189-198.
4. Kohopaa, J. and Ahlstrand, R. (2000) Re-embrittlement behaviour of VVER-440 reactor pressure vessel weld material after annealing, *Int. J. Press. Vess. & Piping* **76**, 575-584.
5. Debarberis, L., von Estorff, U., Crutzen, S., Beers, M., Stamm, H., de Vries, M.I. and Tjoa, G.L. (2000) LYRA and other projects on RPV steel embrittlement Study and mitigation of the AMES Network, *Nucl. Eng. Des.* **195**, 217-226.
6. U.S. NRC Regulatory Guide 1.99, Rev.1 (1977).
7. Suzuki, K. (1998) IAEA report, IWG-LMNPP-98/3.
8. Grosse, M., Denner, V., Böhmert, J. and Mathon, M.H. (2000) Irradiation-induced structural changes in surveillance material of VVER 440-type weld metal, *J. Nucl. Mater.* **277**, 280-287.
9. Koutsky J. and Kocik, J. (1994) Radiaton damage of structural materials, ed. Academia Prague.
10. Phythian, W.J. and English, C.A.(1993) Microstructural evolution in-reactor pressure-vessel steels, *J. Nucl. Mater.* **205**, 162-177.
11. Brauer, G., Liszkay, L., Molnar B. and Krause, R. (1991) Microstructural aspects of neutron embrittlement of reactor pressure-vessel steels - a view from positron-annihilation spectroscopy, *Nucl. Eng. & Desg.* **127**, 47-68.
12. Pareja, R., De Diego, N., De La Cruz R.M. and Del Rio, J. (1993) Postirradiation recovery of a reactor pressure-vessel steel investigated by positron-annihilation and microhardness measurements, *Nucl. Technol.* **104**, 52-63.
13. Lopes Gil, C., De Lima, A.P., Ayres De Campos, N., Fernandez, J.V., Kögel, G., Sperr, P., Triftshäuser W. and Pachur, D.(1989) Neutron-irradiated reactor pressure-vessel steels investigated by positron-annihilation, *J. Nucl. Mater.* **161**, 1-12.
14. Prochazka, I., Novotny, I. and Becvar, F. (1997) Application of maximum-likelihood method to decomposition of positron-lifetime spectra to finite number of components, *Mater. Sci. Forum* **255-257**, 772-774.
15. Valo, M., Krause, R., Saarinen, K., Hautojärvi P. and Hawthorne, R. (1992) ASTM STP 1125, Stoller, Philadelphia.
16. Hartley, J.H., Howell, R.H., Asoka-Kumar, P., Sterne, P.A., Akers, D. and Denison, A. (1999) Positron annihilation studies of fatigue in 304 stainless steel, *Appl. Surf. Sci.* **149**, 204-206.
17. Becvar, F., Cizek, J., Lestak, L., Novotny, I., Prochazka, I. and Sebesta, F. (2000) A high-resolution BaF2 positron-lifetime spectrometer and experience with its long-term exploitation, *Nucl. Instr. Meth.* **A 443**, 557-577.
18. Miller, M.K., Russel, K.F., Kocik, J. and Keilova, E. (2000) Embrittlement of low copper VVER 440 surveillance samples neutron-irradiated to high fluences, *J Nucl. Mater.* **282**, 83-88.
19. Cizek, J., Becvar F. and Prochazka, I. (2000) Three-detector setup for positron-lifetime spectroscopy of solids containing Co-60 radionuclide, *Nucl. Instr. Meth.* **A 450**, 325-337.

20. Cizek, J., Prochazka, I., Kocik, J. and Keilova, E.(2000) Positron lifetime study of reactor pressure vessel steels, *phys. stat. sol. (a)* **178**, 651-662.

21. Van Hoorebeke, L., Fabry, A., van Walle, E., Van de Velde, J., Segers, D. and Dorikens-Vanpraet, L. (1996) A three-detector positron lifetime setup suited for measurements on irradiated steels, *Nucl. Instr. Meth.* **A 371**, 566-571.

22. Ghazi-Wakili, K., Zimmermann, U., Brunner, J., Tipping, P., Waeber, W.B. and Heinrich, F. (1987) Positron-annihilation studies on neutron-irradiated pressure-vessel steels, *phys. stat. sol. (a)* **102**, 153-163.

23. Slugen, V., Segers, D., De Bakker, P.M.A., DeGrave, E., Magula, V., Van Hoecke T. and Van Waeyenberge, B. (1999) Annealing behaviour of reactor pressure-vessel steels studied by positron-annihilation spectroscopy, Mossbauer spectroscopy and transmission electron microscopy, *J. Nucl. Mater.* **274**, 273-286.

24. Slugen and V. Magula, V. (1998) The micro structural study of 15Kh2MFA and 15Kh2NMFA reactor pressure vessel steels using positron-annihilation spectroscopy, Mossbauer spectroscopy and transmission electron microscopy, *Nucl. Eng. Desg.* **186/3**, 323-342.

25. Slugen, V. (2000) IAEA report, SLR10994.

26. Slugen, V., Hascik, J., Gröne, R., Bartik, P., Zeman, A., Kögel, G., Sperr P. and Triftshäuser, W. (2001) Investigation of reactor steels, *Mater. Sci. Forum* **363-365**, 47-51.

27. Magula V. and Janovec, J. (1994) Effect of short-time high-temperature annealing on kinetics of carbidic reactions in 2.7cr-0.6mo-0.3v steel, *Ironmaking and Steelmaking* **21**, 223-228.

28. J. Kocik, J., Keilova, E., Cizek, J. and Prochazka, I. (2000) in: METAL 2000, Proc. of 9th Int. Conf. on Metallurgy (CD-ROM), May 2000 Ostrava, Czech Republic, Tanger Ltd., paper No.719.

29. Lipka, J., Hascik, J., Slugen, V., Kupca, L., Miglierini, M.,Gröne, R., Toth, I., Vitazek, K. and Sitek, J. (1996) In: *Proc. Int. Conf. Vol. 50, ICAME'95*, I. Ortalli (ed.), SIF, Bologna, 161.

30. Cohen, L. (1980) Application of Mössbauer spectroscopy. Volume II. Academic Press, New York.

31. Brauer, G., Matz, W. and Fetzer, Cs. (1990) Experience with neutron-irradiated reactor pressure-vessel steels - a mossbauer study, *Hyperfine Interaction* **56**, 1563-1567.

32. Amaev, A.D., Dragunov, Yu.G., Kryukov, A.M., Lebedev, L.M. and Sokolov, M.A. (1986) Investigation of irradiation embtrittlement of reactor VVER-440 vessel materials. In: *Proc. of IAEA specialists meeting of RPV embrittlement*, Plzeň.

33. Slugen, V. (1999) Microstructural analysis of nuclear reactor pressure vessl steels, In: M.Miglierini and D. Petridis (eds.), *Mössbauer spectroscopy in material science*, Kluwer Academic Publishers, Netherlands, 119-130.

34. De Bakker, P., Slugen, V., De Grave, E., Van Walle, E. and Fabry, A. (1997) Differences between eastern and western-type nuclear reactor pressure vessel steels as probed by Mossbauer spectroscopy, *Hyperfine Interaction* **110**, 11-16.

35. Puska, M.J., Sob, M., Brauer, G. and Korhonen, T. (1994) 1st-principles calculation of positron lifetimes and affinities in perfect and imperfect transition-metal carbides and nitrides, *Phys. Rev.* **B49**, 10947-10957.

36. Gröne, R., Hascik, J., Slugen, V., Lipka, R., Pietryzk, P. and Vitazek, K. (1997) A positron 1D-ACAR spectrometer for the study of Co-60 containing materials, *Nucl. Instr. & Meth. in Phys. Res.* **B 129**, 284-288.

37. Slugen, V., Hascik, J. and Gröne, R. (2000) Angular correlation positron annihilation spectroscopy applied in investigation of neutron irradiated RPV-steels, *International Journal of Applied Electromagnetics and Mechanics* **11**, 39-47.

38. Davies, M., Kryukov,.A., English, C., Nikolaev Y. and Server, W.L. (2000) ASTM STP1366.

39. Böhmert J. and Grosse M. (1998) In: *Proc. of Jahrestagung Kerntechnik 1998*, ed. Inforum Verlag, Bonn, 741.

40. de Bakker, P.M.A., De Grave, E., van Walle E. and Fabry, A. (1996) *Proc. Int. Conf. Vol. 50, ICAME'95*, I. Ortalli (ed.), SIF, Bologna, 145.

41. Sperr, P., Kögel, G., Willutzki P. and Triftshäuser, W. (1997) Pulsing of low energy positron beams, *Applied Surface Science* **116**, 78-81.

42. Bauer-Kugelmann, W., Sperr, P., Kögel G. and Triftshäuser, W. (2001) Latest version of the Munich pulsed low energy positron system, *Mater. Sci. Forum* **363-365**, 529-531.

43. Prochazka, I., Novotny, I. and Becvar, F. (1997) Application of maximum-likelihood method to decomposition of positron-lifetime spectra to finite number of components, *Mater. Sci. Forum* **255-257**, 772-774.

MŐSSBAUER SPECTROSCOPY OF COMMERCIAL GALVANNEALED ZINC COATINGS

M. ZMRZLY[1,2], O. SCHNEEWEISS[1] AND J. FIALA[2]

[1] *Institute of physics of materials, Academy of Science of The Czech Republic, Žižkova 22, 61662 Brno, Czech Republic*
[2] *Institute of chemistry of materials, Faculty of chemistry, Brno University of Technology, Purkyňova 118, 61200 Brno, Czech Republic*

1. Introduction

Corrosion losses amount to 2% of production of the steel. Research in the area of corrosion protection and surface treatment of materials is therefore one of the most frequent topics in contemporary material engineering. The traditional and well-known method of surface treatment technology is the hot-dipping process involving zinc coating [1]. Due to inter-metallic phases resulting from steel and zinc substrates, Mössbauer spectroscopy is one of the convenient methods of phase analysis. However, the sample preparation presents with a problem, because of the top part of the coatings, which is formed by zinc and gamma radiation is not able to penetrate the layers of intermetallics below this zinc layer.

2. Technology

Two similar technologies of surface treatment will be discussed in this paper: hot-dip galvanizing and galvannealing

2.1. HOT-DIP GALVANIZING

Treated surface has to be degreased and pickled. Then it is dipped into the aqueous solution of ammonium chloride and dried in tunnel furnace. Then follows the actual process, the product is dipped into the molten zinc bath containing a minor amounts of Al, Ni, etc. (460°C, in special cases 550°C. Melting point of Zinc: 419,7°C) from a few seconds to a few minutes, according to the steel thickness. After the zinc layer formation and cooling, the surface can be passivated by chromating or phosphating. This technology is used mainly for piece goods treatment; coating the thickness reaches 100 μm [2].

M. Mashlan et al. (eds.), Material Research in Atomic Scale by Mössbauer Spectroscopy, 199–204.

2.2. GALVANNEALING

This process is almost the same as the previous one. It is used for coating steel sheets in continuous production lines. The band of steel is lead through the pre-treatment baths (degreasing, pickling), then through the annealing furnace with controlled atmosphere ($N_2 + H_2$) where the surface is deoxidised. Following this, it continues directly into the molten zinc bath (460°C) usually doped by cca 0.13%wt. of Al. Time for stay in bath is cca five seconds. Almost no intermetallics occur here, apart from the layer of molten zinc on the surface. Then the band passes through air knives that wipe away excess zinc and decrease the thickness of the layer to cca 20 μm. The band continues to the galvannealing furnace, where diffusion of iron and zinc takes place and the intermetallic phases are conceived. By controlling time and temperature in the furnace, we can control the phase composition of the coating.

3. Coating Structure

As we can see in phase diagram [3], there are four intermetallic phases in binary system of iron and zinc. Note the width of concentration ranges and the overlapping of Gamma and Gamma 1 phases.

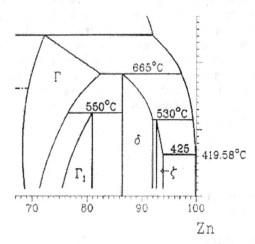

Figure 1. Zinc- rich corner of Fe- Zn binary phase diagram [3]. Concentration in atomic percents of zinc.

The top of the coating is composed of almost pure zinc. It provides corrosion resistance of the coating. Hardness is 70 HV.

First intermetallic phase is Zeta phase $FeZn_{13}$. The mechanical properties are quite good (185 HV) but corrosion resistance is poor. This phase is monoclinic. On metallographic micrographs it can be seen as typical needle-like crystals, usually forming most of the coating thickness.

Delta phase is the hardest component of the coatings (300 HV). It is hexagonal, and it is also present in coatings. It is very brittle [4].

Gamma and Gamma 1 are cubic structures, bcc, resp. fcc. They form a very narrow layer or they can absent totally. Their mechanical properties are probably the best (180 HV) [4].

4. Mössbauer spectra

The intermetallic phase (IMP) have very different spectra. The chart below summarises Mössbauer spectroscopy parameters obtained on laboratory prepared powders [5]. One very narrow doublet of zeta phase betrays only one site of iron in the structure, with high symmetry of nearest surrounding. Changes of relative areas of sites of Gamma and Gamma 1 phase's spectra have helped to resolve the problem of occupation of position in structure clusters that form these phases. There were two possible versions (Johansson, Brandon) obtained by neutron diffraction. Mössbauer spectroscopy confirmed the Johansson version [5].

TABLE 1. Mössbauer spectroscopy parameters of Fe-Zn IMP.

Phase	Site	Quadrupole splitting [mm/s]	Isomer shift [mm/s]
ζ		0.153	0.489
δ	A	0.468	0.508
	B	0.474	0.216
	C	0.275	0.372
Γ	A	0.156	0.365
	B	0.503	0.126
Γ_1	A	0.079	0.470
	B	0.555	-0.042
	C	0.224	0.042

5. Experimental

5.1. AIM

Since the top layer of zinc is impermeable to gamma radiation, it was necessary to find some well-defined method for removing of this layer. It meant to find a convenient etching solution and to estimate its rate of etching. Moreover, it was essential to avoid contamination of the IMP by etching solution or products of etching.

5.2. MATERIAL

Two commercially produced zinc coatings were the subject of the study. The first material was 5.5 mm thickness construction steel coated by hot-dip process. The coating thickness was 80μm. The second material was continuously galvannealed steel sheet of thickness 0.7mm with 20 μm of coating thickness.

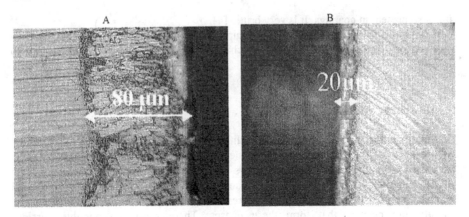

Figure 2. Micrographs of used materials.500x, etchant: 2% CrO_3 A) hot-dipped B) Galvannealed.

5.3. ETCHING

The chromate solution, prepared by dissolving of 200g CrO_3 and 7g Na_2SO_4 .10H_2O in distilled water, was applied. The samples were etched in this solution, followed shortly by rinsing in 2% HCl to remove chromate deposits from the surface and the decrease in coating thickness was measured at metallographical cross-section micrographs and with micrometric thread gauge. However, the measurement with gauge was difficult because of the so-called, "hard zinc" grains which occur always, and reach up to 50μm in diameter. They are caused, technologically, by the oxidation of the molten zinc bath surface.

5.4. RESULTS

The fall in thickness was measured at nine points of the sample. From the slopes of obtained lines was estimated the mean etching rate of chromate solution as being −4.65 μm min^{-1} .

From the measurement of metalographic samples three slopes were identified: zinc layer −10 μm min^{-1}, Zeta phase −3.7 μm min^{-1}, Gamma phase −0.4 μm min^{-1}.

Mössbauer spectroscopy measurements were carried out after removal of the zinc layer. The galvannealed coating was scratched off from the surface of the sample and the powder was measured by transmission Mössbauer spectroscopy. The hot-dipped sample was measured by backscattering Mössbauer spectroscopy.

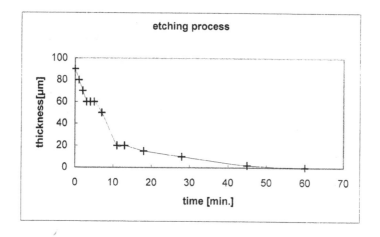

Figure 3. Measurement of decrease of coating thickness during the etching proces.

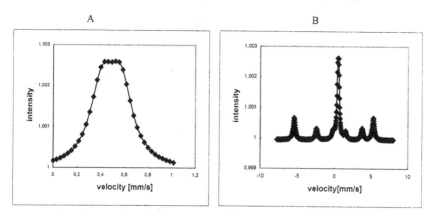

Figure 4. Backscattering Mössbauer spectra of hot-dipped material. Source: ^{57}Co in Rh matrix, cal. against α – Fe. A) narrow doublet of zeta phase (α – Fe background was removed in MS Excell) B) complete spectrum of the sample.

A well-defined zeta phase was identified in the hot-dipped sample. Isomer shift is in agreement with values of [5] (0.489 mm/s). Quadrupole splitting is narrower (0.080 mm/s) than the values obtained by Cook (0.153 mm/s). One reason could be the texture of the coating and the extensive dimensions of monocrystals in the layer.

In the galvannealed sample, zeta phase (IS: 0.511 mm/s, QS: 0.129 mm/s) was also identified, but the dominant component is a doublet with IS: 0.195 mm/s and QS: 0.186 mm/s. It is supposed that the remaining chromate deposits form this site of iron, and rinsing in HCl was probably not sufficient. On the other hand, it can be caused by aluminium doping of the molten zinc bath. This issue remains unresolved to the present day.

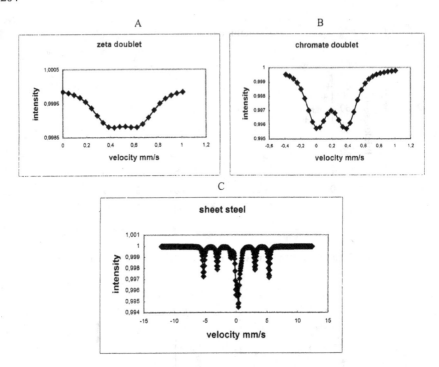

Figure 5. Transmission Mössbauer spectra of galvannealed material. Source: ^{57}Co in Rh matrix, cal. against α – Fe. A) narrow doublet of zeta phase (α – Fe background was removed in MS Excell) B) doublet of chromate deposits (α – Fe background was removed in MS Excell) C) complete spectrum of the sample.

6. Conclusions

The mean etching rate of chromate solution was estimated as –4.65 μm min^{-1}. In both studied materials, zeta phase was identified with narrow quadruple splitting, as is explained by the texture of coatings. The rinsing process in HCl must be improved to avoid chromate deposits on the samples' surfaces. The efficiency of rinsing will be checked by EDX. Research on galvannealed steel sheets will continue, focusing on the deeper layers of inter-metallic phases.

References

1. Zmrzlý, M. (1999) Corrosion protection of steel by electroplated, thermal spayed and hot-dip galvanized zinc coatings, Batchelor thesis, VUT Brno.
2. Zmrzlý, M. (2001) Corrosion protection of steel by diffusion zinc coatings, Diploma thesis, VUT Brno.
3. Mackowiak, J. and Short, N.R. (1979) Metallurgy of galvanised coatings, *Int. Metals Reviews* **1**, 241-269.
4. Leško, A., Kollárová, M. and Parilák, Ľ. (2002) Deformation and fracture of hot dip galvanized steel sheets, *Metallic materials* **40**, 1 –10.
5. Cook, C.D. (1993) Mossbauer Spectroscopy Applied to Magnetism and Materials Science *(Modern Inorganic Chemistry,* Plenum Pub Corp.

NUCLEAR RESONANT SCATTERING OF SYNCHROTRON RADIATION

O. LEUPOLD[1,2], A.I.. CHUMAKOV[1] AND R. RÜFFER[1]

[1]*European Synchrotron Radiation Facility, F-38043 Grenoble, France*
[2]*On leave from:*
Fachbereich Physik, Universität Rostock, D-18055 Rostock, Germany

Abstract

Since its observation in 1985 nuclear resonant scattering of synchrotron radiation has become an excellent tool to study hyperfine interactions as well as dynamical effects in solids. It combines the advantages of both local probe experiments and scattering techniques and gives valuable information on magnetic structures in case of NFS experiments and of the density of phonon states in case of NIS experiments. Experiments benefit from the outstanding beam quality of 3rd generation synchrotron radiation sources, as the small beam size and divergence.

1. Introduction

Since the pioneering work of Gerdau et al.[1], which showed for the first time nuclear resonant scattering of synchrotron radiation (short NRS) this field has dynamically developed from an exotic "playground" - with the need of highly isotopic enriched perfect single crystals - to an established method. Since then the technical development, like dedicated synchrotron radiation sources, high-resolution x-ray optics and fast detectors, have made a rapid development possible. Nowadays, third generation synchrotron radiation sources like the European Synchrotron Radiation Facility (ESRF) in Grenoble/France, the Advanced Photon Source (APS) in Chicago/USA and SPring-8 in Harima/Japan provide high brilliant beams to perform efficiently NRS under extreme sample conditions, like small sample size, grazing incidence geometry or high pressure. A number of recent reviews on the development and on various aspects of the NRS technique can be found in [2].

There exist mainly two branches: nuclear forward scattering (NFS) and nuclear inelastic scattering (NIS). NFS can be considered to be a complementary technique to Mössbauer spectroscopy, since it means in most cases hyperfine spectroscopy, combining the energy resolution of Mössbauer resonances with the outstanding properties of third generation sources. Due to the linear polarization of synchrotron radiation NFS is especially sensitive to the direction of magnetic hyperfine fields. NIS, on the other hand, opens the field of highly element specific investigation of lattice dynamics.

205

M. Mashlan et al. (eds.), Material Research in Atomic Scale by Mössbauer Spectroscopy, 205–216.
© 2003 *Kluwer Academic Publishers. Printed in the Netherlands.*

The two following chapters will give a general introduction to nuclear resonant scattering and to the beamline layout for NRS experiments. Here we will restrict ourselves to NFS utilizing the two nuclei ^{57}Fe and ^{151}Eu and then turn towards selected applications.

2. Nuclear Resonance Scattering Beamlines – Typical Layout

Synchrotron radiation sources and "standard" beam line optics deliver a very intense photon beam with a bandwidth in the order of some eV, which is very large as compared to the extremely narrow widths of nuclear resonances, which are usually in the order of 1 neV to 100 neV in the case of Mössbauer isotopes. So only a very tiny part of the photon intensity in the beam is able to excite a nuclear resonance. More sophisticated optics achieving energy resolutions in the meV regime were designed, in order to reduce the load on the detectors and the timing electronics.

In this work we report on experiments which were performed either at the wiggler station BW4 [3] at DESY or at the undulator based nuclear resonance beamline ID18 [4] at the ESRF. The experimental setup for nuclear forward scattering at the ESRF beamline ID18 [4] is displayed in Fig 1.

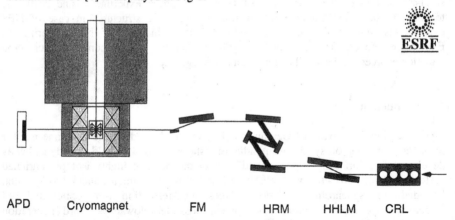

Figure 1. Experimental setup for nuclear forward scattering of synchrotron radiation. CRL: compound refractive lens; HHLM: high heat load monochromator; HRM: high resolution monochromator; FM: focusing monochromator; APD: avalanche photo diode (for details see text).

The x-ray beam delivered from one to three undulators is first slightly focused or collimated by the compound refractive lens CRL [5] to ≈150 µm FWHM, in order to improve the throughput and the energy resolution of the monochromators, as well as to reduce the beam size at the sample. The high heat load monochromator (HHLM) consisting of two cryogenically cooled Si (111) single crystals delivers an energy band of some eV, depending on the X-ray energy. The high-resolution monochromator (HRM) utilizes two Si channel cut crystals in nested configuration as proposed in [6]. For each isotope a special monochromator solution has to be found, sometimes also depending on the experiment, either optimized in flux or in energy resolution. The focusing monochromator (FM) [7] designed with a pair of two Si(1 1 1) single crystals reduces the horizontal beam size to about 200 μm FWHM.. This is achieved by bending the 400 μm thick Si crystal.

The cryomagnetic system with a superconducting split coil magnet can reach magnetic fields up to 6 T in a temperature range from ≈ 2 K to ≈ 400 K. This system has a large sample space (diameter 48 mm) in order to mount high-pressure cells or samples for grazing incidence experiments. The cryostat is mounted on a two circle goniometer for adjusting tilt and glancing angle, respectively. As fast detector serves an avalanche photo diode (APD) [8,9], with a time resolution of below 1 ns. Further reduction of the beam size can be achieved by Kirkpatrick-Baez optics using bent multilayer mirrors. For a review on monochromatization cf. [10], on fast detectors cf. [11]. Detectors and timing electronics have to discriminate the *prompt* non-resonant from the *time-delayed* resonantly scattered photons.

The setup at BW4 is in general comparable. The cryomagnet system is more versatile from the point of view of magnetic fields, since it has two superconducting split pairs in orthogonal directions, which can be operated simultaneously and such offer the possibility to align an external magnetic field in various different orientations with respect to the beam. The larger beam size cannot be reduced by optical elements like CRLs. The only accessible nuclear resonance is that of ^{57}Fe.

3. Nuclear Forward Scattering of Synchrotron Radiation

The first NFS experiments were performed by Hastings et al. [12] and van Bürck et al. [13] on magnetized α-Fe foils. An overview on NFS can be found in recent review articles, as for example in Refs. [14,15], which, however, cover only the ^{57}Fe resonance. In the following we will give a short introduction to the NFS method with special focus on the isotopes ^{57}Fe and ^{151}Eu and their magnetic hyperfine interaction.

3.1. BASICS OF NFS

We will give here a very schematic sketch of NFS. The highly collimated, pulsed synchrotron radiation impinges on the sample; the transmitted and forward scattered radiation is recorded by a fast detector. Since the resonantly scattered photons are reemitted delayed due the lifetime of the nuclear levels, which is much longer than the length of the exciting pulse (typically 10...100 ns as compared to 100 ps), the time evolution of this intensity can be used to discriminate the resonant from the nonresonant intensity. The exciting pulses, which define the "time zero" of the nuclear excitation, are suppressed by electronic means in the coarse of the experiment. At the ESRF in 16-bunch mode the distance between two successive pulses is 176 ns corresponding to the electron bunch distance.

The time behavior of the reemitted photons in forward direction is, in general, not just an exponential decay with the corresponding lifetime of the excited nuclear state. The time decay shows characteristic modulations, which are on one hand caused by the hyperfine interactions (*quantum beats*) and on the other hand by multiple scattering in thick samples (*dynamical beats*).

Like in conventional energy domain Mössbauer spectroscopy [57]Fe is the most widely used isotope in nuclear resonant scattering of synchrotron radiation. Experiments with several isotopes, which are already accessible at synchrotron radiation sources, i.e. [169]Tm [16], [119]Sn [17,18], [83]Kr [19,20], [181]Ta [21], and [151]Eu [22,23] have been treated in a recent review [24]. Others, which are especially important for magnetism, are, e.g. [161]Dy [23,25], or [149]Sm [26].

3.2. NFS AND HYPERFINE INTERACTIONS

3.2.1. *Magnetic hyperfine interaction in α-iron*

As first example we show in Fig. 2 NFS spectra of a magnetized α-[57]Fe foil. The $I_e=3/2 \rightarrow I_g=1/2$ Mössbauer transition in [57]Fe (I_e, I_g are the spin quantum numbers of the excited and ground state, respectively) is a magnetic dipole transition. The magnetic hyperfine interaction completely lifts the degeneracy of the nuclear levels, the selection rules for dipole radiation allow $\Delta m=0, \pm 1$ transitions, i.e. 6 transitions. Spectra were measured at a temperature $T \approx 100$ K in a helium bath cryostat in external fields of 1 T in plane and 3 T perpendicular to the plane of the foil. Figure 2 shows the pronounced dependence of the NFS time patterns on the external magnetic field direction. This is caused both by the angular characteristics of the dipole transition and the linear polarization of the synchrotron radiation, which is depicted in the coordinate system in Fig. 2. We call it σ-polarization, when the electric field vector E is in the storage ring plane. The relevant Mössbauer transitions, the corresponding polarization eigenstates of the transitions and the change of the magnetic quantum number Δm are given in the figure, as well.

In case (a) the external magnetic field is perpendicular to the storage ring plane, only the $\Delta m=0$ transitions are σ-polarized and can be excited. The two possible hyperfine transitions visualized by the grey line in the stick diagram and in the transition scheme give rise to the fast quantum beat modulation superimposed by the slower dynamical beat with its intensity minimum around 90 ns.

In case (b) the external magnetic field is parallel to the polarization of the synchrotron radiation, now the $\Delta m=\pm 1$ transitions have the proper polarization and can be excited. The interference of now four possible transitions with different amplitudes gives a more complicated beat pattern.

In case (c), where H is parallel to k, the eigenpolarization states of the nuclear transitions are left and right circular. All 4 transitions can be excited, because the polarization of the incoming beam can be described as a linear combination of left and right circular polarization. A rather simple quantum beat pattern appears, due to the fact that $\Delta m=+1$ cannot interfere with $\Delta m=-1$ transitions. Furthermore, the γ-quanta from all transitions belonging to the same polarization state have the same energy.

The two cases (a) and (b) in Fig. 2, where the hyperfine field is aligned perpendicular to the beam by the external field, would not be distinguishable in Mössbauer spectroscopy with a conventional unpolarized source. Polarized single line Mössbauer sources are the exception rather than the rule. We want to mention the concept [27] and recent applications utilizing circular polarized [28-30] and linear polarized [31] Mössbauer sources.

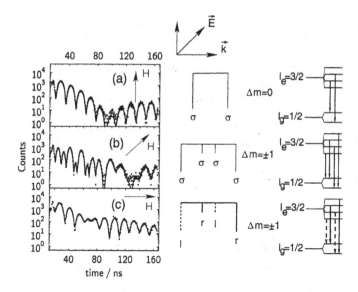

Figure 2. NFS spectra of a 3μm thin α-^{57}Fe foil in external magnetic fields of different directions as indicated. Measurements were done at the nuclear resonance station BW4 at HASYLAB/DESY. *k* is the direction of the photon, *E* the direction of the polarization vector of the beam (σ-polarization). The stick diagrams represent the corresponding Mössbauer spectra, the transition schemes on the right assume a linear polarized source. l and r denote left and right circular polarization, respectively.

3.2.2. *Magnetic and electric hyperfine interactions in Eu compounds*

One very interesting nucleus in the field of magnetism is ^{151}Eu with resonance energy of 21.5417 keV. The $I_e=7/2 \rightarrow I_g=5/2$ Mössbauer transition in ^{151}Eu is a magnetic dipole transition, as well. This results in 18 transitions if a magnetic hyperfine interaction acts on the nucleus.

The numerous spectroscopic applications of this resonance benefit from the two valence states of europium and their large difference in isomer shift and different magnetic properties. The Eu^{2+} ion with the magnetic $4f^7(^8S_{7/2})$ state is an excellent example of pure spin magnetism, therefore, magnetic properties can be easily monitored with the ^{151}Eu resonance. Together with the Eu^{3+} ion with its non-magnetic $4f^6(^7F_0)$ configuration of the 4f shell, europium compounds offer the possibility to study, e.g., temperature and pressure dependent valence transitions and intermediate valence systems. ^{151}Eu Mössbauer spectroscopy was very successful to study these phenomena [32], general reviews on ^{151}Eu Mössbauer spectroscopy, see Refs. [33]. The resonance was first observed in nuclear forward scattering at the ESRF in 1995 [22] and in nuclear inelastic scattering geometry at TRISTAN (KEK, Tsukuba) [23]. A recent review on NRS with this isotope and on instrumentation is given in [24].

As an introductory example the effect of magnetic hyperfine interaction on the ^{151}Eu nuclei is depicted in Fig. 3, where we show experimental results on a polycrystalline sample of EuS, which is ferromagnetically ordered below $T_C=16.5$ K. The experimental setup at ID18 at the ESRF corresponds to that in Fig. 1 given above. By proper choice of the direction of an external magnetic field one can select either $\Delta m=\pm 1$ or $\Delta m=0$

210

transitions, when studying ferromagnetic compounds, and simplify considerably the time spectra as in the case of ^{57}Fe. The time spectra depend in a very pronounced manner on the external magnetic field direction. Note, that without an external field all 18 transitions contribute to the time spectrum (Fig. 3, case(a)), while with external field in both cases (b) and (c) only the $\Delta m=\pm 1$ transitions can be excited. Again in the case that H is parallel to k the eigenpolarization states are left and right circular and a rather simple quantum beat pattern appears, due to the fact that $\Delta m=+1$ can not interfere with $\Delta m=-1$ transitions (see Fig. 3(c)). When H is parallel to the polarization vector E of the synchrotron radiation beam, *all* $\Delta m=\pm 1$ transitions have the same polarization and can interfere (Fig. 3(b)).

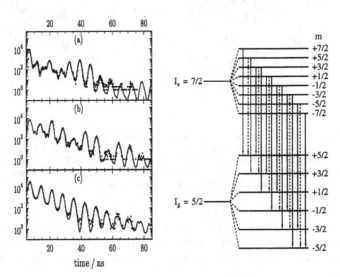

Figure 3. left: NFS spectra of a EuS powder sample measured at ID18/ESRF; (a) T=4K, H_{ext}=0; (b) T=7K, H_{ext}=3T parallel to E; (c) T=4K, H_{ext}=3.66T parallel to k (as defined in fig. 2). right: nuclear level scheme and possible transitions assuming pure magnetic hyperfine interaction; dashed lines: $\Delta m=0$ transitions; full lines: $\Delta m=\pm 1$ transitions

After this first basic experiment several applications at the ^{151}Eu resonance were performed at the ID18 beam line. As mentioned above, the large difference in isomer shift between the Eu^{2+} and the Eu^{3+} valence states makes NFS on europium compounds an almost ideal tool to determine the valence of the Eu atom/ion, by using reference samples of well known isomer shift upstream or downstream the sample under study. This technique was successfully applied at the ^{57}Fe resonance [34,35].
It is worthwhile to note that — in contrast to conventional Mössbauer spectroscopy — a direct determination of the isomer shift is not possible in a single NFS measurement without reference sample, due to the broadband excitation by the synchrotron radiation pulse. In the case of ^{151}Eu there exist two standard compounds, the divalent EuS and the trivalent EuF$_3$. For demonstration of the effect of the isomer shift relative to a reference sample, powder samples of EuS and EuF$_3$ were measured both alone and sandwiched

together [36]. The results are shown in Fig. 4.

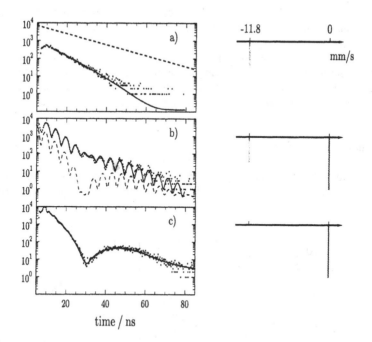

Figure 4. a) Time spectrum of an EuS powder sample. The dashed line corresponds to a decay with the natural life time. b) Time spectrum of an EuS/EuF₃ powder sample "sandwich". The dashed line resembles the shape of the spectrum of an "ideal sandwich" with two samples of the same thickness. c) Time spectrum of an EuF₃ powder sample. The stick diagrams on the right visualize the position of the resonance lines in energy scale. The (small) line broadening in EuF₃ due to quadrupole splitting is neglected in the stick diagram.

Fig. 4a: The time spectrum of a EuS powder sample can be fitted with a single resonance line, assuming a slight thickness inhomogeneity, which smears out the expected Bessel minimum of the dynamical beat structure.

Fig. 4 c: Time spectrum of a EuF₃ powder sample. This spectrum cannot be fitted with a single line only. The Bessel minimum is pronounced and excludes the possibility of a large thickness distribution. The best fit result was obtained assuming a quadrupole interaction parameter of the nuclear ground state, $\frac{1}{2}eQ_gV_{zz} = (1.7\pm0.25)$ mm/s, (where Q_g is the quadrupole moment of the ground state of the ^{151}Eu nucleus) together with an asymmetry parameter of the electric field gradient (EFG) tensor, $\eta=0.8$. The sign of V_{zz}, the main component of the EFG-tensor, can not be determined, again due to the broad bandwidth of the exciting beam compared with the nuclear level width. The unavoidable line broadening of a standard SmF₃ Mössbauer source obscured this splitting of the EuF₃ resonance in conventional Mössbauer spectra [37].

Fig. 4b: The time spectrum of the EuS/EuF₃ sandwich clearly reveals the large isomer shift between the Eu²⁺ and the Eu³⁺ states, exhibiting fast quantum beats with a period of ≈5 ns. The fitted difference in isomer shift $\Delta(IS)=(11.8\pm0.05)$mm/s corresponds to an energy of $8.5 \cdot 10^{-7}$eV$\leftrightarrow$205MHz. With the use of the EuS isomer shift reference the determination of the sign of the quadrupole interaction parameter for EuF$_3$ is possible, in principle. But, at large η the spectrum of EuF$_3$ is almost symmetric, which makes an unambiguous determination difficult. However, the χ^2 of the fits favors the negative sign.

The intensity at the Bessel minimum around 30ns does not reach the zero level, due to the different thickness of the two samples. It only reduces the quantum beat contrast. The dashed line shows the simulation of the time delayed intensity of an "ideal sandwich" with two samples of the same thickness, keeping all other parameters fixed.

4. Examples

In this chapter we show two examples – in grazing incidence geometry and under high pressure, respectively – where NFS benefits from the small beam size and divergence at 3rd generation sources.

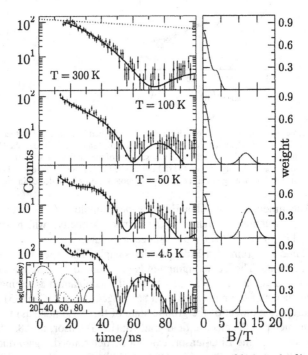

Figure 5. Top: Schematics of NRS in grazing incidence geometry (θ is the angle of incidence) Bottom: The left panel shows time spectra of nuclear resonant grazing-incidence reflection from ultrasmall ^{57}Fe islands on W(1 1 0). The modulation of the intensity is attributed to a perpendicular magnetization of the Fe islands. The solid lines are the results of simulations. For comparison the dashed line in the insert corresponds to magnetization in the sample plane. The right panel displays the probability density for the hyperfine field distribution that was obtained from the simulations (from [38]).

As first example the perpendicular spin orientation in ultra small Fe islands on W(1 1 0) is discussed [38]. Ultra small pseudomorphic Fe islands on an atomically clean W(1 1 0) crystal have been prepared by thermal evaporation of Fe enriched to 95 % in ^{57}Fe. The coverage of the Fe islands was 0.57 which is slightly below the percolation limit. Their average diameter was determined to be 2.0 nm. A coating of five monolayers of Ag prevents the sample from contamination. Figure 5 (left panel) displays the time spectra taken in grazing incidence geometry between RT and 4.5 K. The right panel of figure 5 shows the weight and distribution of the magnetic component. It increases with decreasing temperature. This can be attributed to superparamagnetic relaxations of the magnetic moments. At high temperatures the magnetization of small particles is subject to fast thermal fluctuations so that the effective magnetic hyperfine field averages to zero. The transition from the fast relaxation regime to the magnetically ordered state occurs at about 50 K. However, it is a very broad transition certainly due to the size distribution of the islands. The modulation in the time spectra is characteristic for a perpendicular magnetization. This result is quite remarkable because Fe films on W (110) are known to be magnetized in plane for coverages of more than 0.6.

In the second example we will discuss the pressure induced valence transition in EuNi$_2$Ge$_2$. As discussed before, the valence states in Eu systems determine the isomer shift, which is quite large between Eu^{2+} and Eu^{3+}. The isomer shift shows up as quantum beats in NFS studies, if both charge states coexist in the same sample [22] or if a reference sample with a well defined isomer shift is measured together with the sample under investigation as demonstrated in Fig. 4.

A specially designed high-pressure cell with B$_4$C anvils was used with a sample diameter of about 1 mm. EuNi$_2$Ge$_2$, an intermetallic compound with divalent Eu at ambient pressure and a valence transition around 5 Gpa [39], was studied at pressures up to 10 GPa (cf. Fig. 6). EuF$_3$ (left column) and EuS (right column) were used as isomer shift references. The spectra around 5 GPa indicate – from a strong variation of the spectral shape – a mixed-valent behavior with a distribution of valence states. At higher pressures the trivalent state is approached. Note that spectra at 5.4 GPa were taken with both references, demonstrating that large differences in isomer shifts can be better determined than small shifts. The isomer shift bar diagrams in the middle of Fig. 6 illustrate how the beat frequencies correspond to the isomer shifts between EuNi$_2$Ge$_2$ and the EuF$_3$ and EuS references, a large energy difference corresponds to a fast beating. The valence transition, being already completed at 10 GPa, is depicted in the isomer-shift vs. pressure diagram at the bottom of Fig. 6.

5. Summary

The third generation synchrotron radiation sources have a new impact on the fields of hyperfine spectroscopy, due to the high brilliance and the wide tunability. Especially, the small beam size and the high collimation of the synchrotron radiation beam gives applications in high pressure, surfaces and interfaces, and small crystals new impetus. Polarizing magnetic fields simplify the NFS spectra and can be used to unravel the magnetic ordering type.

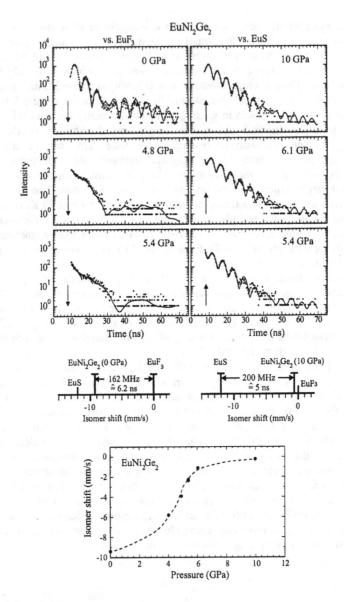

Figure 6. [151]Eu-NFS spectra of EuNi$_2$Ge$_2$ at room temperature as a function of pressure [37]. As reference samples serve EuF$_3$ (left column) and EuS (right column). The observed modulations in the time spectra are due to the isomer shift between EuNi$_2$Ge$_2$ and the reference samples (see bar diagram below the spectra). The isomer shift in the diagram at the bottom is given vs. EuF$_3$, where 1 mm/s corresponds to 17.31 MHz.

Synchrotron radiation is tunable in energy over the entire energy range of Mössbauer nuclei. Although most experiments are still performed at the resonance of the [57]Fe nucleus, which seems to be the ideal probe in solids, other Mössbauer isotopes, especially above 20 keV, like [151]Eu, are nowadays routinely available for nuclear resonance experiments at 3rd generation synchrotron radiation sources. Especially, high-pressure experiments benefit from the higher γ-energy, which causes less absorption losses in the diamond anvil high-pressure cells.

Acknowledgements

One of the authors (O.L.) is especially grateful to Prof. E. Gerdau and the nuclear resonance group at University of Hamburg for numerous fruitful discussions. The financial support of the German BMBF, contracts no. 05643GUA1 and 05 KS1HRA/8 is highly acknowledged.

References

1. Gerdau, E., Rüffer, R., Winkler, H., Tolksdorf, W, Klages, C.P., and Hannon, J.P. (1985) Nuclear bragg-diffraction of synchrotron radiation in yttrium iron-garnet, *Phys. Rev. Lett.* **54**, 835-838.
2. Gerdau, E., and de Waard, H., Eds. (1999,2000) Nuclear Resonant Scattering of Synchrotron Radiation (Part A), *Hyp. Int.* **123/124**, Nuclear Resonant Scattering of Synchrotron Radiation (Part B), *Hyp. Int.* **125**.
3. Leupold, O., Gerdau, E., Rüter, H.D., Meyer-Klaucke,W., Trautwein, A.X., and Winkler, H. (1996) *HASYLAB Annual Report Part I*, DESY, Hamburg, p.73.
4. Rüffer, R., and Chumakov, A.I. (1996) Nuclear resonance beamline at ESRF, *Hyp.Int.* **97/98**, 589-604.
5. Snigirev, A.A., Kohn,V.G., Snigireva, I.I., Lengeler, B. (1996) *Nature* **384**, 49.
6. Ishikawa, T., Yoda, Y., Izumi, K., Suzuki, C.K., Zhang, X.W., Ando, M., and Kikuta, S. (1992) Construction of a precision diffractometer for nuclear bragg scattering at the photon factory, *Rev. Sci. Instrum.* **63**, 1015-1018.
7. Freund, A.K., Comin, F., Hazemann, J., Hustache, P., Jenninger, B., Lieb, K., and Pierre, M. (1998) *Proc. SPIE* **3448**, 1.
8. Kishimoto, S. (1992) High time resolution x-ray measurements with an avalanche photodiode detector, *Rev. Sci. Instruments* **63**, 824-827.
9. Baron, A.Q.R. (1995) Report on the x-ray efficiency and time response of a 1-cm(2) reach through avalanche-diode, *Nucl. Inst. Meth. A*} **352**, 665-667.
10. Toellner, T.S. (2000) Monochromatization of synchrotron radiation for nuclear resonant scattering experiments, *Hyp. Int.* **125**, 3-28.
11. Baron, A.Q.R. (2000) Detectors for nuclear resonant scattering experiments, *Hyp. Int.* **125**, 29-42.
12. Hastings, J.B., Siddons, D.P., van Bürck, U., Hollatz, R., and Bergmann, U. (1991) Mossbauer-spectroscopy using synchrotron radiation, *Phys. Rev. Lett.* **66**, 770-773.
13. van Bürck, U., Siddons, D.P., Hastings, J.B., Bergmann, U., and Hollatz,R. (1992) Nuclear forward scattering of synchrotron radiation, *Phys. Rev. B* **46**, 6207-6211.
14. Smirnov, G.V. (1996) Nuclear resonant scattering of synchrotron radiation, *Hyp. Int.97/98*, 551-588.
15. Smirnov, G.V. (1999) General properties of nuclear resonant scattering, *Hyp. Int* **123/124**, 31-77.
16. Sturhahn, W., Gerdau, E., Hollatz, R., Rüffer, R., Rüter, H.D., and Tolksdorf, W. (1991) Nuclear bragg-diffraction of synchrotron radiation at the 8.41 kev resonance of thulium, *Europhys. Lett.* **14**, 821-825.
17. Alp, E.E., Mooney, T.M., Toellner, T., Sturhahn, W., Witthoff, E., Röhlsberger, R., and Gerdau, E. (1993) Time-resolved nuclear resonant scattering from Sn-119 nuclei using synchrotron radiation, *Phys. Rev. Lett.* **70**, 3351-3354.
18. Kikuta, S. (1994) Studies of nuclear resonant scattering at TRISTAN-AR, *Hyp. Int.* **90**, 335-349.
19. Johnson, D.E., Siddons, D.P., Larese, J.Z., and Hastings, J.B. (1995) Observation of nuclear forward scattering from Kr-83 in bulk and monolayer films, *Phys. Rev. B* **51**, 7909-7911.
20. Baron, A.Q.R., Chumakov, A.I., Ruby, S.L., Arthur, J., Brown, G.S., van Bürck, U., and Smirnov, G.V. (1995) Nuclear resonant scattering of synchrotron-radiation by gaseous krypton, *Phys. Rev. B* **51**, 16384-

216

16387.
21. Chumakov, A.I., Baron, A.Q.R., Arthur, J., Ruby, S.L., Brown, G.S., Smirnov, G.V., van Bürck, U., and Wortmann, G. (1995) Nuclear-scattering of synchrotron-radiation by Ta-181, *Phys. Rev. Lett.* **75**, 549-552.
22. Leupold, O., Pollmann, J., Gerdau, E., Rüter, H.D., Faigel, G., Tegze, M., Bortel, G., Rüffer, R., Chumakov, A.I., and Baron, A.Q.R. (1996) Nuclear resonance scattering of synchrotron radiation at the 21.5 keV resonance of Eu-151, *Europhys. Lett.* **35**, 671-675.
23. Koyama, I., Yoda, Y., Zhang, X.W., Ando, M., and Kikuta, S. (1996) Nuclear resonant excitation of Dy-161 and Eu-151 by synchrotron radiation, *Jpn. J. Appl. Phys.* **35**, 6297-6300.
24. Leupold, O., Chumakov, A.I., Alp, E.E., Sturhahn, W., and Baron, A.Q.R. (1999) Noniron isotopes, *Hyp. Int.* **123/124**, 611-631.
25. Shvyd'ko, Yu.V., Gerken, M., Franz, H., Lucht, M., and Gerdau, E. (2001) Nuclear resonant scattering of synchrotron radiation from Dy-161 at 25.61 keV, *Europhys. Lett.* **56**, 309-315.
26. Röhlsberger, R., Quast, K.W., Toellner, T.S., Lee, P.L., Sturhahn, W., Alp, E.E., and Burkel, E. (2001) Observation of the 22.5-keV resonance in Sm-149 by the nuclear lighthouse effect, *Phys. Rev. Lett.* **87**, 047601.
27. Gonser, U., and Fischer, H. (1981) in *Topics in Current Physics, Mössbauer Spectroscopy II* Springer, Berlin, Vol. 25, p. 99.
28. Szymanski, K., Dobrzynski, L., Prus, B., and Cooper, M.J. (1996) Single line circularly polarised source for Mossbauer spectroscopy, *Nucl. Instr. Methods B* **119**, 438-441.
29. Szymanski, K., Dobrzynski, L., Prus, B., Mal'tsev, Yu., Rogozev, B., and Silin, M. (1998) *Hyp. Int.* **C3**, 265.
30. Szymanski, K., Satula, D., and Dobrzynski, L. (1999) On the validity of a Fe-57 hyperfine magnetic field distribution measured by a monochromatic, circularly polarized Mossbauer source, *J. Phys.: Condensed Matter* **11**, 881-887.
31. Jäschke, J., Rüter, H.D., Gerdau, E., Smirnov, G.V., and Sturhahn, W. (1999) A single line linearly polarized source of 14.4 keV radiation by means of resonant absorption, *Nucl. Instr. Methods B* **155**, 189-198.
32. Nowik, I. (1983) Mossbauer studies of valence fluctuations, *Hyp. Int.* **13**, 89-118.
33. Barton, C.M.P., and Greenwood, N.N. (1973) in Stevens, J.G., and Stevens, V.E. (eds.), *Mössbauer Effect Data Index* Plenum Press, New York, p.395., Grandjean, F., and Long, G.J. (1989) in Long,G.J., and Grandjean, F. (eds.), *Mössbauer Spectroscopy Applied to Inorganic Chemistry*, Plenum Press, New York, Vol. 3, p.513., Harmatz, B., (1976) *Nuclear Data Sheets* **19**, 33.
34. Alp, E.E., Sturhahn, W., and Toellner, T. (1995) Synchrotron mossbauer-spectroscopy of powder samples, *Nucl. Instr. Meth. B* **97**, 526-529.
35. Leupold, O., Grünsteudel, H., Meyer, W., Grünsteudel, H.F., Winkler, H., Mandon, D., Rüter, H.D., Metge, J., Realo, E., Gerdau, E., Trautwein, A.X., and Weiss, R. (1996) *Conference Proceedings "ICAME-95"*, Ortalli,I. (ed.), SIF, Bologna, Vol. 50, p. 857.
36. Pleines, M., Lübbers, R., Strecker, M., Wortmann, G., Leupold, O., Shvyd'ko, Yu.V., Gerdau, E., and Metge,J. (1999) Pressure-induced valence transition in EuNi2Ge2 studied by Eu-151 nuclear forward scattering of synchrotron radiation, *Hyp. Int.* **120/121**, 181-185.
37. Pleines, M. (1998) Diplomarbeit, Universität Paderborn.
38. Röhlsberger, R., Bansmann, J., Senz, V., Jonas, K.L., Bettac, A., Leupold, O., Rüffer, R., Burkel, E., and Meiwes-Broer, K.H. (2001) Perpendicular spin orientation in ultrasmall Fe islands on W(110), *Phys. Rev. Lett.* **86**, 5597-5600.
39. Hesse, H.-J., Lübbers, R., Winzenick, M., Neuling, H.W., and Wortmann, G. (1997) Pressure and temperature dependence of the Eu valence in EuNi2Ge2 and related systems studied by Mossbauer effect, X-ray absorption and X-ray diffraction, *J. Alloys Compounds* **246**, 220-231.

SYNCHROTRON MÖSSBAUER REFLECTOMETRY FOR INVESTIGATION OF HYPERFINE INTERACTIONS IN PERIODICAL MULTILAYERS WITH NANOMETER RESOLUTION

M.A. ANDREEVA[1] AND B. LINDGREN[2]
[1] *M.V. Lomonosov Moscow State University, Department of Physics, 117234, Moscow, Russia*
[2] *Uppsala University, Department of Physics, Box 530, 751 21 Uppsala, Sweden*

1. Introduction

Magnetic multilayers are, nowadays, the object of an undefeated attention in surface physics, the theory of magnetism and the theory of electron transport because they possess unique properties and they are very promising materials for the future "spinelectronics" (or "magnetoelectronics", or "spintronics"). The methods of their experimental investigations are accordingly intensively developing (see e.g. a review introduction in [1]). The most difficult problem is the investigation of the magnetic state of buried layers and interfaces. Spin-polarized neutron reflectivity, Mössbauer spectroscopy, X-ray magnetic circular dichroism, X-ray magnetic-resonant scattering, nuclear probe methods and some others have been continuously improving for such depth selective investigation. In particular, the nanometer depth selectivity of Mössbauer spectroscopy is achieved by inserting ^{57}Fe probe layers at a definite depth in investigated samples [2,3].

Mössbauer reflectometry is a relatively new method for the investigation of ultrathin surface layers and multilayers although the first investigation of the total external reflection of Mössbauer radiation was done in 1963 [4]. The broad application of grazing incidence Mössbauer spectroscopy [5] was restricted by the small intensity of radioactive sources selected at narrow angular range required for such experiments. Main progress in "Mössbauer" reflectometry has been achieved at the nuclear resonant scattering stations of synchrotron radiation [6-7]. Highly collimated synchrotron radiation is well suited for the experiments requiring an angular resolution < 1 arc.min such as reflectometry. The lack of energy resolution required for the common Mössbauer spectrum recording has been overcome at such stations by time gating [8]. The hyperfine splitting of the resonant spectrum is measured in the time representation as quantum beat oscillations [9].

In the total external reflection region (typically 2-4 mrad for 14.4 keV radiation) the penetration depth changes from several nanometers up to several microns with the angle variation [10] allowing to perform depth selective investigations by means of measuring the energy [5] or time spectra of reflectivity (see e.g. Fig. 15 in [6]). For the case of

M. Mashlan et al. (eds.), Material Research in Atomic Scale by Mössbauer Spectroscopy, 217–228.
© 2003 *Kluwer Academic Publishers. Printed in the Netherlands.*

periodical multilayers, the nonresonant prompt and the delayed nuclear resonance reflectivity curves exhibit Bragg peaks, the peaks in the delayed curve being correspondent to the magnetic periodicity of the structure according to the sensitivity of the nuclear resonant scattering to the hyperfine field parameters [11-12]. The energy or time spectra of the Bragg reflectivity are depth selective in the scale of one repetition period [11,13-14] but this circumstance has not been properly understood and used for the investigation of buried layers and interfaces. Here we try to explain the main peculiarities of the method and give some experimental examples.

2. Structure Sensitivity of the Bragg Reflectivity Spectra

At first let us show how the Bragg reflectivity Mössbauer spectra differ from the absorption Mössbauer spectrum (Fig.1).

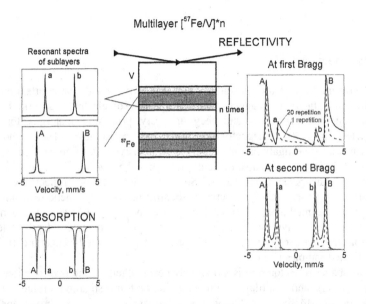

Figure 1. Comparison of absorption Mössbauer spectrum and Bragg reflectivity Mössbauer spectra for the simplest model when two doublet hyperfine contributions characterize the resonant spectrum of the sample, the smaller split doublet being placed in the interface region.

The model is the following one: $[V(1.5 \text{ nm})/^{57}Fe(1.5 \text{ nm})]*n$ multilayer and two hyperfine fields: $\mathbf{B}_1^{hf} = 33$ T and $\mathbf{B}_2^{hf} = 21$ T in ^{57}Fe layers in equal proportion. We assume also for simplicity that the hyperfine field orientations are the same $\mathbf{B}_1^{hf} \parallel \mathbf{B}_2^{hf}$ along the polarization magnetic vector of the linear polarized radiation field so only two lines (2nd and 5th in each magnetic sextet) interact with the radiation. The transmission Mössbauer spectrum at the same assumptions for our sample has four lines (two doublets) with equal intensity irrelative where each hyperfine field is located in the iron layers. But the reflectivity spectrum "feels" very well how the hyperfine fields are distributed in depth through a repetition period. If we place $\mathbf{B}_2^{hf}$ to the interface region than that contribution will be strongly suppressed at the Bragg peak of the first order (see Fig. 1). For the Bragg peak of the second order the ratio of contributions is drastically changed. We calculate here the reflectivity spectrum by means of Parratt recursive equation [15] neglecting the polarization mixing at the multiple reflections.

Notice that such depth selectivity of the Bragg reflections has been effectively used in X-ray magnetic resonance scattering (see e.g. [16] where the reflectivity spectra near $L_{2,3}$ edges are measured for nine low angle Bragg peaks in a La/Fe multilayer and the depth profile of 5d magnetic moments determined).

The explanation of the shown depth selectivity is very simple in the kinematical limit of the Parratt formula. For grazing angles ϑ larger than the critical angle of total reflection we can neglect multiple scattering, replace the normal component of the wave vectors in each layer by their approximate expression $\eta_j = \sqrt{\sin^2 \vartheta + \chi_j} \approx \sin \vartheta + \dfrac{\chi_j}{2 \sin \vartheta}$ and neglect the refraction and absorption for the phase shifts of scattered waves. In such a way we get the amplitude of the reflected waves as the following [17]:

$$R = \frac{1}{4 \sin^2 \vartheta} \sum_{j=1}^{L} (\chi_{j-1} - \chi_j) e^{iQz_{j-1}} \qquad (1)$$

In (1) χ_j is susceptibility of the layer j, $Q = \dfrac{4\pi}{\lambda} \sin \vartheta$, λ is the radiation wavelength, z_j is the depth of the boundary between the layers j-1 and j. Formula (1) determines the reflectivity as a sum of reflected waves by the boundaries between two continuous media. We are going to distinguish the properties (hyperfine interactions) of each layer, so it is convenient to use in (1) the scattering amplitude by a single thin layer

$$r_j = \frac{1}{4 \sin^2 \vartheta} ((\chi_{j-1} - \chi_j) + (\chi_j - \chi_{j+1}) e^{iQd_j}) \approx \frac{iQd_j}{4 \sin^2 \vartheta} \chi_j, \qquad (2)$$

where d_j is its thickness. Than (1) takes the form

$$R = \frac{iQ}{4 \sin^2 \vartheta} \sum_{j=1}^{L} \chi_j d_j e^{iQz_{j-1}} \qquad (3)$$

In the periodical medium the most essential part of (3) includes the sum only over sublayers in a single repetition period. We can call this part the structure factor F

$$F = \frac{iQ}{4 \sin^2 \vartheta} \sum_{k=1}^{K} \chi_k d_k e^{iQ\xi_k}, \qquad (4)$$

where ξ_k is the depth calculated from the top of the repetition period and k numerates the discrete layers in the «unit cell». The sum over all repetition periods is described by the well-known Laue-factor L_n

$$L_n = (1+e^{iQD}+e^{2iQD}+e^{3iQD}+...+e^{(n-1)iQD}) = \frac{1-e^{inQD}}{1-e^{iQD}}, \tag{5}$$

where n is the number of repetitions, D is the period of the structure. At the exact Bragg angle $QD = m2\pi$ (m is an integer number) and $L_n = n$. So in the kinematical limit for periodical structures the amplitude of scattering (3) is represented as

$$R = L_n F \tag{6}$$

and all structure information is contained in the structure factor F. This factor is responsible for the interference of all waves reflected by layers in one repetition period.

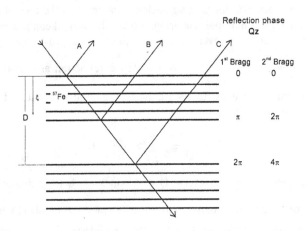

Figure 2. The interference of the reflected waves in one repetition period following from (4). At the Bragg angle the waves reflected by layers separated by the period D (A and C) are added with $n \times 2\pi$ phase difference, for a half of the period distance between layers (A and B) the interference is destructive in the first order Bragg peak.

As follows from Fig.2 the contributions from the interfaces in the reflectivity spectrum to the Bragg peak of the first order are added in anti-phase if they are separated by half of the period. This explains the suppression of the lines a and b observed in Fig.1.

3. Fourier Analysis of the Time Spectra of Reflectivity

The susceptibility of the nuclear resonant medium contains the electronic χ_j^{el} and the nuclear resonant $\chi_j^{nuc}(\omega)$ parts and it is the function of the energy E (or frequency $\omega = E/\hbar$):

$$\chi_j(\omega) = \chi_j^{el} + \chi_j^{nuc}(\omega) = -\frac{\lambda^2}{\pi} N_j^{at} r_0 (Z_j + \Delta f_j' + i\Delta f_j'') - \lambda \mu^{nuc} \sum_l \frac{A_l \frac{\Gamma_l}{2\hbar}}{\omega - \omega_l + \frac{i\Gamma_l}{2\hbar}} \quad (7)$$

where N_j^{at} is the volume density of atoms, r_0 is the radius of the electron, $Z_j + \Delta f'_j$ and $\Delta f''_j$ are the real and imaginary parts of the atomic scattering amplitude, μ^{nuc} is the linear nuclear absorption coefficient: $\mu^{nuc} = \sigma_{res} N_j^{nuc} f^{LM} P$, σ_{res} is the cross-section of the nuclear resonant absorption (for ^{57}Fe $\sigma_{res} = 2.56 \ 10^{-4} \ nm^2$ [9]), N_j^{nuc} is the density of the resonant nuclei, f^{LM} is the probability of the Mössbauer effect and P is the enrichment by the nuclear resonant isotope. So the previous section was actually a consideration of the reflectivity in the energy domain $R = R(\omega)$.

With pulsed synchrotron radiation the measurements are performed typically in the time domain. In order to calculate the time spectrum of reflectivity the Fourier transform of the amplitude $R(\omega)$ should be done:

$$R(t) = \frac{1}{2\pi} \int_{-\infty}^{+\infty} R(\omega) e^{-i\omega t} d\omega \quad (8)$$

Hyperfine splitting of the nuclear resonant levels reveals itself in the time spectra of the nuclear decay as quantum beats between all existing frequencies of the resonant transition. The general picture of the multi-frequency oscillations in the time spectra is quite complicated so for qualitative analysis we simplify this analysis by Fourier transform of the time spectra of the reflectivity $|R(t)|^2$ as it is shown in Fig.3.

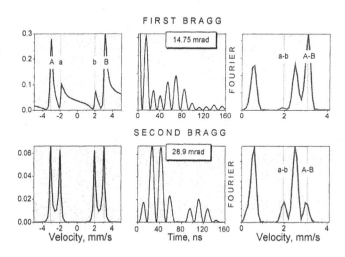

Figure 3. Energy spectra of reflectivity (the same as in Fig.1), corresponding time spectra and their Fourier transform, presented on the same scale as the energy spectra (relative velocity in mm/s) for comparison.

For the example presented in Fig.1, we obtained the same result concerning the structure sensitivity in the time spectra of reflectivity as in the energy spectra of reflectivity. The beat frequency between lines a-b at the Bragg maximum of the first order is suppressed but at the Bragg reflection of the second order the frequencies a-b and A-B give almost equal contributions. Admittedly, the most pronounced in the last case is the interference between cross lines a-B and B-a (line at ~2.5 mm/s in Fourier transform), so the time spectra are more complicated for analysis than the energy spectra even in such a simple example.

4. Standing Wave Effects

As we have shown the depth selective information about the periodical multilayers can be obtained from the reflectivity spectra measured at different Bragg maxima. The structure sensitivity follows directly from the interference of waves reflected in one repetition period according to (4). Now we consider another circumstance, which was not taken into account in the simplest kinematical formula (4-6). These formulas suppose that the shape of the energy or time spectra of reflectivity is completely determined by the structure factor (4), the increase in the number of repetition periods in the multilayer enhances only the intensity of reflectivity but does not change the spectrum shape. Calculations show that in most cases it is a very rough approximation, the spectrum shape changes with the number of repetitions. We get much more accurate result if we take into consideration the refraction and absorption corrections for the phases of reflected waves in the kinematical formula (which are really very essential in the resonant medium):

$$R(\omega) = \frac{iQ}{4\sin^2 \vartheta} \sum_{j=1}^{L} \chi_j(\omega) d_j e^{iQz_j + \frac{i}{\lambda \sin \vartheta} \sum_{k=1}^{j-1} \chi_k(\omega) d_k} \tag{9}$$

Energy dependence for the phases of the adding waves means that the radiation field during its propagation inside periodical multilayer acquires the resonant dependence. This effect was investigated in [18] and called the resonant standing waves.

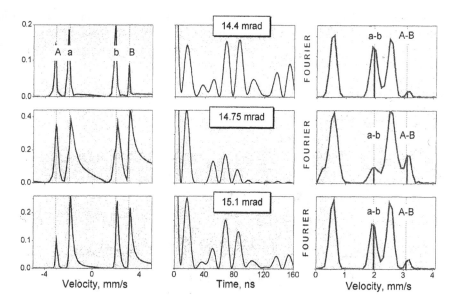

Figure 4. The energy and time spectra of reflectivity and its Fourier transform for several angles in the vicinity of the Bragg angle of the first order. Calculations were done for the model: $[^{57}Fe_{21T}(0.3nm)/^{57}Fe_{33T}(0.3nm)/^{57}Fe_{21T}(0.3nm)/V(2.1nm)]x30$, the exact Bragg angle is 14.75 mrad.

The most interesting consequence of that behavior is that the reflectivity spectra (in the energy and in the time domain) gain dependence on the small shifts of the angle in the vicinity of the Bragg peak. In other words the change in the radiation field configuration with the angle variation can enhance or suppress some contributions to the total spectrum of reflectivity. So the additional depth selectivity appears. This is illustrated in Fig.4 where it is apparent that the interface contributions almost completely suppressed in the time spectrum at the exact Bragg angle appear in the time spectra measured with a small shift of the angle (only ± 0.35 mrad) from the exact Bragg position.

Notice that the effect of the resonant modulation of radiation field appears in the reflectivity spectra only for a large number of repetitions and only when the Bragg reflectivity is considerable enough.

5. Experimental Examples

5.1. $[^{57}Fe/Co]$ MULTILAYERS

The time spectra of the nuclear resonance reflectivity from multilayers $[^{57}Fe(6ML)/Co(3ML)]_{35}/MgO$ and $[^{57}Fe(5ML)/Co(5ML)]_{25}/MgO$ were measured at the Nuclear Resonance Beamline ID18 of the European Synchrotron Radiation Facility (ESRF) [19]. The reflectivity from such multilayers is relatively low ($<10^{-2}$) because the electronic densities of Fe and Co are close. So no visible variation of the spectrum shape with the angle variation was observed. Anyhow the effect of the structure sensitivity at the Bragg maximum is clearly seen (Fig. 5): the contribution from the middle part of [57]Fe in multilayer is enhanced.

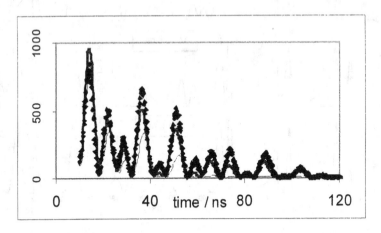

Figure 5. The time spectrum of [[57]Fe(6ML)/Co(3ML)]$_{35}$ measured at the first order Bragg peak (dots). Thick full line is the result of calculations for the case when the smaller hyperfine field (33.8 T) is in the interfaces. Thin line presents the case when the smaller field is in the centre of the Fe-layers.

A comparison of the experimental time spectrum with calculated spectra shows (Fig.5) that the larger hyperfine field (36.4 T) is associated with the centre of [57]Fe layers. This result is in contradiction with the theoretical predictions for Fe/Co multilayers.

5.2. MICROCRYSTALLINE [Cr/[57]Fe]$_{26}$ MULTILAYER

This sample [Cr(1.7nm)/[57]Fe(1.6nm)]$_{26}$ on a glass substrate prepared at the Institute for Physics of Microstructures RAS (Nizhnii Novgorod, Russia) has a relatively large reflectivity (~0.1) so we have observed the effect of the standing wave formation at the Bragg reflection: the shape of the reflectivity spectra essentially changes with the angle variation [20,21]. By attributing the main contribution in the exact Bragg angle spectrum to the central part of the film we have concluded that the orientation of the hyperfine field is different in the interface region as compared to the centre part of iron layers. In the Fourier transform of the time spectra the additional frequencies appear with the angle shift. That was confirmed by the fit of the time spectra [20]. Note that in the fit program we have used the general formula for the reflectivity, based on the 4x4-propagation matrix formalism, taking into account the mixing of the polarization states of the radiation field during multiple reflections [22-24]

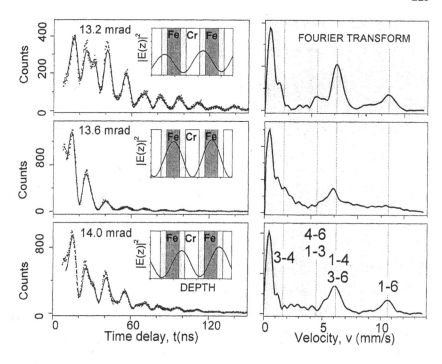

Figure 6. Time spectra of microcrystalline $[Cr(1.7nm)/^{57}Fe(1.6nm)]_{26}$ film measured in the vicinity of the Bragg angle (13.6 mrad) and their Fourier transforms (right side). The frequencies of interference between the magnetic sextet lines (pointed out in the bottom graph) are marked by dashed vertical lines. Inserts in the time spectra show the standing wave pattern inside the multilayer calculated for the case of pure α-Fe time spectrum in iron layers.

5.3. $[^{57}Fe/V]$ MULTILAYERS

Two samples $[^{57}Fe(7)/V(10)]_{29}$ with Pd covering and $[^{57}Fe(10)/V(5)]_{20}$ with V covering were prepared in UHV by sputtering technique and investigated by Mössbauer reflectometry at ESRF [25]. During the time spectra measurements it happened that both of our samples have a spontaneous magnetization along the surface diagonal (crystallographic direction [100]). The time spectra of reflectivity near the Bragg maximum for the sample $[^{57}Fe(10)/V(5)]_{20}$ are presented in Fig.7a. The result of their fit (Fig.7c) shows that the interfaces are not sharp (we see a slight penetration of ^{57}Fe atoms to V layers) and that the lower fields (9.5 T) are connected preferentially with the Fe-on-V interface, but surprisingly, nuclei possessing the relatively large fields (33.9 T) are placed in the V-on-Fe interface. The reason of the interface asymmetry may be attributed to the technology of the sample preparation.

226

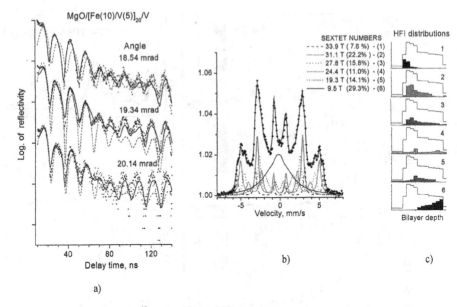

Figure 7. Time spectra of $[^{57}Fe(10ML)/V(5ML)]_{20}$ sample measured near and at the exact Bragg angle 19.34 mrad (a), CEMS spectrum of this sample (b) and the depth distribution (c) in bilayer depth of the different B_{hf}-components as obtained from the fit of the time spectra. Dots are the experimental curves, dashed lines in (a) are for the ideal symmetrical model assuming smaller B_{hf} in the interface region, solid lines are the result of the fit, giving the B_{hf} distribution presented in (c).

The examples show the first application of Bragg reflectivity measurements with synchrotron radiation for the depth selective investigation of multilayers. We see that while CEMS characterises the whole set of hyperfine fields B_{hf} in a sample, Mössbauer reflectometry measurements allow us to get the depth distributions of different B_{hf} across bilayer depth.

Acknowledgments

The work is supported by the Royal Swedish Academy of Sciences, grant of the Swedish Institute and by the Russian Foundation of Basic Research (RFBR), grants No. 01-02-17541. We are grateful to A.I. Chumakov, O. Leupold, R. Rüffer and all ESRF-staff for the possibility to perform the experiments, their essential help with measurements and important discussions. We thank our collaborators L. Häggström, P. Blomquist, B. Kalska, V.G. Semenov, K.A. Prokhorov and N.N. Salashchenko for the sample preparation and help with the experiments.

References

1. Kortright, J.B., Awschalom, D.D., Stöhr, J., Bader, S.D., Idzerda, Y.U., Parkin, S.S.P., Schuller, I.K. and Siegmann, H.-C. (1999) Research frontiers in magnetic materials at soft X-ray synchrotron radiation facilities, *J. Magn. and Magn. Mater.* **207**, 7-44.
2. Kalska, B., Blomquist, P., Häggström, L. and Wäppling R. (2001) A study of the different interfaces in Fe/V superlattices, *Europhysics Letters* **53**, 395-399.
3. Shinjo, T. and Keune, W. (1999) Mössbauer-effect studies of multilayers and interfaces, *J. Magn. and Magn. Mater.* **200**, 598-615.
4. Bernstein, S. and Campbell, E.C. (1963) Nuclear anomalous dispersion in Fe^{57} by the method of total reflection, *Phys. Rev.*, **132**, 1625-1633.
5. Irkaev, S.M., Andreeva, M.A., Belozerskii, G.N., Semenov, V.G. and Grishin O.V. (1993) Grazing Incidence Mössbauer Spectroscopy: New Method for Surface Layers Analysis. I. Instrumentation, *Nucl. Instr. and Meth.* **B74** 545-553. II. Theory of Grazing Incidence Mössbauer Spectra, *ibid.* 554-564; (1995) III. Interpretation of Experimental Data, *ibid.* **B103**, 351-358.
6. Chumakov, A.I., Niesen, L., Nagy, D.L. and Alp, E.E. (1999) Nuclear resonant scattering of synchrotron radiation by multilayer structure, *Hyperfine Interactions* **123/124**, 427-454.
7. Röhlsberger, R. (1999) Nuclear resonant scattering of synchrotron radiation from thin films, *Hyperfine Interactions* **123/124**, 455-479.
8. Rüffer, R. and Chumakov, A.I. (1996) Nuclear resonance beamline at ESRF, *Hyperfine Interactions* **97/98**, 589-604.
9. Smirnov, V.G. (1999) General properties of nuclear resonant scattering, *Hyperfine Interactions* **123/124**, pp. 31-77.
10. Wagner, F.E. (1968) Totalreflexion der rückstoßfreien 8.4 keV –strahlung des Tm^{169}, *Z.für Physik* **210**, 361-379.
11. Toellner, T.S., Sturhahn, W., Röhlsberger, R., Alp, E.E., Sowers and C.H., Fullerton, E.E. (1995) Observation of pure nuclear diffraction from a Fe/Cr antiferromagnetic multilayer, *Phys.Rev.Lett.*, 1995, **74**,. 3475-3478.
12. Bottyán, L., Dekoster, J., Deák, L., Baron, A.Q.R., Degroote, S., Moons, R., Nagy, D.L. and Langouche, G. (1998) Layer magnetization canting in 57Fe/FeSi multilayer observed by synchrotron Mössbauer reflectometry, *Hyperfine Interactions* **113**, 295-301.
13. Andreeva, M.A., Irkaev, S.M., Semenov, V.G., Prokhorov, K.A., Salashchenko, N.N., Chumakov, A.I. and Rüffer, R. (1999) Mössbauer reflectometry of ultrathin multilayer Zr(10нм)/[^{57}Fe(1.6нм)/ Cr(1.7нм)]$_{26}$/Cr(50нм) film using synchrotron radiation, *J. Alloys & Compounds* **286**, 322-332.
14. Andreeva, M.A., Irkaev, S.M., Semenov, V.G., Prokhorov, K.A., Salashchenko, N.N., Chumakov, A.I. and Rüffer, R. (2000) Mössbauer reflectometry of multilayer structure Zr(10nm) / [Cr(1.7 nm) / ^{57}Fe(1.6 nm)]$_{26}$ / Cr(50nm) - comparative measurements in energy and time domains, *Hyperfine Interactions* **126**, 343-348.
15. Parratt, L.G. (1954) Surface studies of solids by total reflection of X-rays, *Phys.Rev.* 95, 359-369.
16. Séve, L., Jaouen, N., Tonnerre, J.M., Raoux, D., Bartolome, F., Arend, M., Felsch, W., Rogalev, A., Goulon, J., Gautier, C. and Bérar, J.F. (1999) Profile of the induced 5f magnetic moments in Ce/Fe and La/Fe multilayers probed by X-ray magnetic-resonant scattering, *Phys.Rev.* B 60, 9662-9674.
17. Hamley, I.W. and Pedersen, J.S. (1994) Analysis of neutron and X-ray reflectivity data. I. Theory, *J Appl.Cryst.* 27, 29-35.
18. Andreeva, M.A. (1999) Space-time properties of the nuclear resonant excitation at the conditions of Bragg reflection from multilayers, *JETP Lett.* 69, 863-868.
19. Lindgren, B., Andreeva, M. A., Häggström, L., Kalska, B., Semenov, V.G., Chumakov, A.I., Leupold, O. and Rüffer, R. (2002) ^{57}Fe/Co multilayers investigated by CEMS and nuclear resonance reflectivity time spectra using grazing incident SR, *Hyperfine interactions*, in press.
20. Andreeva, M.A., Semenov, V.G., Lindgren, B., Häggström, L., Kalska, B., Chumakov, A.I., Leupold, O. and Rüffer, R. (2002) Interface sensitive investigation of ^{57}Fe/Cr superstructure by means of nuclear resonance standing waves in time scale, , *Hyperfine interactions*, in press.
21. Andreeva, M.A., Semenov, V.G., Häggström, L., Lindgren, B., Kalska, B., Chumakov, A.I., Leupold, O., Rüffer, R., Prokhorov, K.A. and Salashchenko, N.N. (2001) Interface selective investigation of ^{57}Fe/Cr multilayer by nuclear resonance Bragg reflectivity in time scale, *The Physics of Metals and Metallography,* **91**, suppl.1, 22-27.
22. Andreeva, M.A. and Rosete, C. (1986) Theory of reflection from a Mössbauer mirror taking account of laminar variation in the parameters of the hyperfine interactions close to the surface, Vestnik Moscovskogo Universiteta, Fizika, **41**, No.3, 65 –71 (English transl. by Allerton press, Inc.).
23. Andreeva, M.A., Irkaev, S.M. and Semenov, V.G. (1994) Secondary radiation emission at Mösbauer total external reflection, *Hyperfine Interactions* **97/98**, 605-623.

24. Röhlsberger, R. (1999) Theory of X-ray grazing incidence eflection in the presence of nuclear resonance excitation, *Hyperfine Interactions* **123/124**, 301-325.
25. Kalska, B., Häggström, L., Lindgren, B., Blomquist, P., Wäppling, R., Andreeva, M.A., Nikitenko, Yu.V., Proglyado, V.V., Aksenov, V.L., Semenov, V.G., Chumakov, A.I., Leupold, O. and Rüffer, R. (2002) Magnetic properties of monocrystal ^{57}Fe/V multilayers investigated by CEMS, nuclear resonance reflectivity in time scale and polarized neutron scattering, *Hyperfine interactions*, in press.

DIFFUSION STUDIES IN ORDERED ALLOYS

Revive old methods with new ideas and new possibilities

M. SLADECEK[1], M. KAISERMAYR[1], B. SEPIOL[1], L. STADLER[1],
G. VOGL[1], C. PAPPAS[2], G.GRÜBEL[3]AND R. RÜFFER[3]
[1] *Institut für Materialphysik der Universität Wien*
Strudlhofgasse 4,1090 Wien, Austria
[2] *Hahn-Meitner-Institut Berlin*
Glienicker Straße 100,D-14109 Berlin, Germany
[3] *ESRF*
BP 220, F-38043 Grenoble Cedex, France

1. Introduction

Diffusion is an important and fundamental phenomenon for the properties and the behavior of metals, alloys, semiconductors, ceramics, glasses and polymers at higher temperatures. It plays an important role in the kinetics of microstructural changes of a material. It is a driving force for nucleation of new phases, recrystallization and phase transformations with a wide use in current technology, e.g. surface hardening, changing of deformation behavior by nucleation, diffusion doping or sintering. Especially the intermetallic alloys attracted attention as suitable materials for high-temperature applications due to their corrosion stability and strength. Many of the physical processes, developed for intermetallics are applicable for diffusion in all crystalline solids. The knowledge of the diffusion behavior of intermetallic alloys is, therefore, of interest for basic material science and for their use in technological applications. [1]

2. Diffusion Studies

2.1. MACROSCOPIC AND MICROSCOPIC DIFFUSION

Diffusion can be studied in a macroscopic range (μm) using the tracer method [1]. This method allows access to the "effective" diffusion. To study the microscopic jump diffusion mechanism, i.e. to gain information about the jump vectors and residence times on different lattice sites, we use scattering methods like Quasielastic Mössbauer spectroscopy (QMS) [2], Quasielastic Neutron Scattering (QNS) [3], Nuclear Resonant Scattering (NRS) [4, 5] or Quasielastic Helium Scattering [6, 7]. Principles and recent results of these methods will be shortly sketched in the following chapter.

229

M. Mashlan et al. (eds.), Material Research in Atomic Scale by Mössbauer Spectroscopy, 229–237.

2.1.1. *Diffusion studies with QNS - The elementary jump in intermetallic alloys with CsCl-structure*

The knowledge of the kinetics on the atomic scale in intermetallic systems permits the control of crucial processes, e.g. different stages of precipitation, and the optimization of macroscopic characteristics of these materials. However, even in intermetallics with structures as simple as the CsCl structure, the mechanism of self diffusion is not yet fully understood. While it is generally believed that diffusion in these alloys happens via jumps to vacant sites in the lattice, the nature of the jumps is not yet clear. There exist two possibilities that are illustrated in Figure 1:

a) An atom A may perform a short jump to a nearest neighbor site (NN). This jump is energetically expensive insofar as a nearest neighbor site in the CsCl structure is a so-called antistructure site, i.e. 'belonging' to the other element B.

b) The A-atom may prefer to stay on its own sublattice (A-sites). In this case the jump distance is considerably longer, the corresponding jump being a jump to a next-nearest neighbor site (NNN) or even further.

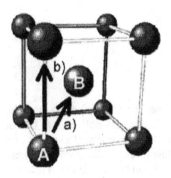

Figure 1. Unit cell of the CsCl structure. Spheres A represent nickel or cobalt atoms, spheres B gallium atoms. The two possible jumps for an A atom are jumps to NN sites (a) which belong to the 'foreign' B sublattice or jumps to more distant but regular NNN sites (b) or even further. Please note that there has to be a vacancy on the target site in order to allow a jump.

It is obvious that the preference for one or the other jump will be principally governed by the ordering energy of the alloy. While intermetallic alloys with low ordering energies like FeAl [8] favor NN jumps, diffusion in compounds with high ordering energies like NiAl [9] is likely to happen via NNN jumps. However, an open question remains: What happens in intermetallic alloys with intermediate ordering energies like NiGa and CoGa ?

Quasielastic neutron scattering allows measuring directly the elementary jump vector on single crystals, i.e. to observe direction, length and frequency of the diffusive jump in these alloys. Ni diffusion in NiGa and Co diffusion in CoGa are ideal for quasielastic neutron scattering studies in terms of scattering cross sections. However, the rather slow diffusion in these alloys (10^{-13} to 10^{-12} m^2/s) represents an experimental challenge because it corresponds to a quasielastic broadening of at best 0.1 μeV.

In both alloys a diffusion process, which involves two distinct time scales, was detected: besides the principal quasielastic line, a second much broader line is present in the signal (Figure 2). These two time scales can be attributed to long residence times on the regular sublattice and short residence times on the 'forbidden' antistructure sites [10].

Figure 3 shows measured line widths if the fast process is taken into account (full circles) or not (open circles). A comparison with calculations (line) shows clearly that diffusion happens via jumps to antistructure sites B and that NNN jumps (to A - sites directly) can be excluded. Further information on the residence time of the diffusing atoms on the antistructure sites is provided by intensities and widths of the two components. These residence times are very long, what indicates unusually high concentrations of thermal antistructure atoms - up to several percent. These antistructure atoms in turn influence the diffusion process by decreasing the degree of order in the system.

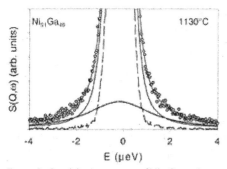

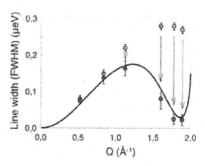

Figure 2. Quasielastic spectrum of $Ni_{51}Ga_{49}$ taken at 1400 K and Q = 1.6 Å^{-1}, fitted with the sum of a narrow and broad Lorentzians, both folded with the resolution (dashed). The broadening is 0.3 μeV for the narrow line and 3 μeV for the broad line (FWHM).

Figure 3. $Co_{64}Ga_{36}$, 1400 K
The maximum of the line width at Q = 1.8 Å^{-1} near a reciprocal <100> lattice point which is obtained from a fit with a single Lorentzian (open circles) can not be explained by a model assuming jumps on the regular sublattice. Jumps to antistructure sites, however, produce an additional broad line. If this line is subtracted from the spectra the resulting narrow line (close circles) shows a minimum at Q=1.9 Å^{-1} as expected for the NN jump model (solid line).

The neutron results on both alloys, NiGa [11] and CoGa [12], point towards a diffusion process which involves more than one defect and which leads to a practically unlimited set of possible jump sequences. An analytical treatment of this scenario is naturally not possible. Monte Carlo simulations, however, taking into account such high defect concentrations, can explain the observed data in alloys that are as different as FeAl or NiGa. This can be achieved by varying interaction energies. Therefore self diffusion at high temperatures in most intermetallic alloys with CsCl-structure can be explained within a single and surprisingly simple model.

2.1.2. *Diffusion studies with nuclear resonant scattering of synchrotron radiation*
Nuclear Resonant Scattering (NRS) has become an established technique for studying diffusion on an atomic scale. The power of the technique for studying diffusion was predicted by the theoretical work of Smirnov and Kohn [4, 5] and demonstrated by several experiments [13 - 16]. At least two conditions have to be fulfilled to perform NRS experiments. The first one is the presence of a Mössbauer isotope in the sample and the second condition is the time structure of the synchrotron radiation adapted to the

lifetime of the used isotope. At the ESRF this time structure originates from 16 electron bunches in the storage ring, that create a short pulse (~100 ps) of synchrotron radiation every 176 ns (the lifetime of the used [57]Fe isotope is 141 ns).The synchrotron radiation pulse, with a resonant energy, creates a coherent collective nuclear state in the sample which may be perturbed or destroyed by diffusion. This leads to an accelerated decay of the resonantly scattered intensity (delayed intensity) with respect to an undisturbed scattering process [4, 5]. Or in a more simplified description: The phase of a wave re-emitted from a diffusing (jumping) atom, changing its position after the excitation by the pulse, will be shifted relative to waves re-emitted by the other atoms. Thus an accelerated decay of the intensity in the coherent channels (forward direction, Bragg directions) can be observed. The delayed intensity is proportional to the intermediate scattering function $I(Q,t)$ [5] which becomes a simple exponential function in the limit of a thin sample:

$$I(\boldsymbol{Q},t) = \exp\left[-\frac{t}{\tau} \sum_{i=1}^{N} N^{-1} \left\{ 1 - \exp(-i\boldsymbol{\bar{Q}}\boldsymbol{\bar{l}}_i) \right\} \right] \qquad (1)$$

where Q is the outgoing wave vector, l_i are jump vectors between lattice sites, τ is the residence time on a lattice site and N is the number of possible jump sites.

The accelerated decay is not only determined by τ but also by the relative orientation of the jump vector to the outgoing wave vector Q, thus enabling studies of the microscopic jump diffusion mechanism.

2.1.3. *Surface and near-surface study of diffusion by nuclear resonant scattering*

To enhance the surface sensitivity of NRS we have combined the techniques of NRS and grazing incidence reflection, the latter being an established technique in X-ray and neutron scattering for studying the structure of thin films.

It has been proven that NRS in grazing incidence geometry provides depth-selectivity for hyperfine parameters [17-20]. We will exploit this depth-selectivity to investigate diffusion phenomena in near surface regions of metallic films of iron. Experiments of this kind became feasible with the advent of 3rd generation synchrotron radiation sources. For a first feasibility study we have chosen a [57]Fe layer on a (001)-MgO substrate [21].

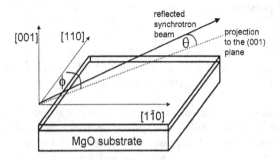

Figure 4. Sketch of the scattering geometry. θ is the angle of incidence and ϕ is the angle between the [110] direction of the iron layer and the projection of the outgoing wave vector of the synchrotron radiation to the (001) plane of the iron layer.

The maximum delayed intensity has been found at an incidence angle $\theta=3.41$ mrad close to the critical angle of total reflection. However, in order to improve the surface sensitivity the incidence angle of 1.66 mrad was chosen in the presented measurements. Spectra were recorded in a temperature range from RT to 1230 K. The paramagnetic spectra above T_c show accelerated intensity decay due to fast diffusion of iron atoms. In order to determine the direction of jump vectors the dependence on the angle ϕ, which is the angle between the in plane [110] direction of iron and the synchrotron radiation beam (Figure 4), has been measured for different ϕ values.

Preliminary results are published in [21]. First attempts to fit the measured spectra using the EFFINO routine written by Spiering et al. [22] show an iron self-diffusion in a 2 nm thick near-surface layer almost two orders of magnitude faster than in bulk material. The activation energy determined from the slope of an Arrhenius plot is about 450 meV. Surprisingly, the best match of Q dependence to the experimental points was achieved with a 2D square-lattice diffusion model in the (001) iron plane with the jump length equal to the lattice parameter of iron.

2.2. NEW DEVELOPMENTS IN DIFFUSION STUDIES

2.2.1. *Probing jump diffusion in crystalline solids with Neutron Spin-Echo spectroscopy (NSE)*

Today the current world-class backscattering spectrometers have practically reached the theoretical limit for the energy resolution that can be expected from this technique (see section 2.1.1). This is a severe restraint for future research in this field because the presently accessible diffusivities limit the field of possible investigations to systems with low to intermediate order which, in addition, have to be studied at very high temperatures, generally close to the melting point.

Due to recent developments in the field of neutron spin-echo spectroscopy we have, for the first time, succeeded to use NSE for investigation of diffusion in the intermetallic alloy NiGa [23].

The advantages of NSE compared to the conventional QNS method are the following:

(i)   NSE will enable measurements in the regime of lower temperatures.

(ii)   NSE will allow one to access regions in the phase diagram that were previously inaccessible.

(iii)   NSE will enable the investigations of highly ordered alloys. A prominent example is the technologically important intermetallic alloy NiAl.

With NSE spectroscopy, the time correlation $S(Q,t)$ is obtained by measuring the polarization of the scattered beam at a given scattering vector Q as a function of the strength of a magnetic precession field [24] Mathematically $S(Q\ t)$, which is also known as incoherent intermediate scattering function, is the $(r \rightarrow Q)$ Fourier transform of the space-time correlation function of diffusing atoms.

234

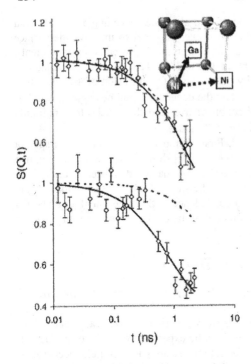

Figure 5. Time correlation as measured on the $Ni_{52.5}Ga_{47.5}$ single crystal at 1400 K. The two spectra were obtained in a simultaneous measurement at $2\Theta = 94°$ (top) and $2\Theta = 130°$ (bottom). The lines represent the time correlation as calculated for NN jumps (solid line) and NNN jumps (dashed line).
Inset:
B2-unit cell. A Ni atom may perform a NN jump into a vacancy on a Ga site (solid arrow) or a NNN jump into a vacancy on the Ni sublattice (dashed arrow).
Figure taken from Kaisermayr et al. [23].

The spectrum in Figure 5 corresponds to a diffusivity of 4.4×10^{-12} m^{2}s^{-1} and shows an effect of about 50%. With some straightforward experimental improvements such as the use of a furnace with a low-magnetic-field heater, even a 10% effect will be resolvable. This means that with the actual setup diffusivities in the 10^{-13} m^{2}s^{-1} range are easily accessible. Fourier times up to 10 ns, as they are provided by current NSE spectrometers in the relevant Q range, will permit us to penetrate the domain of Mössbauer spectroscopy [14], i.e., diffusivities around 10^{-14} m^{2}s^{-1}. In conclusion, we state that neutron spin-echo spectroscopy enables the determination of the diffusive jump as well as the measurement of the different time scales involved in the diffusion process. Compared to experiments on backscattering spectrometers substantial improvements are observed: Measurements at only two appropriately chosen points in reciprocal space are sufficient to allow a clear decision between different jump vectors.

The pronounced difference in the time correlation for different jump vectors in the present experiment shows that this method will make diffusion dynamics on the 1–100 ns scale accessible for neutron scattering, hence providing a powerful tool, which offers a resolution comparable to Mössbauer spectroscopy without being restricted to a single isotope, namely [57]Fe. Information on the mean residence times that diffusing atoms spend on different sublattices is obtained in a very direct manner from individual spectra. This immediate access to different time scales is expected to push the understanding of complicated diffusion mechanisms.

2.2.2. X-ray Photon Correlation Spectroscopy (XPCS) - new method for studying dynamics in solids with coherent X-rays

When coherent light is scattered by a disordered sample the diffraction pattern consists of a highly modulated so-called speckle pattern. This speckle pattern is a pattern of bright and dark spots that look randomly distributed at first glance. But this pattern is in direct relation to the exact spatial distribution of the scattering centers. So, if the sample evolves in time, i.e., the positions of the scattering centers change, also the speckle pattern will change. That means that speckle intensity will fluctuate in time. By calculating the autocorrelation function of one speckle's fluctuating intensity, one gets information about the time scales of the dynamics taking place in the sample.

The invention of the laser made it possible to use the method – called Photon Correlation Spectroscopy (PCS), sometimes also Dynamic Light Scattering – successfully. It became well established during the last 2-3 decades [25]. But PCS lacks of two main things. First, one is restricted to optically transparent materials and second, it is not possible to gain information about processes taking place on atomic scale, due to the long wavelength of several hundreds nanometers.

With the advent of third generation synchrotrons it is possible today to get an X-ray beam which is partially coherent [26] – a completely coherent beam of X-ray photons with • ≈ 1 Å will be available at a hard X-ray Free Electron Laser (FEL) for the first time. Nevertheless one is able today to do PCS with X-rays, the technique is then abbreviated XPCS. With XPCS it is possible to investigate dynamics on atomic scale in optically opaque materials.

In the last years emphasis of XPCS experiments were on dynamics in colloidal systems and glasses [27,28]. Only a few papers dealt with dynamics in crystalline materials [29,30].

Our goal is to establish XPCS for atomistic diffusion. For the following reasons XPCS looks so promising:

(i) XPCS is a non-resonant method, i.e., one is not confined to some special isotope – in principle there is no limitation for the sample's constitution.

(ii) In principle there is hardly an upper limit of accessible time scales for dynamics in the sample. That means that it is possible to investigate processes that relax on a time scale of minutes or even slower, see Figure 6. This is very important when we consider the long-term objective of resolving diffusion of single atoms with XPCS. Often their diffusion coefficient is too small which makes the diffusion mechanism on atomic scale inaccessible for other methods like QMS, QNS, NSR or NSE.

All feasibility studies performed up to now were performed at the TROIKA beamline ID10A at the ESRF, Grenoble.

236

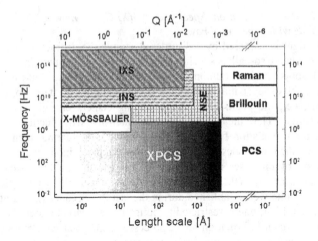

Figure 6. Capabilities of XPCS in comparison to other methods. The y-axis gives the frequency of the dynamics that can be investigated by the several methods. The lower x-axis shows the length scale of objects accessible by the methods and the upper x-axis indicates the length of the corresponding Q-vector. The grey scale gradient in the field of XPCS indicates, that the length scale down to 1 Å will become fully accessible with the FEL.

Acknowledgements

This work was partially supported by project of the Austrian Fonds zur Förderung der wissenschaftlichen Forschung (FWF), contract No. P15421, F & E Vorhaben, Hahn-Meitner Institut (XIFS and Ofenbau). We kindly acknowledge very helpful discussion with L. Deák and U. Pietsch.

References

1. Mehrer, H., (1996) Diffusion in intermetallics, *Mater. Transactions*, JIM **37**, 1259-1280.
2. Vogl, G. and Feldwisch, R. (1998) in *Diffusion in Condensed Matter*, Eds J. Kärger, P. Heitjas and R. Haberlandt, Vieweg Verlag, 40.
3. Hempelmann, R. (1994) Ed. Furrer, A., Proc. Second Summer School on Neutron Scatt., Zuoz, Switzerland
4. Smirnov, G. V. and Kohn, V. G. (1995) Theory of nuclear resonant scattering of synchrotron-radiation in the presence of diffusive motion of nuclei, *Phys. Rev.* B **52**, 3356-3365.
5. Kohn, V. G. and Smirnov, G. V. (1998) Theory of nuclear resonant scattering of synchrotron radiation in the presence of diffusive motion of nuclei. II. *Phys. Rev.* B **57**, 5788-5797.
6. Levi, A.C. Spadacini, R. and Tommei, G.E. (1982) Quantum-theory of atom surface scattering - incommensurate and fluid adsorbates, *Surface Science* **121**, 504-518.
7. Bertino, M.F. , Hofmann, F., Steinhögl , W. and Toennies, J.P. (1996) Quasielastic helium atom scattering measurements of microscopic diffusion of CO on the Ni(110) surface, *J. Chem. Phys.* **105**, 11297-11304.
8. Weinkamer, R., Fratzl, P., Sepiol, B. and Vogl, G.(1999) Monte Carlo simulations of Mössbauer spectra in diffusion investigations, *Phys. Rev.* B **59**, 8622-8625.
9. Mishin, Y. and Farkas,D. (1998) Atomistic simulation of point defects end diffusion in B2 NiAl, *Scr. Mater.* 39, 625-630.
10. Randl, O. G., Sepiol, B., Vogl, G., Feldwisch, R. and Schroeder, K. (1994) Quasi-elastic Mössbauer-

spectroscopy and quasi-elastic neutron-scattering from non-Bravais lattices with differently occupied sublattices, *Phys. Rev.* B **49**, 8768-8773.

11. Kaisermayr, M., Combet, J., Ipser, H., Schicketanz, H., Sepiol, B. and Vogl, G. (2000) Nickel diffusion in B2-NiGa studied with quasielastic neutron scattering, *Phys. Rev.* B **61** 12038-12044.

12. Kaisermayr, M., Combet, J., Ipser, H., Schicketanz, H., Sepiol, B. and Vogl, G., (2001) Determination of the elementary jump of Co in CoGa by quasielastic neutron scattering - art. no. 054303, *Phys. Rev.* B **63** 054303.

13. Sepiol, B., Meyer, A., Vogl, G., Franz, H. and Rüffer, R. (1998) Diffusion in a crystal lattice with nuclear resonant scattering of synchrotron radiation, *Phys.Rev* .B **57**, 10433-10439.

14. Sepiol, B. (1998) *Mat. Res. Symp. Proc.* **527**, 147.

15. Vogl, G. and Sepiol, B. (1999) Diffusion in crystalline materials, Hyperfine Interactions **123/124**, 595-609.

16. Thiess, H., Kaisermayr, M., Sepiol, B., Sladecek, M., Rüffer, R. and Vogl, G. (2001) Nuclear resonant Bragg scattering: Measurement of self-diffusion in intermetallics, *Phys. Rev.* B **64**, 104305.

17. Röhlsberger, R. (1999) Nuclear resonant scattering of synchrotron radiation from thin films, *Hyperfine Interactions* **123/124**, 455-479.

18. Chumakov, A.I., Niesen, L., Nagy, D. L. and Alp, E. E. (1999) Nuclear resonant scattering of synchrotron radiation by multilayer structures, *Hyperfine Interactions* **123/124**, 427-454.

19. Nagy, D. L., Bottyán, L., Deák, L., Dekoster, J., Langouche, G., Semenov, V.G., Spiering, H. and Szilágyi, E. (1999) Synchrotron Mössbauer reflectometry in Materials Science, In: *Mössbauer Spectroscopy in Materials Science*, Eds. Miglierini, M. and Petridis, D. (Kluwer Academic Publishers, Dordrecht, 1999) p. 323-336.

20. Lengeler, B. (1987) Synchrotronstrahlung in der Festkörperforschung, 18. IFF Ferienkurs, Kernforschungsanlage Jülich GmbH.

21. Sladecek, M., Sepiol, B., Kaisermayr, M., Korecki, J., Handke, B., Thiess, H., Leupold, O., Rüffer, R. and Vogl,G. (2002) Enhanced iron self-diffusion in the near-surface region investigated by nuclear resonant scattering, *Surface Science* **507-510**, 124-128.

22. Spiering, H., Deák, L. and Bottyán, L. (2000) EFFINO, *Hyperfine Interactions* **125**, 197-204.

23. Kaisermayr, M., Pappas, C., Sepiol, B. and Vogl, G. (2001) Probing jump diffusion in crystalline solids with neutron spin-echo spectroscopy, *Phys. Rev. Lett.* **87**, 175901.

24. Mezei, F. (1980) *Neutron Spin Echo*, edited by F. Mezei (Springer, Berlin, 1980), 3.

25. Brown W. (1993) *Dynamic Light Scattering*, Clarendon Press.

26. Abernathy D.L. , Grübel, G., Brauer,S., McNulty, I., Stephenson, G. B., Mochrie, S.G.J., Sandy, A.R., Mulders, N. and Sutton M. (1998) Small-angle X-ray scattering using coherent undulator radiation at the ESRF, *J. Synchrotron Rad.* **5**, 37-47.

27. Dierker, S.B., Pindak, R., Fleming, R. M., Robinson, I.K. and Berman, L. (1995) X-ray photon-correlation spectroscopy study of brownian-motion of gold colloids in glycerol, *Phys. Rev. Lett.* **75**, 449-452.

28. Malik, A., Sandy, A.R., Lurio, L.B., Stephenson, G.B., Mochrie, S.G.J., McNulty, I. and Sutton, M. (1998) Coherent X-ray study of fluctuations during domain coarsening, *Phys. Rev. Lett.* **81**, 5832-5835.

29. Brauer, S., Stephenson, G.B., Sutton, M., Brüning, R., Dufresne, E., Mochrie, S.G.J., Grübel, G., Als-Nielsen , J. and Abernathy, D.L., (1995) X-ray-intensity fluctuation spectroscopy observations of critical-dynamics in Fe3Al, *Phys. Rev. Lett.* **74**, 2010-2013.

30. Livet, F., Bley, F., Caudron, R., Geissler, E., Abernathy, D., Detlefs, C., Grübel, G. and Sutton, M. (2001) Kinetic evolution of unmixing in an AlLi alloy using x-ray intensity fluctuation spectroscopy, *Phys. Rev.* E **63**, 036108.

NUCLEAR RESONANCE INELASTIC SCATTERING OF SYNCHROTRON RADIATION IN OXIDES WITH COLOSSAL MAGNETORESISTANCE

A. I. RYKOV [1], K. NOMURA [1], T. MITSUI [2], AND M. SETO [3]

[1] School of Engineering, The University of Tokyo, Hongo 7-3-1, Bunkyo-ku, Hongo 113-8656, Japan
[2] Japan Atomic Energy Research Institute ,Kouto 1-1-1, Mikazuki, Sayo, Hyogo 679-5198, Japan
[3] Research Reactor Institute, Kyoto University, Noda, Kumatori-machi, Sennan-gun, Osaka 590-0494, Japan

Abstract

Phonon anomalies near the critical temperature are studied in a number of itinerant ferromagnets, including (Fe,Co)-based perovskite-related oxides, the ^{57}Fe-doped ruthenates and the manganite $La_{0.7}Sr_{0.3}MnO_3$. The partial ^{57}Fe phonon densities of states (DOS) in these oxides were derived from the resonant nuclear inelastic scattering spectra. The spectra were measured with energy resolution of 2.5meV by detecting the 6.3 keV Fe K_a X-rays following after Mössbauer effect on ^{57}Fe transition excited by monochromatized 14.41 keV synchrotron radiation. The changes in the ^{57}Fe phonon DOS $g(E)$ are observed at cooling across T_c. In the perovskites $Sr_2FeCoO_{6-\delta}$ and $SrBaFeCoO_{6-\delta}$, the lowest energy peak near 15 meV develops below T_c. We attribute this change to the narrowing of the phonon bands at the onset of the transport coherence. The non-metallic brownmillerite $SrCaFeCoO_5$ exhibits a peak in the much lower energy region of $g(E)$. In the $g(E)/E^2$ curve, a strong low-energy deviation from Debye behavior ($g(E)/E^2$= const) appears at 7 meV. This peak enlarges dramatically the vibrational amplitudes for brownmillerites relative to perovskites.

1. Introduction

While the topmost colossal magnetoresistors (CMR) belong to oxides of manganese, a few examples of the CMR-materials are found among perovskites [1,2] and spinels [3], containing iron. In the Mössbauer studies of $Sr_2FeCoO_{6-\delta}$ [2] and Sr_2FeMoO_6 [4,5], both of these oxides showed the double-site spectra, in paramagnetic regime consisting of the equally populated doublet and single-line. Pairs of the oxidation states $Fe^{3+\delta}/Fe^{4+}$ and $Fe^{3+}/Fe^{2+\delta}$ follow from the isomer shifts of doublet / singlet (D+S) spectra in Sr_2FeCoO_6 and Sr_2FeMoO_6, respectively [2,5]. Metallic conductivity is in agreement with the exchange schemes $Fe^{3+}-O-Co^{4+} \leftrightarrow Fe^{4+}-O-Co^{3+}$ and $Fe^{3+}-O-Mo^{5+} \leftrightarrow Fe^{2+}-O-Mo^{6+}$.

The appearance of distorted site, especially in cubic Sr_2FeMoO_6, is reminiscent of the doped manganites, where the Jahn-Teller (JT) distortion is associated with the Mn^{3+}

M. Mashlan et al. (eds.), Material Research in Atomic Scale by Mössbauer Spectroscopy, 239–250.

ions, tending to localize the e_g electrons above T_c. In the Fe-based oxides, the validity of double-exchange (DE) mechanism, at variance to intergranular tunneling mechanism, was advocated [6], however, the charge localization above T_c was not clearly observed. On the contrary, the resistivity $\rho(T)$ in a single crystal of Sr_2FeMoO_6 showed rather metallic behavior both below and above T_c, and only a small hump of the $\rho(T)$ at T_c was found [7]. The broadened $\rho(T)$-humps reported in $SrFe_{1-x}Co_xO_{3-\delta}$ ($T_c \sim$ 150 to 250 K [2]), are also similar to metallic behavior in $SrRuO_3$, a typical ferromagnetic "bad metal", known to show the very strong phonon anomalies in the vicinity of T_c [8].

The lattice vibrations underlying the CMR property in this class of oxides were a subject of extensive studies by Raman, IR, and neutron spectroscopies. We employ here the inelastic absorption of the 14.413 keV X-rays, exciting the ^{57}Fe nuclear level via creating/annihilating phonons, a phenomenon complementary to the Mössbauer effect. Although from the ^{57}Fe spectra only the partial phonon density of states results, i.e. the DOS, composed of the only vibrational states, in which the resonant nuclei are involved, a robust change may occur in the partial DOS at magnetic transitions, as was shown by Chumakov et al. [9] in the case of the 173K-transition in $[Fe(bpp)_2]\,[BF_4]_2$.

The perovskite structure is typically susceptible to soft modes capable to break down the cubic symmetry at cooling. In our focus, however, are the transitions of the metal-to-insulator type, which occur without bare change in crystal symmetry, but show a strong lattice involvement. Superconductivity also obeys this prerequisite. In ferromagnetic $La_{0.7}Ca_{0.3}MnO_3$, the long-range magnetic order breaks down thermally along with extinguishing transport coherence. Complex magneto-crystalline cluster structures form in a broad temperature range where the material transforms progressively from metal to insulator. The field-induced conductivity (CMR-effect) culminates near T_c where the ferromagnetic clusters can efficiently merge into larger domains under control of the external magnetic field [10]. Frequencies of several phonons ν_i change around T_c, where $d\nu_i/dT$ culminate in a fairly similar way, with the most marked culminations of $d\nu_i/dT > 0$ for superconductivity [11,12] and $d\nu_i/dT < 0$ for itinerant ferromagnetism [8].

In the manganites, the effect of magnetic field on the infrared reflectivity spectra (optical CMR) was observed to be equivalent of increasing optical conductivity with decreasing temperature at a matching rate of ~1 K per Tesla ($\mu_B H / k_B \Delta T \approx 0.6$ at T_c in $La_{2/3}Ca_{1/3}MnO_3$[13]). At the same time, clear redistribution of oscillator strengths at cooling was observed within the pairs of the transverse optical (TO) phonons split by JT coupling into doublets at (20,22), (43,46) and (72,76) meV [13]. From extinguishing phonon at ~72 meV, the number of the JT-distorted MnO_6 octahedra was claimed to diminish at cooling through T_c. We are not aware of the optical studies of the phonon anomalies performed up to date on the Fe-based CMR-conductors. Neither the phonon DOS was measured by inelastic neutron or nuclear resonant scattering, yet.

In this work, the measurements of the phonon spectra allow us to detect both the effects of shift of the phonon energies and the DOS redistribution through T_c. Besides the samples of Fe-based oxides, the Fe-doped manganite, ruthenates and cobaltite were studied.

2. Experimental Details

The samples of $Sr_2FeCoO_{6-\delta}$, $SrBaFeCoO_{6-\delta}$, $SrCaFeCoO_{6-\delta}$, $La_{0.67}Sr_{0.33}Mn_{0.99}Fe_{0.01}O_3$, $LaCo_{0.99}Fe_{0.01}O_3$, $LaCo_{0.9}Fe_{0.1}O_3$, $SrRu_{0.99}Fe_{0.01}O_3$ and $BaRu_{0.99}Fe_{0.01}O_3$ were synthesized by a sol-gel method and calcined at 850°C as described in Ref. [14]. Additionally, all samples except $LaCoO_3$ were annealed at 1000°C and then at 570°C under high isostatic pressure (HIP) of 8.8 MPa of oxygen gas. In Ref. [2], the HIP treatment has led to the oxygen content "6-δ" ~5.8 in $Sr_2FeCoO_{6-\delta}$. More oxygen deficient is $SrCaFeCoO_{6-x}$, which retains its brownmillerite structure after the HIP treatment. In other samples, except hexagonal $BaRuO_3$, the perovskite phase was just cubic in our X-ray diffraction patterns.

Nuclear inelastic scattering (NIS) spectra were measured at the beamline BL11XU of the SPring-8 storage ring, operated in the timing mode C (11-bunch train ×29). The assembly of nested asymmetric Si(511) and Si(975) channel-cut monochromators allowed to narrow the bandwidth down to 2.5 meV. The cobaltite samples were measured in another experimental run with the bandwidth of 3 meV. A number of avalanche photodiode detectors (3×3 mm^2) were employed to register the 6.3 keV radiation in a large solid angle. The average beam lifetime was 55 hours with the injection once per 24 hours and the initial current of about 100 mA. Measuring time did not exceed 3 hours for each of the scan.

The idea of present measurement exploits the assumption that the phonon DOS is temperature-independent in the approximation of harmonic lattice, unless the crystal exhibits a structural or electronic phase transition in certain range of temperatures. For the ferromagnetic transition in $Sr_2FeCoO_{5.8}$, this temperature range is known to be broad [2]. Therefore, in addition to the temperatures, which can be affected by the transition (80 K, 150 K, 300 K), we have measured several spectra at temperatures of 450 K and 600 K, which are rather far from the transition.

In contrast to the DOS, the experimental NIS spectra are strongly temperature-dependent. A NIS spectrum is given by the cross-section of the inelastic absorption in function of energy around the resonance. The temperature dependence of the spectra is determined mostly by the occupancy of the phonon states given by the Bose occupation factor N_B for the phonon annihilation part of a spectrum, and by N_B+1 for the phonon creation part of the spectrum. Therefore, the temperature dependence of the phonon annihilation part of a spectrum is especially sharp at low temperatures. That is why the procedure of computation of the phonon DOS cannot be universal for all temperatures. The annihilation part of the NIS spectrum can only be taken into account in the case when its intensity prevails reasonably over the noise of avalanche detectors. A border temperature exists, above which both the creation and the annihilation parts should be taken into account. This borderline value depends sensitively on the content of ^{57}Fe in a sample. The lower is the content of ^{57}Fe, the lower is the resonant counting rate, the higher is the relative contribution of the detector noise, and the higher is the value of temperature below which the annihilation part of the spectrum should be disregarded.

A newly built sample cell allowed varying the temperature from 77 K to 600 K. Letting accurately measured temperature into our DOS-computation program makes coinciding the DOS calculated by two formulas:

$$\text{creation} \quad : \quad g(E) = \frac{E}{E_R} S_1(E) \cdot (1 - \exp(-\frac{E}{k_B T})), \text{E} > 0 \tag{1}$$

$$\begin{array}{c}\text{creation}\\ \underline{\text{annihilation}}\end{array} + \quad : \quad g(E) = \frac{E}{E_R}(S_1(E) + S_1(-E)) \tanh \frac{E}{2k_B T}, \text{E} > 0 \tag{2}$$

The Eqs. (1) and (2) were suggested previously by the authors of Refs. [15] and [16], respectively. In Eq. (1), in use is only the phonon creation branch of single-phonon NIS spectrum $S_1(E)$, $E > 0$, while in Eq.(2), both the creation and the annihilation are valid.

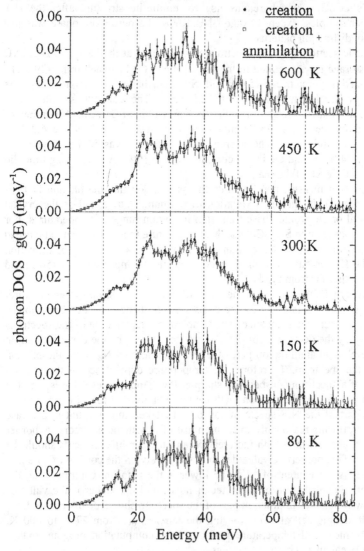

Figure 1. Phonon DOS in $Sr_2FeCoO_{5.8}$ determined from the NIS spectra at indicated temperatures. Closed and open symbols show the DOS calculated with Eqs. (1) and (2), respectively.

The spectra were treated using forward and reverse Fourier transformation and the multiphonon scattering was treated by the log-Fourier method [15,16]. This correction is small at 300 K due to the large Lamb-Mössbauer factors f_{LM} ~0.8 found from the ratio of zero moment M_0 and first moment M_1 of the spectra, $f_{LM} = 1 - E_R(M_0/M_1)$, where E_R =1.937 meV is the recoil energy. Slight asymmetry of the instrumental function was taken into account in calculating M_1 [15]. Another important correction of spectra was related to the subtraction of the noise of avalanche detectors [16]. For this reason, most of the spectra were measured in the extended region, and the background was evaluated between 100 and 110 meV, where both S_1 and multiphonon scattering are negligible. The subtraction of background becomes a major correction especially in the samples containing low abundance of ^{57}Fe. In $Sr_2FeCoO_{5.8}$ ($SrCaFeCoO_5$), the ^{57}Fe content was 20%(50%). In samples with 1% of foreign probe, the ^{57}Fe enrichment was 95%. The detector noise accumulated in spectra of such samples with the measuring time (~6 hours) is a severe bug in the high-energy region (>50 meV), where it is adding to the error, weighted in $g(E)$ by the energy factors of Eqs. (1,2), while the error bars in the region of main DOS peaks are given principally by the standard statistical deviation.

3. Results and Discussion

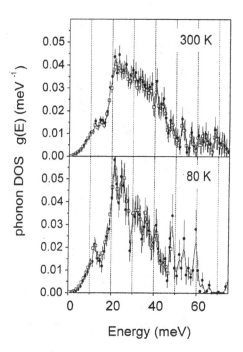

Figure 2. Phonon DOS in $BaSrFeCoO_{6-\delta}$ at indicated temperatures. The ^{57}Fe enrichment is 20%.

Fig.1 show the temperature dependence of the phonon DOS in $Sr_2FeCoO_{5.8}$ at several temperatures above T_c, below T_c, and at T_c (150 K). The DOS features are more pronounced below T_c. Three region of energy may be considered. In the region between 20 and 50 meV, the majority of Fe vibrational states can be related to the bending and torsional modes of the perovskite structure. The higher energy region, around 70 meV, is the stretching mode, involving mainly the vibrations of oxygen, with only a tiny contribution of Fe. The peak at lowest energy (15 meV) is likely to be attributed to the external and acoustic modes, which involve mainly the Sr vibrations relative to the rest of the lattice. This peak shows lower energy when Ba replaces a part of Sr sites (Fig. 2).

Clearly, the soft peak becomes better resolved below T_c in both cases. We therefore attribute this change below T_c to the onset of the transport coherence. The changes of DOS above T_c are not so significant and can be merely attributed to the vibrational anharmonicity, which is clearly seen in the softening of the front edge of the steep climb at 19 meV in the 600K-spectrum (Fig.1). The changes below T_c occur in much narrower temperature range and cannot be attributed to effect of anharmonicity.

In the samples, where the extrinsic probe ^{57}Fe is diluted in the lattice of oxides of Mn, Ru or Co, the NIS spectra would reflect the local vibrational densities of states of ^{57}Fe in a foreign lattice. In such a case, the local DOS is composed of the modes of resonance with the host phonon spectrum [17]. In oxides, where Fe is in the same ligand environment as the host ion, we expect the shift of resonant frequencies with respect to the host frequencies to be smaller than in metals. The DOS would change more significantly if the Fe ions adopt the coordination by oxygen different from that of the host ion. Such a situation is typical in the cuprates because of difference in the preferential oxidation states between Fe and Cu [18], but unlikely in cobaltites, manganites or ruthenates. We observed that the lattice parameters and the NIS spectra of these doped oxides are practically unchanged after treatment under oxygen gas pressure of 8.8 MPa at 570°C.

Of particular interest is $SrRuO_3$ because this "bad metal" shows the outstanding hardening of major phonon modes at cooling through ferromagnetic transition. In the region of phonon DOS observed in our ^{57}Fe spectra, the phonons were reported to harden from 26 to 29, 30.5 to 31.5, 47.5 to 49, and 50.5 to 51.5 meV [8]. Our NIS spectra (Fig.3) and DOS at 300K (Fig.4) show the features at 25, 30 and 35 meV. Although counting rate was weak in the cryostat, one observes in Fig.3 that the mode at 35 meV is enhanced at 80K. This is in agreement with the hardening observed in Ref. [8].

The mode at 26 to 29 meV was assigned in Ref. [8] to the Sr motion mainly. Therefore, the remarkable change in the frequency of this mode at cooling through T_c was unexpected and interpreted as a

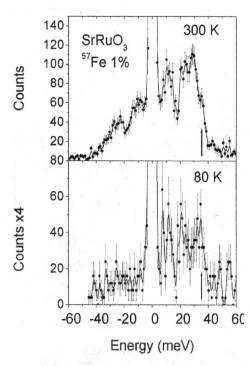

Figure 3. NIS spectra in $SrRu_{0.99}{}^{57}Fe_{0.01}O_3$ at 300 K and at 80 K. The arrow indicates a peak in phonon DOS at 38 meV.

possible indication of complex mechanism of spin-phonon coupling. However, we can see in Fig. 4 that the partial ^{57}Fe DOS is very large in this region. Therefore, if ^{57}Fe reflects faithfully the Ru partial DOS, the change in frequency of this mode is not unusual, because these vibrations contribute to the modulations of the Ru-O-Ru bond angles and bond lengths.

Lanthanum cobaltite is a fascinating material, in which the succession of Co^{3+} spin-state transitions and large drop in resistivity at the 500K-transition was observed [19].

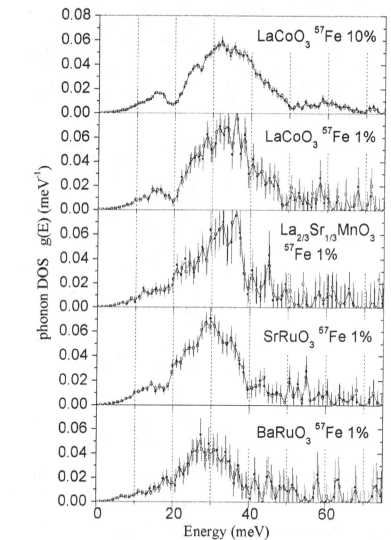

Figure 4. Phonon DOS in LaCo$_{0.9}$^{57}Fe$_{0.1}$O$_3$, LaCo$_{0.99}$^{57}Fe$_{0.01}$O$_3$, La$_{0.67}$Sr$_{0.33}$Co$_{0.99}$^{57}Fe$_{0.01}$O$_3$, SrRu$_{0.99}$^{57}Fe$_{0.01}$O$_3$, and BaRu$_{0.99}$^{57}Fe$_{0.01}$O$_3$, determined from the NIS spectra at 300 K. Closed and open symbols show the DOS calculated with Eqs. (1) and (2), respectively.

We observe in LaCoO₃ that the substitution of 1% or 10% of Co with ⁵⁷Fe leads to the fairly similar ⁵⁷Fe phonon DOS. The increase of doping rate may reasonably produce the broadening only of the intrinsic DOS features in LaCoO₃. Again, except for the main band, the DOS shows the low-energy peak at 15 meV. The main band between 27 meV and 40 meV is accompanied by two shoulders at 24 meV and 44 meV.

The feature at 44 meV is also present in the CMR manganite La₀.₆₇Sr₀.₃₃Mn₀.₉₉⁵⁷Fe₀.₀₁O₃, however, separated better from the sharp cutoff of the main band at 38 meV. Such a cutoff was not observed in the generalized vibrational density of states (GVDOS) obtained with the method of inelastic neutron scattering [20]. It is shown in Fig. 5., that the neutron-weighted GVDOS also reveals the features at 15, 24 and 44 meV. However, the cutoff in the Fe DOS occurs in the interior of the band of GVDOS. It appears that the Mn vibrations and the oxygen vibrations are not mixed in this manganite. The aspects of comparison between partial ⁵⁷Fe DOS and GVDOS have been reviewed recently on the example of quasicrystals, which showed a striking difference between smooth and featureless behavior of GVDOS and strongly peaked at one energy value (27meV) ⁵⁷Fe DOS [21].

Quite similar contrast is observed in our manganite. In the ⁵⁷Fe DOS, the peak at 37 meV is well pronounced. The ratio of the value at 37 meV to the values at 15 meV and 24 meV is much larger in ⁵⁷Fe DOS than in GVDOS. The contributions of different atoms weighted by masses and scattering lengths are present in GVDOS. Since both the ⁵⁷Fe DOS and GVDOS are normalized, we should compare only relative values. From comparison of the ⁵⁷Fe DOS and GVDOS values at 37 meV with those at 15 meV and 24 meV it appears, that the GVDOS-contribution of Sr partial vibrations is nearly half for the 24 meV feature, but larger for the 15 meV feature. It is plausible therefore to attribute the lowest-energy feature to the Van Hove singularity of the acoustic phonon branches, whereas the features at 24 and 37 meV have the character of the Einstein oscillators with different partial contributions of Sr and Fe.

Finally, we compare the DOS

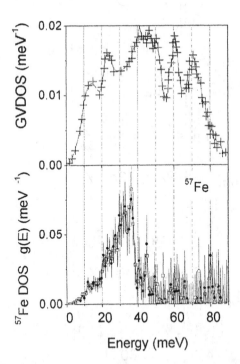

Figure 5. Comparison of the ⁵⁷Fe vibrational DOS at 300 K and the generalized vibrational density of states derived from inelastic scattering of 120 meV neutrons at 15 K [20] in La₂/₃Sr₁/₃MnO₃.

results obtained on the CMR-material $Sr_2FeCoO_{5.8}$ with those for $SrCaFeCoO_5$. In the latter, the "chemical pressure" effect of Ca makes difficult the HIP-insertion of extra-oxygen into brownmillerite structure. Fig.6 shows that the DOS in the brownmillerite is vastly different from that in perovskites. In two structures, we have calculated the temperature dependence of Lamb-Mössbauer factor:

$$f_{LM} = \exp\left(-E_R \int_0^{\infty} \frac{g(E)}{E} \frac{1 + e^{-\frac{E}{k_B T}}}{1 - e^{-\frac{E}{k_B T}}} \, dE \right) \tag{3}$$

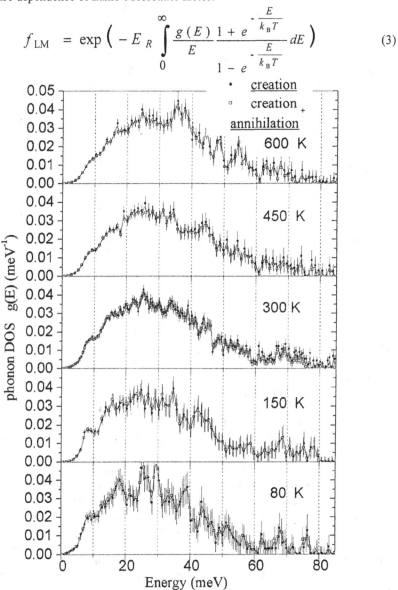

Figure 6. Phonon DOS in the brownmillerite $SrCaFeCoO_5$ at various temperatures.

As shown in Fig. 7, the curves f_{LM} (T) obtained from the ^{57}Fe DOS at various temperatures are well grouped together. The values of f_{LM} (T) calculated directly from the NIS spectral areas may deviate slightly upwards from the curves, especially at the elevated temperatures. The algorithms described previously [15,16] allow normalising the DOS at room temperature and below well to unity. On the other hand, the DOS area is overestimated when the temperature is higher. The situation, when the second renormalization of DOS is required, was mentioned previously [22]. While the integration of spectral areas is sensitive to the high-energy behavior of DOS, affected by the multiphonon scattering, the behavior of f_{LM} from the Eq. (3) is determined by the low-energy behavior of DOS only. Therefore, the results for f_{LM}, presented in Fig.7, are highly reliable even if the multiphonon scattering subtraction was incomplete.

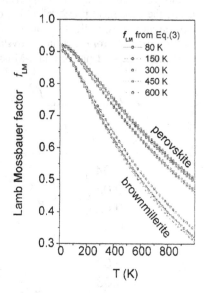

Figure 7. Temperature dependence of the Lamb-Mössbauer factor in perovskite $Sr_2FeCoO_{5.8}$ and brownmillerite $SrCaFeCoO_5$. The curves show f_{LM} are obtained using Eq.(3) from DOS at various temperatures.

Our results confirm the early observation [23] of strong influence on the phonon DOS of the oxygen content in perovskite-derived structures. The DOS was reported previously [23] for several discrete structures derived from perovskite $SrFeO_{3-\delta}$ by ordering of oxygen vacancies into various superstructures. The brownmillerite is the most oxygen-deficient superstructure in this series. From Eq. (3), we show that the DOS of the brownmillerite generates much smaller Lamb-Mössbauer factors than the DOS of perovskite for the closely related compositions. The perovskite $Sr_2FeCoO_{5.8}$ shows much larger factors f_{LM} than the brownmillerite $SrCaFeCoO_5$ (Fig. 7.)

There are two iron sites in brownmillerite, which could have, in principle, quite different factors f_{LM}. It is most plausible to attribute the reduction of the total f_{LM} factor of the brownmillerite to the small f_{LM} factor of the tetrahedral Fe (III) site, which shows the strong vibrational anisotropy in layered oxides [24]. While iron populates both the tetrahedral and octahedral sites in most of the brownmillerites, it goes only into the tetrahedral site in the brownmillerite-related cuprates [24,25], formed by replacement of the chessboard layer FeO_2 of a brownmillerite $Sr_2Fe_2O_5$ with the bilayer CuO_2-Y-CuO_2. In these cuprates, the brownmillerite-inborn tetrahedral Fe (III)-site was isolated in Mössbauer spectra and the anisotropy of the Fe (III) vibrations was observed by the Goldanskii-Karyagin effect [24]. The vibrational ellipsoids of the tetrahedral Fe (III) are oblate, expanded in the plane, in which the oxygen vacancies are ordered. The in-plane bonding is not so tight as compared to the out-of-plane bonding, therefore, the average f_{LM} factors drops due to their reduction for the in-plane directions.

In the octahedral sites, the f_{LM} factors can be also reduced compared to the f_{LM} factors of the original symmetric octahedron in perovskite. The vibrational states of both tetra- and octa-Fe sites are involved into $g(E)$ in the expression under integration of Eq. (3). In the limit of high temperatures, this expression is well approximated by $g(E)/E^2$. The typical curves $g(E)/E^2$ are compared in Fig.8 for a perovskite and a brownmillerite. The low-energy part of $g(E)/E^2$ in the brownmillerite exhibits a peak at 7 meV, which results in the reduction of f_{LM} and in the enlargement of the vibrational amplitudes $\langle \Delta x^2 \rangle = -\ln(f_{LM})/k^2$. It is tempting to identify this feature with the so-called "Boson Peak", observed in this energy range in glasses and disordered systems. On the other hand, such a peak exists also in the

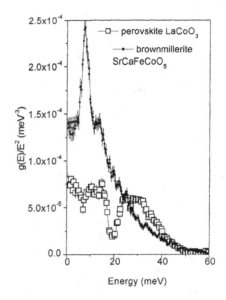

Figure 8. Low-energy part of g(E)/E² in LaCoO₃ (⁵⁷Fe 10%) and in SrCaFeCoO₅.

brownmillerite $Sr_2Fe_2O_5$ [23], which is entirely free of cationic disorder. The analogy with the well-known "Boson Peak" could be still meaningful if one finds in the brownmillerite an assignment of this peak to some medium range vibrational configurations with particularly soft constants. Such configurations may involve both the tetrahedral and the octahedral iron.

4. Concluding Remarks

Nuclear resonant inelastic scattering was used to study the electronic phase transitions in oxides with large magnetoresistance. The changes in the phonon density of states were observed in the NIS spectra, associated with the magnetic ordering transitions. Phonon anomalies, known to occur near T_c in Raman spectra of these oxides, appear in the partial ⁵⁷Fe DOS mainly as sharpening of the phonon bands and density redistributions between them. Similarly to the neutron inelastic scattering on powder samples, instead of phonon softening or hardening, intensity changes were observed below T_c. This can be hardly explained by the lack of experimental resolution. Rather the phonon frequency changes are dissimilar in centre and in high-k points of Brillouin zone. Thus, the optically observed hardenings in the Γ-point can appear as a bare shape change in both the neutron CVDOS and in our ⁵⁷Fe partial DOS.

In the brownmillerite SrCaFeCoO₅, an interesting anomaly in the low-energy part of $g(E)/E^2$ curves was observed. The peak at 7 meV turns out to be a common feature of brownmillerites, at variance with parent perovskites, behaving closer to Debye model.

250

References

1. Kobayashi K.-I., Kimura T., Sawada H., Terakura K., and Tokura Y. (1998) Room-temperature magnetoresistance in an oxide material with an ordered double-perovskite structure, *Nature* 395, 677-680.
2. Maignan A., Martin C., Nguyen N., and Raveau B. (2001) Magnetoresistance in the ferromagnetic metallic perovskite $SrFe_{1-x}Co_xO_3$, *Solid State Sciences* 3, 57-63.
3. Ramirez A.P., Cava R.J., and Krajewski J. (1997) Colossal magnetoresistance in Cr-based chalcogenide spinels, *Nature* 386, 156-158.
4. Algarabel P.A., Morellon L., De Teresa J.M., Blasco J., Garcia J., Ibarra M.R., Hernandez T., Plazaola F., Barandiaran J.M. (2001), Mössbauer spectroscopy in Sr_2FeMoO_6 double perovskite, *Journal of magnetism and magnetic materials* 226-230, 1089-1091.
5. Ballcells Ll., Navarro J., Bibes M., Roig A., Martinez B., and Fontcuberta J., (2001), Cationic ordering control of magnetization in Sr_2FeMoO_6. *Applied Physics Letters* 78, 781-783.
6. Navarro J., Frontera C, Balcells Ll., Martinez B., and Fontcuberta J. (2001) Raising the Curie temperature in Sr_2FeMoO_6 double perovskite by electron doping, *Physical Review B* 64, 092411, 1-4.
7. Moritomo Y., Xu S., Machida A., Akimoto T., Nishibori E., Takata M., Sakata M. ,Ohoyama K., (2001) Crystal and magnetic structure of conducting double perovskite $Sr_2Fe\,MoO_6$, *J. Phys. Soc. Jpn.* 69 1723.
8. Iliev M.N., Litvinchuk A.P., Lee H.-G.,Chen C.L., Dezaneti M.L., Chu C.W., Ivanov V.G., Abrashev M.V. and Popov V.N.(1999) Raman spectroscopy of $SrRuO_3$ near the paramagnetic-to-ferromagnetic phase transition, *Physical Review B* 59, 364-368.
9. Chumakov A., and Rüffer R. (1998) Nuclear inelastic scattering, *Hyperfine Interactions* 113, 59-79.
10. Watahiki M., Metoki N, Suzuki J.-I., Oikawa K.-I., Nie J., Tachiki M., and Yamada Y.(2001) Small-angle neutron scattering study on ferromagnetic correlation in $(La,Tb)_{2/3}Ca_{1/3}MnO_3$, *J. Phys. Soc. Jpn.*70, 1090.
11. Pyka N., Reichardt W., Pintschovius L., Engel G., Rossat-Mignod J., and Henry J. Y. (1993) Superconductivity-induced phonon softening in $YBa_2Cu_3O_7$ observed by inelastic neutron scattering, *Physical Review Letters* 70, 1457-1460.
12. Limonov M.F., Rykov A.I., Tajima S., and Yamanaka A. (2000) Superconductivity-induced effects on phononic and electronic Raman scattering in twin-free $YBa_2Cu_3O_{7-x}$ single crystals, *Physical Review B* 61, 12412-12419.
13. Boris A.V., Kovaleva N.N., Bazenov A.V., van Bentum P.J.M., Rasing Th., Cheong S.-W., Samoilov A.V., and Yeh N.-C. (1999) Infrared studies of a $La_{0.67}Ca_{0.33}MnO_3$ single crystal: Optical magnetoconductivity in a half-metallic ferromagnet, *Physical Review B* 59, R697-R700.
14. Nomura K., Tokumitsu K., Hayakawa T., and Homonnay Z. (2000) The influence of mechanical treatment on the absorption of CO_2 by perovskite oxides, *J. Radioanal. Nucl. Chem.* 246, 69-77.
15. Kohn V.G. and Chumakov A.I. (2000) DOS: Evaluation of phonon density of states from nuclear resonant inelastic absorption, *Hyperfine Interaction* 125, 205-221.
16. Sturhahn W. (2000) CONUSS and PHOENIX: Evaluation of nuclear resonant scattering data, *Hyperfine Interactions* 125 149-172.
17. Seto M., Kobayashi Y., Kitao S., Haruki R., Mitsui T., Yoda Y, Nasu S., and Kikuta S. (2000) Local vibrational densities of states of dilute Fe atoms in Al and Cu metals, *Physical Review B*61 11420-11424.
18. Rykov A., Caignaert V., Nguyen N., Maignan A., Suard E., and Raveau B. (1993) The complex distribution of iron in the $(Y,Ca)Ba_2(Cu,Fe)_3O_{6+y}$ cuprate. *Physica C*205 63-77.
19. Asai K., Yoneda A., Yokokura O., Tranquada J. M., Shirane G., and Kohn K. (1998) Two spin-state transitions in $LaCoO_3$, *Journal of the Physical Society of Japan* 67 290-296.
20. Reichardt W., Bennington S.M. (1999) *Lattice vibrations in $La_{1-x}Sr_xMnO_3$*, ISIS Experimental Report, Rutherford Appleton Laboratory.
21. Brand R.A., Coddens G., Chumakov A. I., Calvayrac Y. (1999) Partial phonon density of states of Fe in an icosahedral quasicrystal $Al_{62}Cu_{25.5}{}^{57}Fe_{12.5}$ by inelastic nuclear-resonant absorption of 14.41-keV synchrotron radiation, *Physical Review B* 59 R14145-R14148.
22. Chumakov A.I., Rüffer R., Baron A.Q.R., Grünsteudel H., and Grünsteudel H.F. (1996) Temperature dependence of nuclear inelastic absorption of synchrotron radiation in α-57Fe, *Phys. Rev. B* 54, R9596.
23. Sturhahn W., Toellner T.S., Alp E.E., Zhang X., Ando M., Yoda Y., Kikuta S., and Seto M. (1995) Phonon density of states measured by inelastic resonant scattering, *PhysicalReview Letters* 74 3832-3835.
24. Rykov A., Caignaert V., and Raveau B. (1994) Quadrupole interactions and vibrational anisotropy of tetrahedral Fe(III) in 123 derivative $LnSr_2Cu_2Ga_{1-x}Fe_x$ (Ln=Y,Ho) *J. Solid State Chem.* 109, 295-306.
25. Rykov A., Caignaert V., Van Tendeloo G., Greneche J. M., Studer F., Nguyen N., Ducouret A., Bonville P., and Raveau B. (1994) Structural aspects and antiferromagnetic ordering in the 123-derivative $LnSr_2Cu_2Ga_{1-x}Fe_x$ (Ln=Y,Ho) *Journal of Solid State Chemistry* 113, 94-108.

PYRITE: LINKING MÖSSBAUER SPECTROSCOPY TO MINERAL MAGNETISM

E.A. FERROW[1], B.A. SJÖBERG[2] AND M. MANNERSTRAND[1]
[1]Department of Mineralogy & Petrology, Lund University, Sölvegatan 13, SE-223 62-Lund, Sweden
[2]Swedish Museum of Natural History, Box 50007, SE-104 05 Stockholm Sweden

Abstract

Gold-bearing pyrite ores are refractory and must be pre-treated to break down the sulphides by oxidation. This is done usually by roasting, bacterial oxidation and smelting. Moreover, ores with high sulphur content require pre-treatment to prevent excessive chlorine consumption, an important source of pollution. The pyrite waste created in mining operation presents serious problems on the environmental impact of acid mine drainage. In order to improve separation of pyrite from other metals as well as for the development of new strategies to inhibit oxidation of pyrite, it is necessary to understand the thermal, chemical, magnetic and biological implications during alteration of pyrite.

The oxidation of pyrite in air was studied using Mössbauer spectroscopy and mineral magnetic methods. A pyrite concretion from biogenic limestone shows a very weak natural remanent magnetization. Heating of the pyrite produced α-hematite as the end product. Intermediate mineral phases created during heating depend mainly on temperature, heating rate, grain size and the atmosphere in the oven. The most magnetic phases occur about 500 °C as determined from magnetic susceptibility and hysteresis measurements. The components of the compound hysteresis were iron sulphates and polymorphs of hematite as determined by Mössbauer spectroscopy. Heating of powder of pyrite produced higher concentration of pyrrhotite than grains of mm size as a result of more thorough oxidation as a result of increased ratio of iron to sulphur produced by degassing during heating.

Since pyrite and its oxidation products are all Fe-bearing phases, combining Mössbauer spectroscopy with rock magnetic methods provides information to monitor the oxidation of pyrite in air and identify the different phases produced and their relation to different experimental parameters.

M. Mashlan et al. (eds.), Material Research in Atomic Scale by Mössbauer Spectroscopy, 251–259.

1. Introduction

Pyrite (FeS_2) is the most common iron sulphide mineral. Aggregates of pyrite are found in magmatic, sedimentary and metamorphic rocks as well as in sediments, including pyritized fossils with the soft-parts preserved [1]. The precipitation of pyrite grains to a concretion in the Limhamn limestone takes place with the involvement of biological processes in the biogene rock [2].

The presence of pyrite and other sulphides in palaeomagnetic samples may, during heating, react in a variety of ways. The magnetic properties are, therefore, different at intermediate steps during heating and depending on local conditions and ageing, pyrite may indeed be an important source of later generations of magnetic minerals in rocks. The magnetic properties of pyrite and its oxidation products, including pyrrhotite and hematite, are unique and can be used in conjunction with Mössbauer spectroscopy to identify their presence. In this study, we intend to combine Mössbauer spectroscopy with rock magnetic methods to characterize the oxidation products, taking advantage of the thermo-chemical and magnetic phase transitions that occur. In addition to the pyrite powder and grain samples used by Ferrow et al. [2], single grains and cylinder shaped samples will be used.

2. Material

The pyrite concretion used in this study comes from a limestone quarry in Limhamn, Southern Sweden [2]. The pyrite concretion investigated was found during active quarrying and has not been affected by weathering. Ten samples with a diameter of 10 mm and a length of 8.5 mm were taken from the concretion with a water-chilled diamond tipped drill. Cylinder-shaped samples were used for measurements of remanent magnetization and magnetic susceptibility. Single grains with a weight of ca. 9 mg were used for hysteresis measurements. From one crushed sample, grains of a size of less than 1 mm were hand picked and the rest were ball-milled to powder for use in the grain and powder samples of the thermo-magnetic measurements.

3. Methods

The natural remanent magnetization (NRM) was measured with a 2-G Enterprises magnetometer model 755-R. A Newport Instrument type A magnet was used for magnetization of cylinder-shaped samples. They were put in an insert holder of Perspex in the JR 5 spinner magnetometer and the isothermal remanent magnetizations (IRM) were measured. The alternating-field demagnetization was carried out in a laboratory-built alternating current device. The coercivity of remanence (Hcr) was calculated from the values measured in reversed fields. An Schönstedt Thermal Specimen Demagnetizer (TSD-1) was used for most thermal experiments [2]. The magnetic susceptibility changes with temperature of grains with a size of less than 1 mm and powder samples were registered on a CS-2 thermal device interconnected with a KLY-2 Kappabridge [3].

An alternating-force magnetometer, MicroMag 2900, was used to determine the hysteresis parameters on single grains. Saturation magnetization (Mrs), saturation remanence (Mr) and coercive force (Hc) were measured.

Thermo-chemical changes and magnetic phase transitions were studied on samples heated in air in the TSD-1 oven. Single grains and a cylinder-shaped sample were heated at pre-selected temperatures up to 700 °C for a certain heating time. The samples were measured when they had cooled to room temperature.

In the furnace of the CS-2 device, the grain sample and the powder sample were heated at a rate of about 9 °C/min. The magnetic susceptibility was registered during the temperature changes with the KLY-2 Kappabridge. The measurements were carried out to a maximum temperature of 700 °C and subsequent cooling to room temperature.

Mössbauer spectra were recorded at room temperature on a constant-acceleration Mössbauer spectrometer with a ^{57}Co/Rh source. The Mössbauer set up are those described by Ferrow et al. [4]. Velocities were calibrated using Fe foil with a thickness of 25 µm. Thin absorber tablets were prepared by mixing the sample with a transoptic material and by pressing the mixture at 80 °C. The spectra were computer-fitted using Recoil, a commercially available Mössbauer spectral analysis software package.

4. Results and Discussion

4.1. REMANENT MAGNETIZATION

The NRM of the ten cylinder samples falls in the range of $(0.06 - 0.60) \times 10^{-3}$ Am^{-1}. The arithmetic mean value was 0.18×10^{-3} Am^{-1}.

The IRM of the progressive step-wise acquisition and subsequent alternating field demagnetization was measured after each step. First the pyrite sample was measured unheated, then it was heated to 440 °C, cooled to room temperature and the measurement sequences were repeated. The saturation remanent magnetization of the unheated sample is stable with time, less than one percent of the saturation remanent magnetization was lost in 1 hr and 20 min. When heated the sample lost 5 per cent of the remanent magnetization during the same time. The sample was stored in the magnetometer and shielded from the external field. The acquisition curves of the IRM reach saturation magnetization at about 0.15 T for the unheated sample and at about 0.25 T for heated the sample (Figs. 1a & 1b, respectively). The forms of the acquisition and demagnetization curves suggest that the remanence originate from the same magnetic phase. The demagnetization field required to reduce saturation remanent magnetization by half was about 25 mT. The stability against alternating demagnetization fields is similar. However, the arithmetic mean saturation remanence value, Mrs is 27.1×10^{-3} Am^{-1} for the unheated and 55 Am^{-1} for the heated samples, showing the high concentration of the magnetic phase in the sample heated at 440 °C.

The normalized direct current demagnetization of the unheated and heated samples shows a coercivity of remanence of $-$ 42 mT for both cases. The observation confirms the presence of a phase with very similar magnetic properties in the sample unheated and heated.

The corresponding room temperature Mössbauer spectra are also plotted in fig. 1. The figure shows a dominant doublet with Mössbauer parameters of low spin Fe^{2+} in octahedral coordination, assigned to iron in FeS_2 and an additional doublet of high spin Fe^{2+} in octahedral coordination, assigned to iron sulphate anhydrite ($FeSO_4$). No magnetic sextet is visible in both cases, indicating that either the concentration of the magnetic phases was below the detection limit of Mössbauer spectroscopy, or the magnetic signals originate from spin transition.

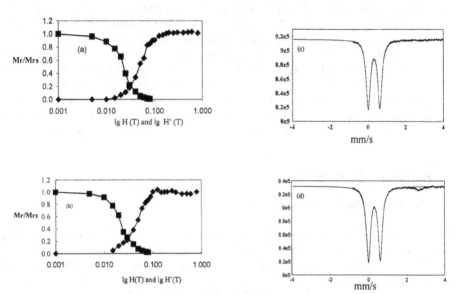

Figure 1. Progressive step-wise acquisition (diamonds) and subsequent alternating field demagnetization (squares) of the IRM for unheated (a) and after heating to 440 °C (b) and their Mössbauer spectra (c, d). The magnetic field (H) and the AF demagnetisation field (H') are shown on a logarithmic scale.

4.2. MAGNETIC SUSCEPTIBILITY

Thermo-magnetic curves of grain and powder samples from the pyrite concretion were heated in the furnace of the CS-2 device, interconnected with a Kappabridge II. The results are plotted in Fig. 2, together with the corresponding Mössbauer spectra. The curves consist of a heating phase from room temperature to 700 $^{\Box}$C and a cooling phase back to room temperature. The Mössbauer spectra were taken at the end of the cycle.

In fig. 2a, the magnetic susceptibility of the grain sample shows a small and gradual increase with temperature between 200 and 400 °C during the heating phase. A noticeable increase starts at 400 °C and terminates a little over 565 °C. Ferrow et al. [2] have shown that pyrite grains heated in these temperature ranges oxidize to α-Fe_2O_3, ε-Fe_2O_3, β-Fe_2O_3, $Fe_2(SO_4)_3$ and $FeSO_4$. It is possible that the susceptibility changes observed at 547, 565, 577 and 605 °C could be correlated to the appearance and disappearance of these phases. However, we have no data on the magnetic properties of these phases other than that of α-Fe_2O_3 to pinpoint every magnetic transition registered in fig. 2. During the cooling phase an increase in the magnetic susceptibility indicate a number of transitions to ferromagnetic states. The prominent one corresponds to monoclinic pyrrhotite with a Curie temperature of 320 °C.

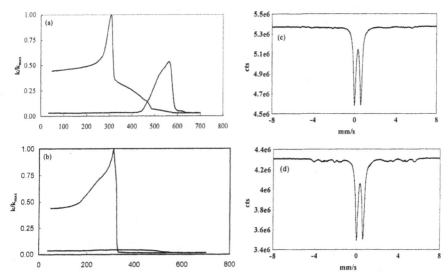

Figure 2. Thermo-magnetic curves of the grain (a) and the powder (b) samples of the pyrite concretion and their corresponding Mössbauer spectra (c and d, respectively). The ratio is the magnetic susceptibility (κ) to the maximum susceptibility value, κ_{max}.

The heating and cooling feature of the powder sample differs from that of the grain (2b). During the heating phase, the susceptibility increase slightly between 195 and 260 °C, stabilizes between 260 and 390 °C and finally falls between 390 and 585 °C. During the cooling phase, the low susceptibility values registered at the highest temperatures are approximately kept low until the creation of a new magnetic phase occur between 330 and 311 °C assigned to pyrrhotite. The decrease of the susceptibility is rectilinear within certain temperature ranges and differs from the smooth susceptibility changes observed in the grain sample. The asymmetry is assigned to the presence of both the monoclinic and hexagonal pyrrhotite polymorphs [5]. Moreover, the intensity of the magnetic phases in the powder sample is higher than that in the grains sample. The Mössbauer spectra show a paramagnetic doublet and magnetic sextets assigned to pyrite and pyrrhotite, respectively (Figs. 2c and 2d). At least two sextets are identified in the pyrrhotite spectrum of the powder sample, supporting the data from the magnetic susceptibility study. Moreover, the intensity of the pyrrhotite lines is more pronounced in the powder sample.

4.3. MAGNETIC HYSTERESIS

The hysteresis parameters of both the time and temperature series were measured on single grains.

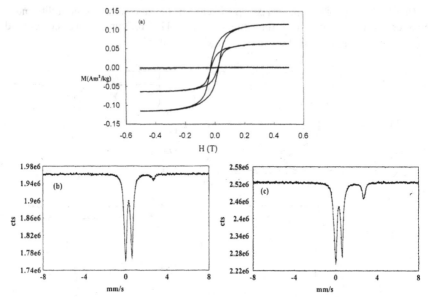

Figure 3. Hysteresis curves of samples heated at 370 °C showing well-developed loops for samples heated for 18 (lower loop) and 45 hours (upper loop) (a). Note the increase of Ms with prolonged heating times. The corresponding Mössbauer spectra show an increase in the concentration of FeSO₄ (b, c).

In the time series, the temperature was kept constant at 370 °C while the run times varied from 1 to 45 hours (Fig. 3a). Samples heated less than 18 hours indicate a heterogeneous material with respect to oxidation. Generally, the amount of ferromagnetic material increases with increasing heating times. The hysteresis loops are well developed for the samples heated for 18 and 45 hours, reflecting the increase in the concentration of the new phase formed during prolonged heating. The coercive force, Hc, has increased from an indeterminate value at 1 hour to 26 mT at 45 hours.

Representative Mössbauer spectra for run durations of 18 and 45 hours are plotted in Fig. 3b & 3c. For samples heated for less than 18 hours, only a single pyrite doublet could be observed. But with prolonged heating the figure shows that pyrite was oxidized to iron sulfate anhydride. No magnetic sextets could, however, be recognized, suggesting that the magnetic signals in fig. 3 could be due to the oxidation of pyrite to iron sulfate anhydride.

In the temperature series, the samples were heated in the oven for one hour to temperatures ranging between 200 and 685 °C. The results of the 500 and 560 °C runs and their corresponding Mössbauer spectra are shown in fig. 4. After heating to 500 °C, the value of the saturation magnetization indicates the creation of a strongly magnetic phase (Fig. 4a). The coercive force, Hc, at this temperature has a value of 19 mT. Further heating of pyrite to 560 °C has resulted in the formation of a phase with higher coercive force but weaker magnetization (Fig. 4b). The Ms is less than 1 per cent of the value measured for the sample heated to 500 °C while the value of Hc is twice as much, indicating a more or less thorough alteration of the original mineral (FeS_2) to the high coercivity phase (α-Fe_2O_3).

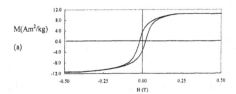

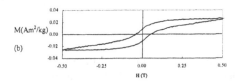

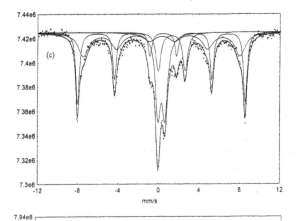

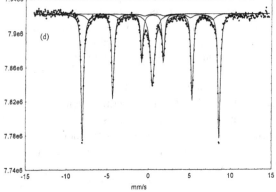

Figure 4. Hysteresis curves of pyrite samples heated for one hour at 500 (a) and 560 °C (b) and the corresponding Mössbauer spectra (c, d). The maximum Ms was obtained at 500 °C. At 560 °C, the Hc has increased and Ms decreased. Mössbauer spectra show an increase in α-Fe$_2$O$_3$ as the temperature increases.

The Mössbauer spectrum for the sample heated at 500 °C shows the presence of FeS_2, $FeSO_4$, α-Fe_2O_3 and ε-Fe_2O_3 as the dominant phases. The corresponding hysteresis loop can therefore be assigned to the presence of the two Fe_2O_3 polymorphs. The Mössbauer spectrum for the 560 °C run shows clearly that the dominant phase present is α-Fe_2O_3, collaborating the hysteresis diagram in fig. 4b. Mössbauer spectra show only α-Fe_2O_3 above 650 °C, as the paramagnetic β-Fe_2O_3 phase at this stage also has been altered to α-Fe_2O_3.

5. Summary

Pyrite concretions in the Limhamn limestone were occasionally found during quarrying, probably formed by mineralization of iron sulphides in local pockets in the biogenic limestone. The Mössbauer data confirm pyrite as the single Fe-bearing mineral of the concretion. Pyrite has a low-spin form of Fe^{2+} and the mineral is diamagnetic.

The NRM is very weak. The IRM characteristics confirm the presence of a remanence carrying magnetic mineral. Heated to 440 °C, the cylinder sample shows normalized IRM characteristics close to the values of the unheated sample. The saturation magnetization is much higher and the metallic lustre of the sample surface has altered to a reddish colour after heating.

Heated continuously from room temperature, the magnetic susceptibility increases in the grain sample but decreases in the powder sample at about 400 °C. The intermediate magnetic phases between pyrite and hematite are obvious with the grain sample. The Mössbauer data shows that a pyrite doublet dominates both the grain and powder samples with a small contribution from pyrrhotite sextets. In the powder sample two sextets could be observed, suggesting the presence of the monoclinic and hexagonal phases.

Mössbauer spectra of the single grain samples revealed an alteration of low spin mineral (FeS_2) to a high spin mineral ($FeSO_4$) after 9 h heating at 370 °C. The amount of $FeSO_4$ increases with prolonged heating time. With the particle sizes used, a heating time of 18 hours and more results in a well-developed hysteresis loop. Saturation magnetization continues to increase, showing increasing amounts of $FeSO_4$ during extended heating.

Increasing heating temperatures of the single grain samples result in a maximum saturation magnetization at 500 °C. The thermo-magnetic alteration products are, according to the Mössbauer spectra, $FeSO_4$, $Fe_2(SO_4)_3$, α-Fe_2O_3 and ε-Fe_2O_3. As the temperature is raised to 560 °C, the α-Fe_2O_3 becomes dominant while the other phases decrease. At 500 °C, an Hc of 19 mT indicate small amounts of α-Fe_2O_3 but at 560 °C, the Hc is more than twice as large, indicating a significant increase of α-Fe_2O_3. The Ms and Mrs have decreased to less than 1 per cent of the value measured of the sample heated to 500 °C.

Acknowledgments

The paper was supported by fund provided by the Swedish Research Council (VR) to EAF. The authors thank chemical engineer Åke Truedsson, Scancem Ltd. for providing the pyrite concretion.

References

1. Briggs, D.E.G., Raiswell, R., Botrell, S.H., Hatfield, D. and Bartels, C. (1995) Controls of the pyritization of the exceptionally preserved fossils: an analysis of the Lower Devonian Hunsrück slate of Germany. *Am. J. Sci.* **295**, 282-308.
2. Ferrow, E., Sjöberg B. and Mannerstrand, M. (2002) Reaction kinetics and oxidation mechanisms of pyrite heated at high temperatures in air: a Mössbauer spectroscopy study. *Eur. J. Mineral.* Accepted.
3. Orlický, O. (1994) Study and detection of magnetic minerals by means of the measurement of their low-field susceptibility changes induced by temperature, *Geologica Carpathica*, **45**, 113-119.
4. Ferrow, E.A., Kalinowski, B.E., Veblen, D.R. and Schweda, P. (1999) Alteration products of experimentally weathered biotite studied by high-resolution TEM and Mössbauer spectroscopy. *Eur. J. Mineral.*, **11**, 999-1010.
5. O'Reilly, W. (1984) *Rock and Mineral Magnetism*, Blackie, Glasgow and London, 220.

SYNTHESIS OF IRON SULFIDES: A MÖSSBAUER STUDY

N.I. CHISTYAKOVA[1], V.S. RUSAKOV[1],
S.V. KOZERENKO[2] AND V.V.FADEEV[2]
[1]Lomonosov Moscow State University, Leninskie gory, 119992 Moscow,
Russia
[2]Vernadsky Institute of Geochemistry and Analytical Chemistry Russian
Academy of Sciences,19, Kosygin St.,117975 Moscow, Russia

The kinetics of iron sulfide crystallization from X-ray amorphous water bearing precipitates (hydrotroilite) was investigated by ^{57}Fe Mössbauer spectroscopy. Synthesis was carried out in the temperature range $20^0C<T<200^0C$. The investigated samples were exposed to aging from several days to several years. It was shown that the reaction proceeded in different ways depending on redox conditions. Hydrotroilite was converted to pyrite in the presence of elemental sulfur. In more reducing conditions without elemental sulfur, however, a sequence of metastable sulfide phases including hydrotroilite, mackinawite, greigite, pyrite and marcasite were formed. In aqueous solutions unsaturated by sulfide sulfur a precursor phase - tochilinite was synthesized. In this study Mössbauer investigation of mackinawite and tochilinite as intermediate phases in the process of hydrotroilite aging are carried out.

1. Introduction

Interest in sulfide minerals is connected with their wide abundance in nature. Sulfide forming reactions are considered an important factor in the global carbon, sulfur and iron cycles in the Earth crust [1]. Iron sulfide sedimentation is considered a factor in H_2S binding, which has governed sulfate and oxygen content in World Ocean waters throughout terrestrial history [2].

In this paper, the processes of iron sulfide crystallization (pyrite FeS_2, mackinawite FeS_{1-x}, tochilinite $2FeS \cdot 1.67Fe(OH)_2$) from amorphous water-bearing precipitates (hydrotroilite) were investigated using ^{57}Fe Mössbauer spectroscopy. This method permits tracking of sulfide formation reactions and provides data on structure transformation.

The distinctive particularity of the formation of iron sulfide in temperatures above 20^0C and below 300^0C is the difficulty to crystallize the thermodynamically stable phases (the pyrite, the pyrrhotite) due to high nucleation barriers. Under these conditions and instead of stable phases, their metastable analogues are formed. In the initial stage hydroxide-sulfides are formed as precursor phases, which gradually transform to sulfides. In the solution unsaturated by sulfur ion species tochilinite is

261

M. Mashlan et al. (eds.), Material Research in Atomic Scale by Mössbauer Spectroscopy, 261–270.

formed as an intermediate phase. Tochilinite is known as a mixed layered hydroxide-sulfide $2FeS \cdot 1.67(Mg,Fe)(OH)_2$, consisting of alternating mackinawite-like tetrahedral FeS nets and $(Mg,Fe)(OH)_2$ brucite-like octahedral layers.

2. Experimental Details

Synthesis was carried out within the temperature interval $20^0C < T < 200^0C$ at pH 6.3-7.0 and the concentration of sulfur ion species to near saturation under two sets of conditions - in the presence and in the absence of excess of elemental sulfur in the mineral forming media. In this temperature range as our investigations have shown the reaction velocity increases with the increase of synthesis temperature and the sequence of forming sulfides remains the same. Therefore, the data obtained for room temperature synthesis (when processes of phase formation are proceeding slowly) are presented in this paper for an indepth study of iron sulfide crystallization processes. Samples aged from several days to several years. The sulfide precipitates studied were initially amorphous as shown by X-ray diffraction.

Tochilinite in the compositional range of pure iron-bearing up to magnesial of Fe:Mg $\approx 5:1$ was synthesized in aqueous solutions unsaturated by sulfide sulfur. Experimental procedures for tochilinites synthesis have been presented [3,4].

Mössbauer investigations were carried out at room temperature on the spectrometer operated in constant acceleration mode and equipped with [57]Co-source in Rh-matrix. Both superposition of parameter distribution functions and curve fitting were used for the analysis of Mössbauer spectra.

3. Results and Discussion

3.1. KINETICS OF IRON SULFIDE CRYSTALLIZATION WITH ELEMENTAL SULFUR IN THE SOLUTION

Mössbauer spectra of both the starting substance and the end product showed asymmetric quadrupole doublets; the initial spectrum had the distinctive broadening characteristics of amorphous materials [5]. The distribution of Mössbauer line shift δ and quadrupole shift ε have been calculated assuming a linear correlation between δ and ε. It is distinctive that the correlation degree of these parameters $d\delta/d\varepsilon$ is about 6×10^{-2} for all spectra. We determined Mössbauer line and quadrupole shift distribution functions $p(\delta)$ and $p(\varepsilon)$ for the initial and the aged products (*Fig.1*), the changes in the average values of the shift $\bar{\delta}$ and $\bar{\varepsilon}$ with aging are shown in *Fig.2*. Essential changes in the features in the initial stage is not observed. The analysis has shown substantial changes in these parameters at aging times of about 1200 hours. These results are also confirmed by analyzing the distribution $p(\varepsilon)$ characteristics (*Fig.2*): full width at half maximum of the distribution $(\Gamma_{p(\varepsilon)})$; excess coefficient (β). Clear correlation between temporary dependencies of different Mössbauer characteristics and the ratio of sulfur ion species and iron atom numbers n_S/n_{Fe}, obtained by chemical phase analysis [6], is observed (*Fig.2*). As to hydrogen content in sulfide precipitate ΣH, in the initial stage the intensive content decrease of electro neutral particles of adsorbed water is fixed. Observable changes do not occur up to three months of syntheses, the contents of OH⁻ and HS⁻-groups are intensively reduced later (*Fig.2*). This means that spectrum transformation is stipulated by changes in the anion sublattice (the substitution of OH⁻ and HS⁻-groups by S^{2-} ions).

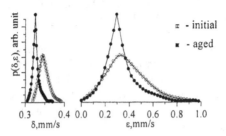

Figure 1. Mössbauer line and quadrupole shift distribution functions $p(\delta)$ and $p(\varepsilon)$ for the initial and the aged (7200 hours) in an excess of elemental sulfur products at room temperature.

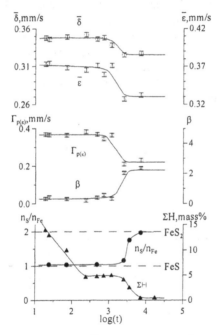

Figure 2. Average values of Mössbauer line shift $\bar{\delta}$ and quadrupole shift $\bar{\varepsilon}$; full width at half maximum $\Gamma_{p(\varepsilon)}$ and excess coefficient β for the distribution $p(\varepsilon)$; a ratio n_S/n_{Fe} of sulfur ion species and iron atom number and total content of hydrogen ΣH in absorbed water, OH⁻- and HS⁻-groups in the synthesis products as dependence on synthesis time.

No intermediate phases were found and pyrite (FeS_2) was the final product of aging (7200 hours). Mössbauer investigations have shown that Fe^{2+} ions in hydrotroilite are in the low spin state and presumably exist in pyrite-like nearest surroundings [5,6]. Our investigations demonstrate that the main factors in pyrite crystallization are sediment aging conditions, temperature and time interval. We ascertained that the reaction depends on redox conditions. When there was an excess of elemental sulfur in the mineral forming media the following scheme was realized: hydrotroilite is altered to pyrite.

3.2. KINETICS OF IRON SULFIDE CRYSTALLIZATION WITHOUT ELEMENTAL SULFUR IN THE SOLUTION

In the absence of an excess of elemental sulfur in the mineral forming media several metastable sulfides formed in succession: hydrotroilite and mackinawite, greigite, pyrite and marcasite (*Fig.3*). Quadrupole shift $p(\varepsilon)$ and hyperfine field $p(H)$ distribution functions for samples from this series are shown on *Fig.4*. According to X-ray phase analysis, the first sample from this series (10 days of aging) was mackinawite. The analysis of distribution $p(\varepsilon)$ for this sample showed that it was a mixture of hydrotroilite and mackinawite. Our investigations of mackinawites synthesized at different temperature range revealed that their Mössbauer spectra were different (see 3.3.1). The spectrum of mackinawite, synthesized at room temperature, was similar to hydrotroilite. We conclude that the first sample was probably a mixture of two different mackinawites and hydrotroilite. On $p(\varepsilon)$ distribution function for the 35-day sample, a peak corresponding to small particles of greigite [7], was observed The total intensity of greigite (magneto-ordering and superparamagnetic parts) was 45.5±1.2%. Probably a small amount of these superparamagnetic particles were present in the 10-day sample too. For the 120-day sample the increase in greigite content up to 60.3±1.2% was

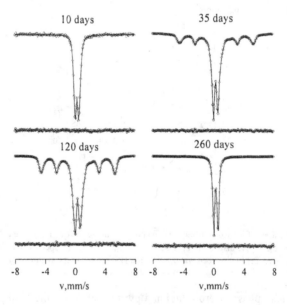

Figure 3. Mössbauer spectra of samples aged in the absence of an excess of elemental sulfur at room temperature.

observed. Herewith the intensity of the spectrum of small greigite particles increased. At 260 days, greigite and mackinawite practically disappeared. X-ray phase analysis showed that the sample was a mixture of pyrite and marcasite, minerals with similar Mössbauer parameters. Average values of $\bar{\delta}$ and $\bar{\epsilon}$, full width at half maximum of the distribution and positive correlation $\Delta\delta/\Delta\epsilon = 0.036\pm0.013$ showed that pyrite and marcasite mixture to be the end product of the synthesis. The processes of phase formation in this system could be divided into three stages, the decrease of hydrotroilite and mackinawite contents with greigite formation, further formation of greigite and pyrite and marcasite appearance from greigite, which gradually disappears.

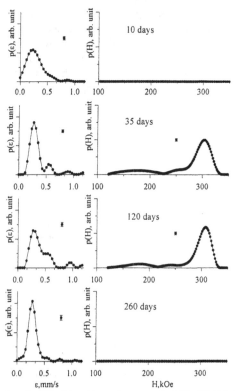

Figure 4. Mössbauer quadrupole shift p(ε) and hyperfine magnetic field p(H) distribution functions for the samples aged in the absence of an excess of elemental sulfur at room temperature.

3.3. MACKINAWITE AND TOCHILINITE INVESTIGATIONS

The interest of researchers in mackinawite and tochilinite stems from the widespread natural occurrence of these minerals (in sea deposits and ores) and in the Universe (carbonaceous chondrites and cosmic dust) [8-10]. Mössbauer investigation of mackinawite and tochilinite are also very interesting to the point where they are an intermediate phase in the process of hydrotroilite aging. Tochilinite is a hydroxide-sulfide where the FeS tetrahedral network layers of mackinawite-like composition alternate with $(Fe,Mg)(OH)_2$ brucite-like octahedral layers [8, 10, 11]. Tochilinite has only recently been synthesized in laboratory conditions [3], and Mössbauer spectra were recorded for natural magnesial tochilinites only. The Mössbauer data on synthetic mackinawite and natural tochilinite are different, and the corresponding data on synthetic tochilinite are altogether absent.

3.3.1. *Mackinawite*

Mackinawite was synthesized in an alkaline medium (pH 13.0) at 160-200°C. As in [7,12], the spectra of the synthetic mackinawites studied in this work contain a single narrow resonance line (*Fig. 5*). Interpreting the spectra in terms of various models shows, however, that they are quadrupole doublets with small splitting values rather than singlets. In some spectra, the Zeeman sextets due to magnetite impurities are observed in addition to the doublets. In the mackinawite unit cell containing two formula units (sp. gr. P4/nmn, $a_m = b_m = 3.679$ Å, $c_m = 5.047$ Å [1]), iron atoms are known to be situated in one crystallographic site and have a tetrahedral environment of sulfur atoms. This explains the presence of a single partial quadrupole doublet in the spectrum.

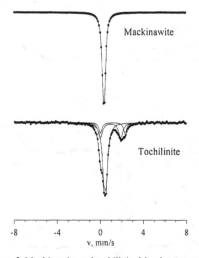

Figure 5. Mackinawite and tochilinite Mössbauer spectra.

In the spectra of various mackinawite samples, the Mössbauer line shifts δ vary from 0.37 to 0.39 mm/s, which is closest to the values reported by Bertaut *et al.* [12], and the quadrupole shifts ε (2ε≡Δ) of the spectrum components vary from 0.05 to 0.07 mm/s. These parameter variations are likely to be caused by different nonstoichiometry degrees of the obtained samples. The values δ and ε are in the regions of hyperfine interaction parameter values characteristic to the Mössbauer spectra of high-spin Fe^{3+} or low-spin Fe^{2+} - ions [13]. The neutron diffraction data evidence shows that the iron ions in mackinawite do not have magnetic moments [12], which leads us to conclude that these ions occur in the divalent low-spin state, which agrees with the results obtained by Vaughan et al. [7].

Studying the iron sulfide synthesis shows that Mössbauer spectra of mackinawites synthesized at room temperatures and at 160-200°C are different. The spectra of mackinawites, synthesized at room temperature, were similar to hydrotroilite spectrum (Fig.3). Maximum values of Mössbauer line shift δ_{max} and quadrupole shift ε_{max}, obtained from p(ε) distribution function (Fig.4), are equal to 0.324±0.005 mm/s and 0.229±0.027 mm/s. X-ray phase analysis showed that the investigated sample was mackinawite. The analysis of distribution p(ε) for this sample showed that the content of high temperature mackinawite (the spectrum with narrow quadrupole doublet) was only 5.8±0.3%. It is possible to expect that the sample was a mixture of small amount of high temperature mackinawite, hydrotroilite and low temperature mackinawite, in which structure OH⁻-groups are presented. The presence of OH⁻-groups in the structure of this mackinawite results in broadening of the quadrupole doublet.

3.3.2. Tochilinite

The synthesis of tochilinite was conducted in alkaline media (pH 12.0-13.0) at temperatures of 80-140°C and fixed concentration of sulfur ion species [3]. The relative amount of Mg atoms in the initial mixture n_{Mg} varied from 0 to 25% (n_{Mg} is the percent ratio of the number of Mg atoms to the total number of cations). A characteristic Mössbauer spectrum of synthetic iron-tochilinite without the subspectra of magnetite (magnetite is present in some samples as a concomitant phase) is shown in Fig. 5. In tochilinite, the mackinawite and brucite-like layers alternate. It is therefore natural to expect that its spectrum should be a superposition of the subspectra of these two layers. The spectrum of tochilinite (Fig. 5) is a superposition of three partial quadrupole doublets. The hyperfine parameters of one doublet (δ_1 and ε_1) are close to the values of the mackinawite spectrum and correspond to Fe^{2+} ions in the low-spin state. The presence of this doublet is due to the presence of the sulfide layer in tochilinite. The appearance of the two other quadrupole doublets is likely to be caused by the presence of iron in brucite layers; the hyperfine parameters of these partial spectra (δ_2, ε_2 and δ_3, ε_3) correspond to Fe^{2+} ions in the high-spin state [13]. The doublet hyperfine parameters of the sulfide layer vary in the ranges 0.43-0.46 mm/s for the shift δ_1 and 0.08-0.11 mm/s for the quadrupole shift ε_1. These data are in agreement with the results obtained for natural magnesial tochilinites [14].

In the subspectra of the brucite-like layer the hyperfine parameters are δ_2 = 1.13-1.18 mm/s, ε_2 = 1.15-1.29 mm/s, δ_3 = 0.88-0.97 mm/s, and ε_3 = 0.87-0.95 mm/s. In synthesized tochilinites, the brucite-like layers were composed of (Fe, Mg)(OH)$_2$. In brucite, Mg(OH)$_2$, and ferrous hydroxide, Fe(OH)$_2$, divalent Fe^{2+} cations are situated in equivalent crystallographic sites and have an octahedral environment of oxygen atoms. It is, therefore, reasonable to expect that ^{57}Fe nuclei present in brucite-like layers should give a single subspectrum. For instance, the Mössbauer spectrum of ferrous hydroxide, Fe(OH)$_2$, studied by Miymoto [15] at room temperature contains a single quadrupole doublet with δ = 1.05 ± 0.02 mm/s and ε = 1.49 ± 0.01 mm/s. Note that the mean values of δ_2 and δ_3 are close to the δ value obtained by Miymoto [15]. It is, therefore likely that conjugation of brucite and mackinawite layers results in the distribution of Fe^{2+} ions over substantially different positions resulting in the appearance of two distinct partial quadrupole doublets (*Fig. 5*). The appearance of nonequivalent iron sites in the brucite-like layer corresponds to the observation by Organova [11] that, in the brucite-like layer of magnesial tochilinite, the Fe and Mg atoms tend to be distributed over different brucite sites. Nonequivalent iron sites also appear in the mackinawite layer because of its conjugation with brucite. This, however, only causes broadening (≥ 0.1 mm/s) of the subspectrum lines.

A comparison of the hyperfine interaction parameters of the Mössbauer spectra of mackinawites and the corresponding subspectra of tochilinites shows that iron ions in the sulfide tochilinite layer are characterized by the Mössbauer line and the quadrupole shift values of 0.06-0.07 and 0.02-0.06 mm/s, larger than those observed for mackinawite. This occurs, probably, due to the tochilinite structure formed in the conjugation of layers via compression of the brucite and expansion of the mackinawite layer. According to Organova [11], the structure of tochilinite is described by a common crystal lattice for its sulfide and brucite parts. The brucite and mackinawite layers are oriented so that the tochilinite unit cell parameters are related to those of the brucite-like and mackinawite layers as $a_t \cong \sqrt{3}\, a_{br} \cong \sqrt{2}\, a_m$ and $b_t \cong 5a_{br} \cong 3\sqrt{2}\, a_m$. These relations and the unit cell parameters of mackinawite [1] and ferrous hydroxide can be used to estimate the relative change in the mackinawite unit cell volume, $\Delta V_m/V_m$, caused by the sulfide and brucite layers conjugation. The calculation value is 0.064 at c_m constant. The S^{2-} and O^{2-} ions have identical outer electronic shells. This makes it possible to estimate the change in $\Delta\delta$ at the Mössbauer line shift of ^{57}Fe nuclei in the sulfide layer of tochilinite as compared to mackinawite, based on the calculations of the dependence of chemical shift on the distance between oxygen in tetrahedrally coordinated iron ions [16]. The obtained value $\Delta\delta$ is 0.053 mm/s and agrees well with the experimental data. We arrive at the conclusion that the increase in the Mössbauer line shift value δ for Fe^{2+} ions in the sulfide layer of tochilinite against that in mackinawite is mainly caused by expansion of the base of the mackinawite unit cell in the conjugation of mackinawite and brucite-like layers. The increase in quadrupole shift ε appears to be caused by substantial distortion of the nearest tetrahedral environment of Fe atoms as a result of the tochilinite structure formation. The hyperfine parameters of the spectrum of ^{57}Fe nuclei in tochilinite may also differ from those of the spectrum of mackinawite because the sulfide layer is deficient in iron [8, 11]. This deficiency somewhat changes the Fe-S distance and distorts the tetrahedra around the iron atoms, which contributes to the observed changes in the shift δ and the quadrupole shift ε.

Tochilinites were synthesized from mixtures with different magnesium contents n_{Mg}. The total relative intensity of the subspectra corresponding to Fe^{2+} ions in brucite layers, $I_{\Sigma2,3}$ decreases as n_{Mg} increases. This trend is well illustrated in *Fig. 6*, where the relative intensities of both quadrupole doublets, I_2 and I_3, and their total relative intensity $I_{\Sigma2,3}$ are plotted as functions of n_{Mg}. The figure shows that the total relative intensity decreases with increasing n_{Mg} mainly due to the change in intensity I_3 of one of the doublets.

Suppose that in tochilinite, Mg^{2+} ions are presented in brucite-like layers only. Then, at the given $n_{br} : n_m$ ratio between the numbers n_{br} of conjugate brucite and n_m of mackinawite layers, we can calculate the ratio between the number of iron ions in brucite layers and the total number of cations in tochilinite depending on the relative content of Mg atoms, n_{Mg} (dashed lines in *Fig. 6*). It is seen that the experimental dependences of total relative intensity $I_{\Sigma2,3}$ and relative intensity I_3 on n_{Mg} have virtually the same slopes as the calculated dependences. It follows that the Mg atoms occur in brucite layers of tochilinite only and preferably occupy positions characterized by smaller values (δ_3 and ε_3) of ^{57}Fe subspectrum hyperfine parameters. Conversely, Fe atoms in the brucite layer preferably occupy the sites with larger Mössbauer spectrum hyperfine parameters, which are closest to the δ and ε parameters of the spectrum of ^{57}Fe nuclei in ferrous hydroxide, $Fe(OH)_2$ [15]. Note that at $n_{Mg} = 0$ and Fe atoms occupying all possible sites in the brucite layers of tochilinite, the $I_3/(I_2+I_3)$ ratio equals 0.58 ± 0.05. This agrees with the results by Organova [11], where it is established, that 0.60 of brucite layer sites is preferably occupied by the Mg atoms in magnesium tochilinite.

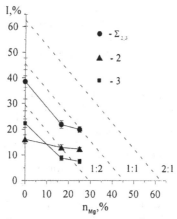

Figure 6. Relative intensities of two subspectra corresponding to Fe^{2+} ions in brucite-like layers and their total relative intensity as functions of n_{Mg}. Dashed lines are obtained by calculating the ratios of the number of iron ions in brucite layers to the total number of cations in tochilinite at various $n_{br}:n_m$ ratios between the numbers of conjugate brucite and mackinawite layers.

It is also seen from *Fig. 6* that the experimental $I_{\Sigma2,3}$ values are closest to those calculated for $n_{br}:n_m = 1:1$. It follows that, in the synthesized tochilinites, conjugation involves equal numbers of mackinawite and brucite layers. This is in agreement with the electron diffraction data on tochilinite synthesized under similar conditions [3]. Some deviation of the experimental values from the calculation results can be explained by a higher Mössbauer effect probability f for the ^{57}Fe nuclei in mackinawite against that in brucite. Indeed, the presence of strong covalent interactions in mackinawite [11] as distinguished from ferrous hydroxide, can be attributed to the fact that the bonds between the Fe and S atoms are more rigid than those between the Fe and O atoms and, as a consequence, the probability f is higher (e.g., see [16]).

4. Conclusion

The alteration of hydrotroilite to pyrite was realized when elemental sulfur was present in the solution. In more reducing conditions without elemental sulfur, a sequence of metastable sulfides formed, including hydrotroilite, mackinawite, greigite, pyrite and marcasite.

The Mössbauer study of synthetic tochilinite performed for the first time has demonstrated that the spectrum of ^{57}Fe nuclei represents a superposition of three quadrupole doublets, one of which corresponds to the Fe^{2+} ion sites in the sulfide layer, and the other two to nonequivalent sites occupied by the high-spin Fe^{2+} ions in the brucite layer.

An increase of the Mössbauer line shift for the subspectrum of sulfur layer of tochilinite against that for mackinawite has been shown to result from the mackinawite unit cell expansion caused by its conjugation with the brucite unit cell.

Magnesium atoms in tochilinite have been found to occur only in the brucite layer, largely in one of the two observed nonequivalent sites.

The suggestion that, in the structure of tochilinite, conjugation involves equal numbers of mackinawite and brucite layers has been substantiated.

This work was supported by RFBR, Grant 00-05-65372.

References

1. Berner, R.A. (1962) A tetragonal iron sulfide, *Science* **137**, 1669.
2. Holland, H.D. (1978) *The Chemistry of the Atmosphere and Ocean,* Wiley and Sons, New York.
3. Kozerenko, S.V., Organova, N.I., Fadeev, V.V., Magazina, L.O., Kolpakova, N.N. and Kopneva, L.A. (1996) Tochilinite produced in laboratory, *Lunar and Planetary Scince XXVII* **2**, 695-696.
4. Moroz, L.V., Kozerenko, S.V. and Fadeev, V.V. (1997) The reflectance spectrum of synthetic tochilinite, *Lunar and Planetary Scince XXVIII*, **2**, 983-984.
5. Rusakov, V.S., Khramov, D.A., Chistyakova, N.I., Kozerenko, S.V. and Fadeev, V.V. (1994) Mössbauer study of the Crystallization process of amorphous water-containing iron sulfide, *Phys. Stat. Sol. (a)* **144**, K45-K48.
6. Kozerenko, S.V., Khramov, D.A., Fadeev, V.V., Kalinichenko, A.M., Marov, I.N., Evtikova, G.A. and Rusakov, V.S. (1996) Pyrite formation mechanisms in aqueous solutions at low T and P, *Geochemistry International* **33**, 16-31.
7. Vaughan, D.J. and Ridout, M.S. (1971) Mössbauer studies of some sulphide minerals, *J. Inorg. Nucl. Chem.* **33**, 741-746.
8. Organova, N.I., Drits, V.A. and Dmitrik, A.L. (1974) Selected area electron diffraction study of a type II "Valleriite-like" mineral, *Am. Mineral.* **59**, 190-200.
9. Vaughan, D.J. and Craig, J. (1978) *Mineral Chemistry of Metal Sulfides,* Cambridge Univ., Cambridge.
10. Mackinnon, I.D.R. and Zolensky, M.E. (1984) Proposed structures for poorly characterized phases in C2M carbonaceous chondrite meteorites, *Nature* **309**, 240-242.
11. Organova, N.I. (1989) *Crystal Chemistry of Incommensurable and Modulated Mixed-Layer Minerals (in Russian),* Nauka, Moscow.
12. Bertaut, E.F., Burlet, P. and Chappert, J., (1965) Sur l'absence d'ordre magnetique dans la forme quadratique de FeS, *Solid State Commun.* **3**, 335-338.
13. Reiff, W.M. (1973) Mixed oxidation states and averages electronic environments in iron compounds, in I.J. Gruverman and C.W. Seidel (eds), *Mossbauer Effect Methodology,* L. Plenum Press, New York, pp. 89-105.
14. Burns, R.G. and Fisher, D.S. (1994) Nanophase mixed-valence iron minerals in meteorites identified by cryogenic Mössbauer spectriscope, *Hyperfine Interactions* **91**, 571-576.
15. Miymoto, H. (1976) The magnetic properties of Fe(OH)₂, *Mater. Res. Bull.* **11**, 329-336.
16. Nikolaev, V.I. and Rusakov, V.S. (1985) *Mössbauerovskie issledovaniya ferritov,* Mosk. Gos. Univ., Moscow.

MÖSSBAUER SPECTROSCOPY IN STUDYING THE THERMALLY INDUCED OXIDATION OF Fe^{2+} CATIONS IN IRON-BEARING SILICATE MINERALS

Examples of applications with almandine, pyrope and olivine

M. MASHLAN[1], R. ZBORIL[2] AND K. BARCOVA[3]

[1]*Department of Experimental Physics, Palacky University, Olomouc, Czech Republic*
[2]*Department of Inorganic and Physical Chemistry, Palacky University, Olomouc, Czech Republic*
[3]*Institute of Physics, Technical University of Ostrava, Ostrava, Czech Republic*

1. Introduction

The oxidation mechanisms of Fe^{2+} in different iron-bearing minerals depend strongly on the *external conditions*, including the heating temperature, mineral crystallinity, pressure conditions and oxidation atmosphere. However, slight differences in the *internal features* (the total content of iron, distribution of iron in non-equivalent sites, chemical environment and coordination of iron) can also significantly influence the oxidation mechanism. Thus, different oxidation routes can occur depending on the structural ordering and external conditions.

The thermal treatment of Fe-bearing minerals under the *low-energy conditions* can lead to the oxidation of Fe^{2+} cations in defined sites of the mineral framework [1] as well as to the change of the electronic equilibrium between Fe^{2+} and Fe^{3+} ions coexisting in a crystal lattice [2]; however, the general mineral structure is preserved. *Higher thermal activation* causes the destruction of mineral structure related with the Fe^{2+} oxidation. Thus, the explanation of the mechanism of these destructive oxidations includes the understanding of the intrinsical Fe^{2+}→Fe^{3+} oxidation (air-oxygen oxidation, internal redox process) as well as the determination of iron ions distribution in the structures of the conversion products. Therefore the knowledge of the general reaction mechanism, including the identification and quantification of reaction products, is necessary to resolve the oxidation mechanism.

Generally, oxidative conversions of Fe^{2+}-bearing minerals are most often related with the formation of different Fe^{3+}-bearing oxides. Among them two main groups can be underlined: iron(III) oxides and spinels.

The α-Fe$_2$O$_3$ with the hexagonal corrundum structure, γ-Fe$_2$O$_3$ with the cubic spinel structure, rare β-Fe$_2$O$_3$ with the cubic bixbyite structure, orthorhombic ε-Fe$_2$O$_3$ and

M. Mashlan et al. (eds.), Material Research in Atomic Scale by Mössbauer Spectroscopy, 271–284.

amorphous Fe_2O_3 can be found in the mixture of oxidation products [3]. The reliable identification and quantification of different *iron(III) oxide polymorphs* is a relatively difficult problem, because the thermally induced isochemical structural transformations of metastable iron oxide forms (amorphous-, β-, γ-, ϵ-Fe_2O_3) to α-Fe_2O_3 usually accompany the primary oxidation processes. Moreover, iron(III) oxides are often formed as ultrafine particles with an imperfect crystallinity. The cation doping and/or substitution of iron in the crystal lattice of ferric oxides also plays also an indispensable role from the viewpoint of their credible identification and structural and magnetic characterization.

The *spinels* are a numerous group of mixed oxides of general formula AB_2O_4. They are based on a face centred, cubic close packed oxide lattice with the B occupying octahedral sites and the A tetrahedral sites. The unit cell contains 32 oxide ions. The cations occupy one eighth of 64 tetrahedral A sites and one half of 32 octahedral B sites. The stability of the spinel structure depends on the cations occupying the A and B sites. Variation of the cations at these sites produces an astonishingly wide variety of magnetic and electric properties, which are a consequence of chemical properties of the constituent metals, mostly of those occupying the B site [4]. The wide variety of incorporated metal cations and imperfect crystallinity complicates the identification and characterization of spinels formed during the oxidation processes.

Various methods are used to identify and characterize products of thermally induced oxidations in minerals. Typically applied experimental techniques could be classified into several main groups including spectroscopy, diffractometry, methods of thermal analysis, magnetic measurements and microscopic techniques. *Spectroscopic methods* produce spectra, which arise as result of interaction of electromagnetic radiation with matter. The type of interaction depends upon the wavelength of the radiation. The most widely applied techniques are infrared, Mössbauer and ultraviolet-visible spectroscopy. A second group of methods belongs to the field of *diffractometry* and involves interaction of X-rays, electrons or neutrons with atoms of a solid. X-ray diffraction is by far the most widely used. *Thermal analysis* techniques use the specific behavior of a solid upon heating. The most common of these techniques are differential thermal analysis, differential scanning calorimetry and thermal gravimetric analysis. For the magnetic characterization and the study of thermally induced magnetic phase transitions, the *molar magnetic susceptibility and magnetization measurement* are usually applied. The group of *microscopic methods* includes optical microscopy, transmission and scanning electron microscopy, scanning tunnelling microscopy and atomic scale microscopy. As a result, the combined application of the above techniques provides the complex information about particle size and morphology, chemical composition, crystal structure, extent of metal substitution, and electronic and magnetic characteristics of the oxidation products (oxides, spinels).

Among all these techniques the position of *^{57}Fe Mössbauer spectroscopy* is very specific due to its selectivity for iron with no interference from other elements. Its great sensitivity to the local environment of the iron atoms in the crystal lattice makes from it an excellent tool for the study of the electronic density, its symmetry and magnetic properties of the ^{57}Fe nucleus incorporated in the structures of both iron-bearing minerals and their oxidation products, iron(III) oxides and spinels. Through hyperfine parameters, Mössbauer spectra provide important information, which is applicable to oxidation studies in minerals including the valence state of iron, the number of iron atoms in non-equivalent sites of a crystal lattice, the type of iron coordination, the level of ordering and stoichiometry, the degree of cation substitution, magnetic ordering and the magnetic transition temperature.

In this paper ^{57}Fe Mössbauer spectroscopy is presented as the indispensable guide at the study of thermally induced oxidations of Fe^{2+} cations in iron-bearing minerals. The advantages and the singularity of this method are demonstrated with examples of the thermal oxidations of three Fe-bearing natural silicates: two garnets (almandine, pyrope) and olivine.

2. Materials and Experimental Methods

Purple-red crystals of almandine (Czech Republic), red crystals of pyrope (Czech Republic) and green crystals of olivine (Norway) were purified with HCl and oxalate acid, in order to remove secondary minerals such as carbonates and Fe oxyhydroxides. The crystals were washed in distilled water and then dried at 100 °C. Electron microprobe analysis (EDAX 9900 + Phillips Scanning Electron Microscope, type 535 M) and XRF analysis (Energy Disperse Spectrometer Spectro X-LAB) were performed with the aim to analyse chemical composition and homogeneity of the used mineral grains. Chemical zoning or inhomogeneities were not observed. The chemical compositions were found using the electron microprobe analysis and Mössbauer spectroscopy (determination of Fe^{2+}/Fe^{3+} ratio)

$$(Fe_{2.31}Mg_{0.18}Mn_{0.07}Ca_{0.07})Al_{1.96}Si_{3.25}O_{12},$$
$$(Mg_{2.03}Fe_{0.72}Mn_{0.02}Ca_{0.23})(Al_{1.78}Ti_{0.05}Cr_{0.12}Fe_{0.08})Si_{3.06}O_{12},$$
$$(Mg_{1.82}Fe_{0.17}Ni_{0.005}Cr_{0.005})SiO_4$$

for almandine, pyrope, and olivine, respectively.

Crystals were crushed in an agate mill to a powder with the particle size of 1.5-3 μm. The powdered samples of almandine, pyrope and olivine were heated in isothermal conditions at temperatures 750, 1000, and 1200 °C, respectively. Silicates were thermally treated for such periods of time as necessary for the comprehensive monitoring of their oxidation process (1-360 hours depending on the type of mineral and calcination temperature).

The transmission ^{57}Fe Mössbauer spectra of 512 channels were collected using a Mössbauer spectrometer in constant acceleration mode with a ^{57}Co(Rh) source. Measurements were carried out in a temperature range of 12 to 300 K using a cryostat with closed He-cycle (Janis Research Company). Isomer shift values are related to α-Fe. The phase composition of samples was monitored by XRD using a Seifert-FPM equipment with CuKα radiation and conventional θ-2θ geometry. Si was used as an external calibration standard. The individual phases were identified from XRD patterns by means of a PDF2 database.

3. The Crystal Chemistry of Garnets

Garnets are orthosilicates with isolated $(SiO_4)^{4-}$ groups. Their general structural formula is $X_3Y_2Z_3O_{12}$, the space group $Ia\overline{3}d$ and the basic cell contains eight formula units. X corresponds to dodecahedral coordination of the metal in position $24c$ (X = Mg^{2+}, Ca^{2+}, Fe^{2+}, or Mn^{2+}), Y corresponds to octahedral coordination in position $16a$ (Y = Al^{3+}, Fe^{3+}, Cr^{3+}) and Z to tetrahedral coordination in position $24d$ (Z=Si^{4+}, with minor amounts of Fe^{3+}, Ti^{4+}). Natural garnets are classified into two main groups - the pyrope-almandine-spessartine series: $(Mg^{2+}, Fe^{2+}, Mn^{2+})_3Al_2Si_3O_{12}$ and the uvarovite-grossular-andradite series: $Ca_3(Al^{3+}, Fe^{3+}, Cr^{3+})_2Si_3O_{12}$. However, a minor content of Fe^{3+} ions in the Y position may be found in natural garnets of the pyrope-almandine-spessartine series.

4. The Crystal Chemistry of Olivine

Olivines have the orthorhombic crystal structure with a space group *Pnma*. The two nearly octahedral sites M1 and M2 have different point symmetries ($\overline{1}$ and *m*) and geometrical arrangements including the nonequivalent M-O distances (d_{M1-O}=2.095 Å, d_{M2-O}=2.131 Å). Cations are generally distributed disorderly over M1 and M2 sites. The ordering of cations in the olivine structure is strongly controlled by the thermal history of the sample. According to the chemical composition, olivines can be classified into four main groups: Mg-olivines (forsterite, Mg_2SiO_4), Mg-Fe olivines (chrysolite, $(Mg,Fe)_2SiO_4$), Fe–olivines (fayalite, Fe_2SiO_4) and Fe–Mn olivines.

5. Results and Discussion – Mössbauer Studies of Thermal Oxidations in Silicates

5.1. ISOCHEMICAL STRUCTURAL TRANSFORMATIONS OF IRON(III) OXIDES DURING THERMAL TREATMENT OF *ALMANDINE AT 750 °C*

The room temperature (RT) Mössbauer spectrum of non-heated almandine consists of a doublet with hyperfine parameters: δ=1.29 mm/s, ΔE_Q=3.51 mm/s and Γ=0.33 mm/s, which originates from Fe^{2+} ions in the dodecahedral $24c$ position of the garnet structure. The thermal behavior of Fe^{2+} was investigated at the minimum oxidation temperature of 750 °C with the aim to identify the primary oxidation phase and to consider its secondary transformation mechanism.

RT Mössbauer spectra of almandine heated at 750 °C for different periods of time are shown in figure 1. The spectra consists of a Fe^{2+} doublet of the initial almandine and a broad Fe^{3+} central doublet with δ=(0.34-0.35) mm/s, ΔE_Q=(0.88-1.05) mm/s and Γ=(0.55-0.74) mm/s corresponding to superparamagnetic nanoparticles of γ-Fe_2O_3. This assignment was based on the XRD identification in combination with the low temperature Mössbauer measurements, where the γ-Fe_2O_3 spectrum shows behavior typical for superparamagnetic relaxation. With the increased heating time, a percentage of the γ-Fe_2O_3 superparamagnetic doublet increases in the RT Mössbauer spectra and, moreover, the sextet with parameters typical for hematite ($\delta = 0.36$ mm/s, $\varepsilon = -0.21$ mm/s, B = 51.8 T and $\Gamma = 0.30$ mm/s) appears in the spectrum of almandine heated for 66 hours (see fig.1). These results indicate that Fe^{2+} cations are oxidized to thermally stable α-Fe_2O_3 via metastable γ-Fe_2O_3, which is the primary oxidation phase.

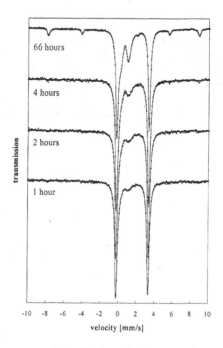

Figure 1. RT Mössbauer spectra of almandine heated in air at 750 °C for different periods of time.

Taking into account the possible formation of ε-Fe_2O_3 as the intermediate during isochemical structural transformation of γ-Fe_2O_3 nanoparticles to hematite [3,5], the presence of this rare polymorph in the heated almandine was investigated. Due to the low oxidation stages of the almandine samples heated at 750 °C, the possibility of the formation of this rare phase should not be foreclosed. To identify eventual traces of ε-Fe_2O_3 in heated samples, iron(III) oxides were separated from the residual non-transformed almandine. This separation was based on the higher density of garnet particles and therefore on their higher sedimentation rate in the water. RT Mössbauer spectrum of iron(III) oxides isolated from the sample heated at 750 °C for 4 hours and results of its mathematical processing are shown in figure 2 and table 1, respectively. In addition to γ-Fe$_2$O$_3$ nanoparticles and α-Fe$_2$O$_3$ subspectra, the spectral lines corresponding to ε-Fe$_2$O$_3$ were unambiguously identified. Four non-equivalent iron sites (three octahedral and one tetrahedral) in the ε-Fe$_2$O$_3$ structure [3,5] were fitted by three sextets as a result of a strong overlap of two "octahedral subspectra". Therefore, the following model describing the transformation mechanism of iron ions during thermal treatment of almandine at 750 °C can be suggested:

$$Fe^{2+} (24c) \xrightarrow{\quad O_2 (air) \quad} \gamma\text{-Fe}_2O_3 \to \varepsilon\text{-Fe}_2O_3 \to \alpha\text{-Fe}_2O_3 \qquad (1).$$

In summary, the example of thermal oxidation of almandine at 750 °C clearly demonstrates the significant contribution of Mössbauer spectroscopy to identification of the primary products of oxidation of Fe-bearing minerals and to the study of the secondary thermally induced isochemical transformations of Fe$_2$O$_3$ polymorphs. Generally this fact is useful in the cases where the initial mineral contains only small amounts of Fe^{2+} and thus, the complicity of other techniques in the study of the oxidation process is strongly limited.

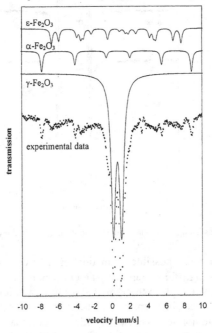

Figure 2. RT Mössbauer spectrum of iron(III) oxides isolated from the almandine sample heated at 750 °C for 4 hours.

TABLE 1. Results of fitting of the RT Mössbauer spectrum of iron(III) oxides isolated from the almandine sample heated at 750 °C for 4 hours.

almandine (750 °C / 4 hours)	δ [mm/s]	$\Delta E_Q(\varepsilon)$ [mm/s]	B [T]	RA [%]
γ-Fe_2O_3 nanoparticles	0.35	0.99	-	73.4
α-Fe_2O_3	0.36	-0.21	51.8	10.2
ε-Fe_2O_3 octahedral sites Fe_1 and Fe_2	0.38	-0.12	44.6	7.2
ε-Fe_2O_3 octahedral site Fe_3	0.39	-0.03	39.5	4.8
ε-Fe_2O_3 tetrahedral site Fe_4	0.23	-0.09	26.6	4.4

δ - isomer shift, $\Delta E_Q(\varepsilon)$ – quadrupole splitting (shift), B – hyperfine magnetic field, RA – relative area

5.2. THERMAL BEHAVIOR OF *PYROPE AT 1000 °C* – THE NONDESTRUCTIVE OXIDATION OF Fe^{2+} IN 24c POSITION OF THE GARNET STRUCTURE

RT Mössbauer spectrum of the initial pyrope contains an asymmetric doublet with δ=1.28 mm/s, ΔE_Q=3.54 mm/s, Γ=0.33 mm/s, and RA=90.0% and a doublet with δ=0.32 mm/s, ΔE_Q=0.25 mm/s, Γ=0.30 mm/s, and RA=10.0%, which originate from Fe^{2+} in the 24c site and Fe^{3+} in the 16a site of the pyrope structure, respectively. The asymmetry of Fe^{2+} doublet is commonly observed in the RT Mössbauer spectra of garnets and can be explained as arising out of paramagnetic relaxation of Fe^{2+} in the dodecahedral position [6,7]. XRD pattern shows exclusively the lines corresponding to the cubic pyrope structure (the space group $Ia\overline{3}d$, lattice parameter a_0=1.14802(101) nm) without any indications of the presence of other phases.

To consider the oxidation in its initial stage, the samples were thermally treated at 1000 °C for different periods of time and analyzed mainly using Mössbauer spectroscopy and XRD. The slow course of the oxidation process is manifested by the gradual decrease of the spectral area corresponding to Fe^{2+} component in RT Mössbauer spectra of samples heated for 1- 360 hours (Table 2). Generally, iron ions appear in the thermally treated pyrope powders in four non-equivalent structural and oxidation states at 1000 °C. Hyperfine parameters of a Fe^{2+} doublet unambiguously correspond to the non-oxidized ferrous ions in dodecahedral position 24c of the pyrope structure. Their content gradually decreases due to the occurring oxidation. The other two paramagnetic doublets in the central part of RT Mössbauer spectra show hyperfine parameters confirming Fe^{3+} state. The doublet with the lower value of quadrupole splitting parameter (0.25 mm/s) corresponds to the ferric ions incorporated in the octahedral 16a position of the original pyrope structure; their content remains constant during thermal treatment. The second Fe^{3+} doublet with the significantly higher value of quadrupole splitting parameter (0.70 mm/s) should be assigned to Fe^{3+} ions in dodecahedral 24c position.

TABLE 2. Hyperfine parameters of RT Mössbauer specta of pyrope heated in air at 1000 °C for 1-360 hours.

Fe phase	parameters	1 h	3 h	15 h	120 h	360 h
Fe^{2+} (24c)	δ [mm/s]	1.27	1.27	1.27	1.26	1.26
	ΔE$_Q$ [mm/s]	3.53	3.53	3.53	3.51	3.50
	Γ [mm/s]	0.32	0.31	0.29	0.34	0.34
	RA [%]	65.8	62.5	48.1	11.5	3.2
Fe^{3+} (16a)	δ [mm/s]	0.32	0.32	0.29	0.30	0.29
	ΔE$_Q$ [mm/s]	0.24	0.22	0.23	0.25	0.22
	Γ [mm/s]	0.29	0.30	0.30	0.30	0.29
	RA [%]	9.5	9.8	8.8	9.1	9.3
Fe^{3+} (24c)	δ [mm/s]	0.24	0.27	0.23	0.26	0.26
	ΔE$_Q$ [mm/s]	0.69	0.68	0.70	0.67	0.72
	Γ [mm/s]	0.50	0.50	0.50	0.54	0.71
	RA [%]	11.7	12.3	18.0	32.8	36.8
α-Fe$_2$O$_3$	δ [mm/s]	0.36	0.35	0.37	0.37	0.38
	ε [mm/s]	-0.22	-0.23	-0.22	-0.22	-0.21
	B [mm/s]	50.5	50.3	50.4	49.9	50.3
	Γ [mm/s]	0.38	0.39	0.37	0.38	0.44
	RA [%]	13.0	15.4	25.1	46.6	50.7

The higher value of the quadrupole splitting parameter of 0.70 mm/s reflects the lower symmetry of the iron environment (in comparison with Fe^{3+} in 16a position) [8] and the increasing spectrum area with time indicates the gradual oxidation of Fe^{2+} ions in 24c position. The only magnetically ordered component in RT Mössbauer spectra of pyrope samples thermally treated at 1000°C presents hyperfine parameters typical for hematite (δ=0.34÷0.36 mm/s, ΔE$_Q$=-0.19÷-0.21 mm/s, B=49.9÷51.0 T, Γ=0.4÷0.52 mm/s). Its relative area increases with time as in the case of Fe^{3+} ions in 24c position. It seems that thermally induced air-oxidation of Fe^{2+} ions in the pyrope structure results in the partial stabilization of Fe^{3+} ions in 24c position and in the incorporation of the redundant ferric ions in the hematite structure. In this redox system, hematite serves as the compensator of the charge change of iron atoms in 24c position as well as the host compound for the reduced atoms of the air-oxygen as could be demonstrated by the following equation:

$$6Fe^{2+} (24c) + 3/2O_2 \rightarrow 4Fe^{3+} (24c) + \alpha\text{-}Fe_2O_3 \qquad (2)$$

The course of the oxidation process at 1000 °C including the changes in distribution of iron ions can be clearly demonstrated by the representative Mössbauer spectra in figure 3.

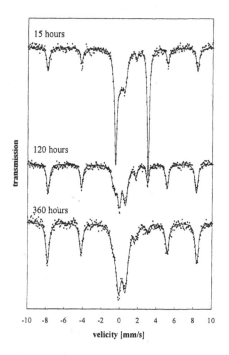

Figure 3. RT Mössbauer spectra of pyrope heated at 1000°C for 15, 120 and 360 hours.

The alternative that the central doublet in the RT Mössbauer spectra (assigned to Fe^{3+} in 24c) could correspond to the superparamagnetic nanoparticles of ferric oxide (as in the case of almandine transformation at 750°C) was excluded on the basis of the Mössbauer measurements at 20K (fig. 4), where this phase remains paramagnetic and its isomer shift parameter at 20 K (0.38 mm/s) is significantly lower than typical for hematite and other ferric oxides.

The most interesting fact related with the transformation mechanism at 1000 °C is the non-destructive character of the oxidation process. The almost complete oxidation of Fe^{2+} in 24c position and their partial extraction from pyrope and transformation to hematite don't result in the decomposition of the garnet structure. For illustration, XRD patterns of the initial powdered pyrope ($Fe^{3+}/\Sigma Fe$=10%) and the pyrope heated at 1000 °C for 360 hours ($Fe^{3+}/\Sigma Fe$=97%) differs only by the presence of hematite, however the positions of the lines corresponding to the garnet structure remain almost the same and the found lattice parameter varies only imperceptibly.

We can conclude that Mössbauer spectroscopy seems to be the exclusive technique for monitoring the oxidation in the defined position of the mineral structure, as in the case of pyrope heated at 1000 °C.

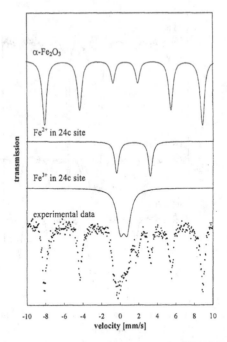

Figure 4. (20K) Mössbauer spectrum of pyrope heated at 1000°C for 120 hours.

5.3. IDENTIFICATION OF PRIMARY OXIDATION PHASE DURING THERMAL DECOMPOSITION OF *Mg-Fe OLIVINE AT 1200 °C* – MAGNESIOFERRITE OR HEMATITE?

Identification of the primary product of thermally induced Fe^{2+} oxidation in the cases where the mineral contains only a small fraction of iron belongs to the difficult problems of physics and chemistry of minerals. The low content of iron restricts the use of the usual identification techniques. Moreover, the identification process can be complicated by the simultaneous (side and parallel) reactions, which occur at higher temperatures. Let us demonstrate these problems with the example of thermal decomposition of Mg-Fe olivine (content of iron: 6.5%wt) at 1200 °C.

RT Mössbauer spectrum of Mg-Fe olivine heated at 1200 °C for 70 hours (fig. 5) contains two Fe^{2+} doublets corresponding to the M1 and M2 sites in the olivine structure [9] (δ=1.15 mm/s, ΔE_Q=3.02; δ=1.16 mm/s, ΔE_Q =2.15 mm/s) and two Fe^{3+} sextets (δ = 0.28 mm/s, ϵ = 0.01 mm/s, B = 46.2 T; δ = 0.33 mm/s, ϵ = 0.01 mm/s, B = 48.7 T). The found Mössbauer parameters of magnetic components correspond well with those reported for Fe^{3+}

in tetrahedral and octahedral sites of the spinel structure of magnesioferrite ($MgFe_2O_4$) [10]. It would seem that magnesioferrite would be the direct product of oxidative decomposition of olivine at 1200 °C. However, RT Mössbauer spectra of samples heated for a shorter time reveal the joint presence of magnesioferrite and hematite (fig. 5, table 3). The relative spectral areas reflect the transformation of ferric ions from the hematite to magnesioferrite structure with the increasing time of thermal treatment. These data show that the hematite is the primary oxidation product and $MgFe_2O_4$ forms as the transformation phase.

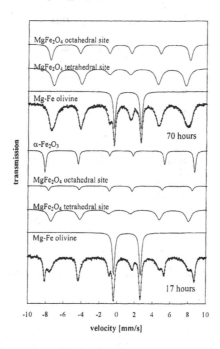

Figure 5. RT Mössbauer spectra of Mg-Fe olivine heated in air at 1200 °C for 17 and 70 hours.

The detailed explanation of the formation mechanism of both hematite and magnesioferrite is related with the XRD identification of non-iron phases originating from the thermal decomposition of Mg-Fe olivine. XRD analysis of the original mineral shows exclusively the diffraction lines characteristic for Mg-Fe olivines (PDF database card number: 31-795). In the all heated samples, the shift of positions of diffraction lines toward pure forsterite (Mg_2SiO_4, 34-189) was observed. Moreover, the new low-intensive lines corresponding to enstatite ($MgSiO_3$, 19-768), magnesioferrite ($MgFe_2O_4$, 17-464), cristobalite (SiO_2, 11-695) and hematite (Fe_2O_3, 33-664) appeared in XRD patterns of samples heated for 1 and 17 hours. After 70 hours, the lines of hematite and enstatite vanished, while the line intensities of magnesioferrite and cristobalite increased.

Thus, the combination of XRD and Mössbauer spectroscopy allows us to suggest the following general mechanism of oxidation of the Fe^{2+} cations in Mg-Fe olivines with the low content of iron:

$$Mg_{2-x}Fe_xSiO_4 + x/4O_2 \rightarrow (1-x)Mg_2SiO_4 + xMgSiO_3 + x/2Fe_2O_3 \tag{3},$$

where x corresponds to the molar content of iron (in this study, x=0.17).

TABLE 3. Hyperfine parameters of RT Mössbauer spectra of Mg-Fe olivine heated at 1200 °C for 1, 17 and 70 hours.

Fe phase	parameters	1 h	17 h	70 h
Fe²⁺ in olivine	δ [mm/s]	1.15	1.15	1.15
	ΔE_Q [mm/s]	3.03	3.01	3.01
	Γ [mm/s]	0.30	0.31	0.32
	RA [%]	41.5	26.3	19.7
Fe³⁺ in MgFe₂O₄ tetrahedral site	δ [mm/s]	0.28	0.26	0.27
	ε [mm/s]	-0.02	0.0	-0.01
	B [mm/s]	47.0	47.2	46.4
	Γ [mm/s]	0.51	0.6	0.61
	RA [%]	14.9	33.0	51.3
Fe³⁺ in MgFe₂O₄ octahedral site	δ [mm/s]	0.34	0.36	0.34
	ε [mm/s]	0.03	0.01	0.01
	B [mm/s]	49.5	49.6	48.9
	Γ [mm/s]	0.4	0.41	0.51
	RA [%]	8.7	18.6	29.0
Fe³⁺ in α-Fe₂O₃	δ [mm/s]	0.37	0.37	-
	ε [mm/s]	-0.20	-0.19	-
	B [mm/s]	51.9	51.8	-
	Γ [mm/s]	0.27	0.27	-
	RA [%]	34.9	22.1	-

* Fe^{2+} ions in M1 and M2 positions of olivine structure were fitted by one doublet due to the low content of Fe^{2+} ions in M2 position resulting in the strong overlap of both sites

It seems that decomposition of the olivine structure is triggered by the oxidation of Fe^{2+} cations and hematite is formed as the primary oxidation phase. The oxidation is accompanied by the reconstruction of the silicate structure resulting in the formation of the thermally stable pure forsterite (Mg_2SiO_4). Both Mössbauer and XRD data reveal that magnesioferrite is the product of the secondary reaction of hematite with enstatite according to the scheme (4). The possibility of this solid-state reaction at 1300 °C was also mentioned by Slovenec et al. [11]

$$MgSiO_3 + Fe_2O_3 \rightarrow MgFe_2O_4 + SiO_2 \tag{4}.$$

The example of the thermal study of Mg-Fe olivine at 1200 °C points out the great potency of the combination of Mössbauer spectroscopy and XRD in looking for the mechanism of thermally induced solid-state reactions. The general contribution of Mössbauer spectroscopy comes from the identification and quantification of Fe-bearing products.

6. Conclusion

Mössbauer spectroscopy was used for the identification of the primary oxidation phase during thermal treatment of almandine at 750 °C and for the study of the secondary polymorphic transformations of Fe_2O_3. Thus, superparamagnetic nanoparticles of maghemite were found as the direct product of oxidation of Fe^{2+} ions in the almandine structure, and ε-Fe_2O_3 was identified as the intermediate product during structural change of maghemite to hematite.

The application of Mössbauer spectroscopy for the study of the oxidation mechanism in the defined site of the mineral structure was demonstrated with the example of thermal study of pyrope at 1000 °C. The possibility of the air-oxidation of Fe^{2+} ions in the dodecahedral 24c position accompanied by the formation of hematite was proved.

The significant contribution of the Mössbauer spectroscopy for the investigation of the mechanism of thermally induced oxidative decompositions was illustrated with olivine at 1200 °C. Two-step formation mechanism of the spinel structure magnesioferrite was suggested on the basis of combined Mössbauer and XRD analysis.

Acknowledgements

Financial support from the Grant Agency of the Czech Republic under projects 202/00/0982, 202/00/D091 is gratefully acknowledged.

References

1. Waerenborgh, J.C., Figueiredo, M.O., Cabral, J.M.P. and Pereira L.C.J. (1994) Powder XRD structure refinements and Fe-57 Mossbauer-effect study of synthetic $Zn_{1-x}Fe_xAl_2O_4$ ($0 < x \leq 1$) spinels annealed at different temperatures, *Physics and Chemistry of Minerals* **21**, 460-468.
2. Schmidbauer, E., Kunzmann, T., Fehr, T. and Hochleitner, R. (2000) Electrical resistivity and Fe-57 Mossbauer spectra of Fe-bearing calcic amphiboles, *Physics and Chemistry of Minerals* **27**, 347-356.
3. Zboril, R., Mashlan, M. and Petridis, D. (2002) Iron(III) oxides from thermal processes-synthesis, structural and magnetic properties, Mössbauer spectroscopy characterization, and applications, *Chem. Mater.*, **14**, No.3, 969-982.
4. Vertes, A. and Homonnay, Z. (1997) *Mössbauer Spectroscopy of Sophisticated Oxides*, Akademiai Kiado, Budapest, p. 88.

5. Tronc, E., Chaneac, C. and Jolivet, P. (1998) Structural and magnetic characterization of epsilon-Fe2O3, *J. Solid State Chem.* **139**, 93-104.

6. Amthauer, G., Annersten, H. and Hafner, S.S. (1976) The Mössbauer spectrum of ^{57}Fe in silicate garnets, *Zeit. Kristallogr.* **143**,14-55.

7. Murad, E. and Wagner, F.E. (1987) The mossbauer spectrum of almandine, *Physics and chemistry of Minerals* **14**, 264-269.

8. Novak, G.A. and Gibbs, G.V. (1971) The crystal chemistry of the silicate garnets, *Amer. Mineral.* **56**, 791-825.

9. Ericsson, T., Amcoff, O. and Kalinowski, M. (1999) Cation preferences in thio-olivines (Fe1-xMgx)(2)SiS4, x <= 0.30, studied by Mossbauer spectroscopy at room temperature, *Neues Jahrbuch Fur Mineralogie-Monatshefte* **11**, 518-528.

10. Harchand, K.S., Taneja, S.P., Raj, D. and Sharma, P. (1989) Mössbauer Studies of Coal Ash, *Fuel Process. Technol.* **21**, 19-24.

11. Slovenec, D., Popovic, S. and Tadej N. (1997) Heating products of glauconitic materials, *Neues Jahrbuch Fur Mineralogie-Abhandlungen* **171**, 323-339.

MÖSSBAUER DIFFRACTOMETRY

Concepts, Instrumentation and Measurements

B. FULTZ AND J.Y.Y. LIN
California Institute of Technology, Pasadena, CA 91125 USA

1. Introduction

The usual incoherent scattering of Mössbauer spectrometry is not useful for Mössbauer diffractometry. The basis for Mössbauer diffractometry is the interference of coherent waves, as with the other three methods for diffraction studies on materials (x-ray, electron, and neutron) [1]. Mössbauer diffractometry uses coherent nuclear resonant scattering of γ-ray photons. Bragg diffraction peaks then occur by the constructive interference of γ-ray wavelets scattered by a periodic crystal of nuclei. Mössbauer diffraction was first observed in studies on single crystals, in which a remarkable "nuclear speed-up" effect [2-5] enhances the diffracted intensity at the Bragg angles. It is difficult to use these "dynamical diffraction patterns" to obtain information about the crystal structure. Our work has focused on "kinematical," or single-scattering diffraction, because of its potential for structural studies on materials at the atomic scale [6-10].

Here we explain the basic concepts of Mössbauer diffractometry in the language of kinematical diffraction. The quantized nature of the nuclear polarizations, and their interferences with x-ray diffraction are inherently different from x-ray diffractometry. More significantly, the spectroscopic capabilities of the Mössbauer effect provide unique but powerful capabilities of Mössbauer diffractometry, not available to x-ray, electron or neutron diffractometries. We describe a Mössbauer diffractometer for measurements on ^{57}Fe-enriched polycrystalline samples (called a "powder" diffractometer). Some characteristic data are presented, and potential applications are discussed.

2. Principles

2.1. SPATIAL CORRELATION FUNCTIONS

In diffraction, a coherent wavelet is emitted from each scatterer, in this case a nucleus of ^{57}Fe that was resonantly excited by an incident γ-ray plane wave. The nuclear excitation is a collective one, a "nuclear exciton" shared by all coherent nuclei. The wavelets emitted from all of these nuclei interfere constructively at the position of the detector to produce a Bragg peak. For each scattering center at r_j (a ^{57}Fe nucleus, for example), the factor $\exp(-i\Delta k \cdot r_j)$ describes the relative phase of its wave, ψ_j, with respect to the waves $\{\psi_i\}$ scattered from centers at other $\{r_i\}$. By working with the relative phases of the

285

M. Mashlan et al. (eds.), Material Research in Atomic Scale by Mössbauer Spectroscopy, 285–295.
© 2003 *Kluwer Academic Publishers. Printed in the Netherlands.*

scattered $\{\psi_i\}$, we can ignore the long distances between the sample and source or detector, and we can concentrate on the angles of emitted rays. The Bragg angle, θ, is included in Δk, the vector difference between the effective plane wave at the detector and the incident plane wave: $\Delta k \equiv k - k_0$, as $\Delta k = 4\pi \sin\theta / \lambda$, where λ is the γ-ray wavelength. The interference of all $\{\psi_i\}$ gives the full scattered wave, $\psi(\Delta k)$:

$$\psi(\Delta k) = \sum_{r_j} f(r_j) e^{-i\Delta k \cdot r_j} . \qquad (1)$$

The intensity, $I(\Delta k)$, is the real function $|\psi(\Delta k)|^2$:

$$I(\Delta k) = \sum_{r_j} f(r_j) e^{-i\Delta k \cdot r_j} \sum_{r_k} f^*(r_k) e^{i\Delta k \cdot r_k} . \qquad (2)$$

The difference in position of pairs of atoms, $r \equiv r_j - r_k$, proves important for the intensity:

$$I(\Delta k) = \sum_r \left(\sum_{r_k} f(r - r_k) f^*(r_k) \right) e^{-i\Delta k \cdot r} , \qquad (3)$$

$$I(\Delta k) = \sum_r P(r) e^{-i\Delta k \cdot r} , \qquad (4)$$

where we have defined the Patterson function $P(r)$ as the term in braces in (3). This $P(r)$ is an autocorrelation function of the positions of the scattering centers. It is the fundamental real-space quantity obtained by Fourier-transformation of the measured diffraction intensity [1]. The quantity $f(r_j)$ is the form factor for all coherent scattering processes. In Mössbauer diffractometry $f(r_j)$ would be the sum of a coherent nuclear resonant scattering component $f_M(r_j)$, plus a coherent x-ray scattering component, $f_X(r_j)$, since wave amplitude is scattered by both processes from an atom at r_j.

2.2. INTERFERENCE

When the nuclear and electronic (x-ray) scattering have comparable amplitudes, the total form factor, $f_T(r_j) = f_M(r_j) + f_X(r_j)$, has important cross-terms when the product $f_T f_T^*$ is taken in (3). It is useful to group terms with different interferences between the x-ray or Mössbauer scattering processes, giving the total Patterson function as a sum of three types of Patterson functions [8]:

$P_{XX}(r)$, the usual Patterson function for x-ray diffraction,

$P_{MX}(r)$ and $P_{XM}(r)$, the Patterson functions associated with the interference between x-ray and Mössbauer scattered wavelets, and

$P_{MM}(r)$, the Patterson function from the coherent nuclear resonant scattering processes.

The mixed Patterson functions $P_{MX}(r)$ and $P_{XM}(r)$ arise from interference between x-ray scattering at one atom and Mössbauer scattering at the same or another atom. This interference is seen most easily in the energy-dependence of the diffracted intensity. The intensity for a single Mössbauer scattering process has the well-known Lorentzian cross-section:

$$\sigma_j(E) = \frac{1}{2\pi} \frac{\Gamma}{E_j^2 + (\Gamma/2)^2} \quad , \tag{5}$$

but for Mössbauer diffractometry we are less concerned with $\sigma_j(E)$, and more concerned with the diffracted wave amplitude, $\psi(\Delta k)$. The cross-section of (5) can easily be shown to equal $f_j f_j^*$, where:

$$f_j(E) = \sqrt{\frac{2}{\pi\Gamma}} \frac{1}{1 - i(2E_j/\Gamma)} \quad . \tag{6}$$

Since $f_j(E)$ is less familiar to Mössbauer spectroscopists, its real and imaginary parts are plotted in the complex plane as dots in Fig. 1, with separations of dots having equal units of $2E/\Gamma$.

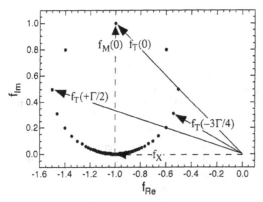

Figure 1. Real and imaginary parts of $f_j(E)$, showing coherent interference of x-ray Rayleigh scattering and Mössbauer scattering for three different energies about the Mössbauer resonance. Dashed arrows are the form factors of individual scattering processes (x-ray and on-resonance Mössbauer). Solid arrows are coherent sums.

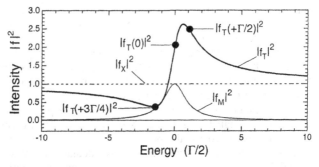

Figure 2. Form factor intensities for x-ray Rayleigh scattering alone, $|f_X|^2$, Mössbauer scattering alone, $|f_M|^2$, and the coherent sum of the two $|f_T|^2$. Points are the intensities of the three form factors, f_T, shown in Fig. 1.

For Mössbauer isotopes such as ^{57}Fe, the real and imaginary parts of $f_M(E)$ change strongly over an energy range of neV, whereas the x-ray form factor, f_X, does not. These different energy dependencies can be used to understand how interference between x-ray Rayleigh scattering and Mössbauer scattering causes an asymmetry in the Mössbauer energy spectrum of a Bragg diffraction. Figure 1 shows the x-ray form factor, f_X, as a constant vector pointing along the real axis towards -1 (dashed arrow). The sign indicates that the scattered x-ray is out-of-phase with respect to the phase of the incident photon (which occurs because the 14.4 keV γ-rays drive the x-ray scattering at a frequency well above the K-shell resonance of the atomic electrons). One $f_M(E)$ is drawn as a dashed arrow in Fig. 1. The waves scattered by the Mössbauer effect are added coherently to the x-ray scattering as a vector sum. The total diffracted wave is labelled as $f_T(E)$ for three values of E, $f_T(\Gamma/2)$, $f_T(0)$, $f_T(-3\Gamma/4)$. The total intensity, $f_T f_T^*$, is shown in Fig. 2. Notice how the intensity of the scattering is strongest for incident energies above the Mössbauer resonance, since both the x-ray and Mössbauer scatterings are approximately in phase. On the other hand, below the Mössbauer resonance the x-ray and Mössbauer scatterings interfere more destructively. For the case shown in Figs. 1 and 2, the form factors, f_M and f_X, are equal (approximately the case for peak 1 of the ferromagnetic sextet from a sample of bcc ^{57}Fe in the (200) diffraction condition). The x-ray intensity and Mössbauer intensity without interference are also shown in Fig. 2.

2.3. POLARIZATION

Photon polarization effects are well-known in Mössbauer spectrometry, but the situation in Mössbauer diffractometry is considerably more complex. For a particular photon polarization and orientation of the hyperfine magnetic field with respect to the photon wavevector, different amplitudes are associated with the nuclear absorption processes, where the angular momentum changes by $\Delta m = -1, 0, +1$. Similar issues occur for the nuclear emission. Further complexity occurs in many samples where the amplitude and polarization of the scattered photons also depend on coherent interference between nuclear transitions with different Δm, and depend on the interference between Mössbauer and x-ray scattering (which has a different polarization behavior). The details of this analysis are beyond the scope of this paper [9,10]. Fortunately, the main issues affecting the intensities of Mössbauer diffraction peaks from polycrystalline samples are understood. It is practical to design experiments to control the intensities of different diffractions from ferromagnetic samples by the application of magnetic fields.

For the case of no interference between nuclear transitions of different angular momenta (different Δm), the polarization factors have been calculated exactly for three orientation distributions of hyperfine magnetic fields [9,10]. The calculation required averaging the Mössbauer polarization factor, $p(h,u) = h \cdot u$, and $|p(h,u)|^2$ over all orientations of the nuclear hyperfine axis (h is the photon magnetic polarization vector, and u is a spherical unit vector associated with $\Delta m = -1, 0, 1$). This was performed by angular momentum algebra of the rotation matrices $D^l_{mm'}(\alpha\beta\gamma)$ [11]. The orientation distribution of hyperfine magnetic fields was expanded in spherical harmonics, and general expressions were obtained for the polarization factor for nuclear magnetic dipole radiation. The polarization factor versus angle was calculated for all coherent processes including the nuclear–x-ray interference contribution. A general solution for arbitrary HMF orientation distributions is accessible with this expansion of the HMF distribution in spherical harmonics, because only terms with index $l \leq 4$ give non-vanishing diffraction intensities for magnetic dipole radiation [9,10].

The interference between nuclear transitions of different angular momenta (different Δm) has been worked out for isotropic HMF distributions [12], but not yet in general. Nevertheless, the angular averages over the HMF distributions tend to wash out differences in the angular-dependence of the intensities of diffraction peaks [9]. It is probably acceptable for most samples to assume an isotropic orientation distribution for hyperfine magnetic fields, for which an exact solution is available.

2.4. ABSORPTION

The absorption or incoherent scattering in the sample from x-ray or Mössbauer scattering of the incident or outgoing diffracted beams affects differently the peaks in a Mössbauer diffraction pattern. This can be calculated with a multislice method assuming incoherent scattering [13], but can be estimated in the case of diffraction from the surfaces of samples having a thickness greater than 2 or 3 extinction lengths. In this case the intensity of diffraction peaks is proportional to [4,14]:

$$I \propto \frac{d\sigma/d\Omega}{\mu} . \qquad (7)$$

In cases where Mössbauer scattering dominates over x-ray scattering, we can ignore x-ray and interference contributions. With the cross-section $d\sigma/d\Omega$ proportional to the averaged square of the polarization factor, $<|p|^2>$, and the absorption coefficient μ obtained from the optical theorem and proportional to the average polarization factor in the forward direction, $<p>$, we calculated the effects of applied magnetic fields on the intensities of Mössbauer diffraction peaks [9]. For diffraction angles 2θ of $63°$ and $89°$, we found that (7) was approximately the same for all three nuclear transitions, $\Delta m = -1, 0, +1$ in cases of an isotropic orientation distribution of hyperfine fields, and for a distribution biased to lie in the plane of the sample [9,10]. With an applied magnetic field orienting the hyperfine axes vertically out of the plane of the diffractometer, however, it was found that the intensity for $\Delta m = 0$ was approximately twice as large as for $\Delta m = \pm 1$, enhancing diffractions when the Doppler drive was tuned to peaks 2 and 5 of the magnetic sextet of ^{57}Fe.

2.5. SPECTROSCOPIC DIFFRACTION

Mössbauer diffractometry is more interesting, and arguably more useful, when there is a distribution of nuclear energy levels in the sample, owing to differences in the hyperfine fields at different nuclei. The first issue of importance is understanding the phase and amplitude of the photons scattered from each nucleus. This requires knowledge of the hyperfine field distribution in the sample. Fortunately there are several ways that this distribution can be obtained from conventional Mössbauer spectra [15], and some of them are discussed in this NATO ARW. With knowledge of the hyperfine field distribution, the effective form factor for coherent scattering can be determined if we assume a spatial distribution for the different nuclei (so we can obtain the $P(r)$ of (3) and (4)). The $f_M(E)$, having a form similar to (6) with angular and polarization factors, and $f_X(E)$ are then summed numerically for each sample. Intrinsic in the calculation is the spatial distribution of the different atoms in the sample. The complexity of the factors described in sections 2.1–2.4 means that a numerical computation will generally be required to interpret the intensity of Mössbauer diffraction peaks. Nevertheless, the ultimate information available from Mössbauer diffractometry, not available from either Mössbauer spectrometry or any other diffraction method is: *the Patterson function of ^{57}Fe atoms in selected chemical environments.*

Some general guidelines can be offered for the design of Mössbauer diffraction experiments. Mössbauer diffraction is much easier to observe when the spectral intensity is localized in a narrow energy range. In cases where the Mössbauer spectrum comprises a broad distribution of peaks, coherent scattering can occur for only a fraction of the available nuclei. Although it is well-known that Mössbauer spectra become more difficult to measure as their intensity is distributed into different peaks, the problem is more serious for Mössbauer diffraction. Coherent intensities scale with the square of the number of nuclei that scatter with the same phase, so the suppression in the diffracted intensity is generally more severe for Mössbauer diffractometry than for Mössbauer spectrometry. The lost intensity occurs as incoherent scattering, and unfortunately contributes to the background of the diffraction pattern. Enrichment in [57]Fe will always be a requirement for Mössbauer diffraction experiments, but it is interesting to speculate on what fraction of atoms must be [57]Fe for diffraction peaks to be observed successfully. Our experience with count rates suggests that sufficient coherent scattering exists to make practical measurements on samples with 10-20% [57]Fe. This requires, however, a background countrate lower than has yet been achieved.

Another concern for the interpretation of Mössbauer diffraction patterns has been whether typical samples, enriched in [57]Fe, are in the kinematical limit for diffraction. Although dynamical effects were not evident through speedup of the nuclear decay and subsequent broadening of the diffraction peaks, this may not be a sufficiently sensitive criterion. Ongoing calculations of dynamical diffraction with the CONUSS package [16] have shown that the samples studied to date have indeed been in the kinematical limit, having coherent lengths of 0.1 μm or less [17]. Furthermore, kinematical theory reproduces the observed trends, as shown in section 4.

3. Experimental Methods

A Mössbauer diffractometer (Fig. 3) looks much like an x-ray diffractometer, the exception being the radioisotope source instead of an x-ray tube. The single most important component of a Mössbauer diffractometer is its detector. Our first experiments were enabled by an Inel CPS-120 position-sensitive detector, but it had serious technical limitations [6,18]. Our present Xe-filled Bruker area detector provided a reduction of 2 orders of magnitude in data acquisition time by increasing the geometrical efficiency to about 1 steradian, and efficiency for 14 keV photon detection to about 80%. Aluminum filters and detector bias voltage can be used to provide some energy discrimination, but this is crude – the Bruker detector is not designed for energy resolution. The extraneous photons it detects generate an unfortunate contribution to the background.

A new solid-state area detector has been developed for an application in x-ray astronomy, the High Energy Focusing Telescope, HEFT, where energy resolution and low noise are crucial [19]. The detector is a pixelated CdZnTe solid state detector of 1.3×2.4 cm area, comprising 26×48 pixels, each 0.5×0.5 mm. The detector array is bonded to an electronics card, and each pixel has its own chain of pulse electronics. A Wilkinson analog-digital converter is multiplexed between all detector pixels (limiting the countrate to approximately 1 kHz, but this is more than sufficient for Mössbauer diffractometry). Associated logic electronics provide an alert for each detected photon, which is identified in a digital output of energy and pixel. Detector efficiency at 14 keV is greater than 95 %. The exciting new capability is energy resolution. With detector cooling to − 20°C, the energy resolution is 350 eV at 14 keV, which allows a considerable reduction in background. With such resolution it is even possible to discriminate against Compton-scattered 14.4 keV photons.

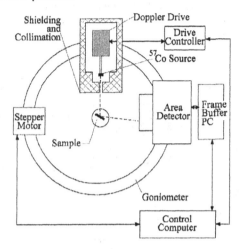

Figure 3. Schematic of a Mössbauer diffractometer.

4. Measurements and Interpretation

The unique *chemical environment selectivity* of Mössbauer diffractometry differs from the *chemical selectivity* often obtained by the anomalous scattering of x-rays. The well-known chemical selectivity modifies the scattering factor of a particular species of atom with respect to other species. Chemical *environment* selectivity picks out different chemical environments of the *same species of atom*. We demonstrated this experimentally by measuring diffraction patterns of DO_3-ordered Fe_3Al (Fig. 4a) [7]. The important point to notice about the DO_3 structure in Fig. 4a is the spatial periodicity of the two Fe sites. One of them (white, denoted 8(c)), has 4 Al neighbors (and 4 Fe neighbors). Its long-range periodicity is simple cubic, so the diffraction pattern from these sites should allow the (100)-type diffractions that are forbidden for bcc structures. The other site for Fe (striped, denoted 4(b)) has 0 Al neighbors. It is arranged in space as an fcc structure with a doubled unit cell. The diffraction pattern from this site should include diffractions in the (1/2 1/2 1/2) family, which are forbidden for the bcc and simple cubic structures. We found these unique diffraction peaks in Mössbauer diffraction patterns [7].

We measured diffraction patterns as in Fig. 4d over all Doppler shift energies in a sample of Fe_3Al. The areas under the Bragg peaks were integrated, giving coherent intensity spectra, some of which are shown in Fig. 4e (described in a preliminary report [20]). Note how the bcc fundamental peaks (211) and (222) show strong left-right asymmetry owing to interference with coherent x-ray scattering (unlike the incoherent spectrum of 1b). There is less asymmetry for the superlattice diffractions (300) and (5/2 3/2 3/2), since the x-ray scattering is much weaker. The solid curve was calculated from a simulation of the ordered alloy. Atom positions on a crystal were varied to produce the best fit to x-ray diffraction patterns and Mössbauer spectra of Fig. 4b. Calculations from these atom arrangements were performed using all interferences and polarization factors.

Roughly speaking, environments from ^{57}Fe atoms with 0, 3, and 4 Al neighbors contribute to the (300) intensity, whereas only the 0 Al environment contributes to the (5/2 3/2 3/2). A more detailed report is in progress.

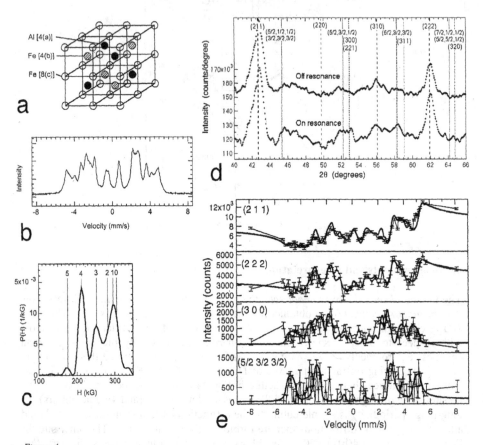

Figure 4.
a/ Fe$_3$Al with the DO$_3$ structure.

b/ Conversion electron Mössbauer spectrum of ^{57}Fe$_3$Al sample with high degree of DO$_3$ order.

c/ HMF distribution of (b) obtained with the method of Le Caër and Dubois [15]. Labels at top denote number of Al atoms in the first-neighbor shell of ^{57}Fe.

d/ Diffraction patterns for tuning of the Doppler drive on resonance near +3.5 mm/s and for +8.3 mm/s (off resonance). Some background variations are detector artefacts, but are consistent for the two diffraction patterns.

e/ Measured and calculated intensity spectra under the diffraction peaks (211), (222), (300), (5/2 3/2 3/2) from diffraction patterns acquired at a different constant velocities.

5. Future Science for Mössbauer Diffractometry

5.1. STRUCTURE

A most appropriate type of science for Mössbauer diffractometry uses the spectroscopic selectivity of local chemical environments about ^{57}Fe atoms. Chemical environments in ordered materials are possible to study this way, but well-ordered crystals are better studied by other methods. Mössbauer diffractometry will likely prove more important for measurements on materials with some degree of disorder. In particular, the technique is well suited for identifying correlations between defect sites in materials.

• Point defects in crystals interact by long-range elastic fields. We expect these interactions will cause some non-randomness in arrangements of point defects, but little is known of such structural characteristics.

• The earliest stages of chemical disorder-order transitions are poorly understood. Even with modern methods of diffractometry, microscopy, and spectrometry, it is often unclear if ordering begins homogeneously or heterogeneously, for example.

5.2. DYNAMICS

An alternative use of a Mössbauer diffractometer is for a new type of "quasielastic" scattering measurements on relaxation processes on the time scale of 10^{-9} to 10^{-7} s. Studies of "relaxation processes" have been performed for many years by conventional Mössbauer spectrometry, but all employed incoherent scattering. With a Mössbauer diffractometer, however, we can measure the spatial coherence of the relaxation processes. The measurements would be akin to coherent quasielastic neutron scattering [21], which has found applications, for example, in studies of diffusion [22], rotational dynamics [23], and motions of groups of atoms [24]. Possible applications for coherent inelastic Mössbauer scattering could be:

• Diffusional processes in solids with time scales of order 10^{-8} seconds have been studied by Mössbauer spectrometry for many years, e.g., [25-27]. Directions of wave coherence during the jump of an individual atom have been used to determine the vector direction of the diffusive jump [27]. With a Mössbauer diffractometer, such measurements could be performed without single crystals by measuring the energy spectra of different Bragg peaks. Furthermore, by examining fundamental and superlattice diffractions in some ordered structures, it should be possible to distinguish atom jumps between and within different sublattices.

• In all previous work on diffusion, interpretations of the data were in the incoherent approximation for spatial coordinates because only one nucleus was involved in each scattering. The diffusive motions were assigned to individual atoms, not collective motions. The motions of individual atoms produce a broad intensity as a function of Δk. Measurements with a Mössbauer spectrometer could distinguish this from correlated movements of groups of atoms, which would modulate the intensity in Δk.

• Some materials have valence instabilities where temperature causes Fe^{2+} and Fe^{3+} to mix between different crystallographic sites. In principle, Mössbauer diffraction patterns could measure the dynamics of valence at particular crystallographic sites.

• Mössbauer spectrometry is often used for measuring the superparamagnetic collapse of magnetically-ordered states in iron oxide nanoparticles, for example [28]. The Δk information provided by Mössbauer diffractometry could prove useful for identifying

294

correlation lengths over which the spins move together.

6. Conclusions

The understanding of the intensities in Mössbauer diffraction patterns is sufficiently complete to allow quantitative measurements on the structure of materials. In favorable cases it is possible to combine diffraction and spectroscopic data to measure the Patterson function of ^{57}Fe atoms in different chemical environments. Furthermore, it may be possible to use a Mössbauer diffractometer to measure coherent quasielastic scattering from cooperative motions of atoms, valences, or spins. The experimental technique still needs development. In particular, a detector with good energy resolution would make practical a much broader range of measurements on materials.

Acknowledgments

The authors acknowledge help with the experimental effort from Drs. U. Kriplani and M.W. Regehr. This work was supported by the National Science Foundation grant DMR-0204920.

References

1. Fultz, Brent and Howe, James M. (2001) *Transmission Electron Microscopy and Diffractometry of Materials*, Springer-Verlag, Heidelberg. ISBN 3-540-67841-7.
2. van Bürck, U., Smirnov, G.V., Mössbauer, R.L., Parak, F. and Semioschkina, N.A. (1978) "Suppression of nuclear inelastic channels in nuclear resonance and electronic scattering of γ-quanta for different hyperfine transitions in perfect ^{57}Fe single crystals", *J. Phys. C: Solid State* 11, 2305.
3. Kagan, Yu., Afanas'ev A.M., and Perstnev, I.P. (1968) "Theory of Resonance Bragg Scattering of γ Quanta by Regular Crystals", *Sov. Phys. JETP* 27, 819.
4. Hannon J.P., and Trammell, G.T. (1969) "Mössbauer Diffraction. II. Dynamical Theory of Mössbauer Optics", *Phys. Rev.* 186, 306.
5. Shvyd'ko, Yu.V. and Smirnov, G.V. (1989) "Experimental Study of Time and Frequency Properties of Collective Nuclear Excitations in a Single Crystal", *J. Phys.: Condens. Matter* 1, 10563.
6. Stephens, T.A., Keune, W., and Fultz, B. (1994) "Mössbauer Effect Diffraction from Polycrystalline ^{57}Fe" *Hyperfine Interact.* 92, 1095.
7. Stephens T.A. and B. Fultz, B. (1997) "Chemical environment selectivity in Mössbauer diffraction from ^{57}Fe$_3$Al", *Phys. Rev. Lett.* 78, 366.
8. Fultz, B. and Stephens, T.A. (1998) "Mössbauer Diffraction and Interference Studies of Polycrystalline Metals and Alloys", *Hyperfine Interact.* 113, 199-217.
9. Kriplani, U., Lin, J.Y.Y., Regehr, M.W., and Fultz, B. (2002) "Intensities of Mössbauer Diffractions from Polycrystalline ^{57}Fe" *Phys. Rev. B* 65, 024405.
10. Lin, J.Y.Y., Kriplani, U. Regehr, M. and Fultz, B. "Polarization Factors for ^{57}Fe Mössbauer Diffractions from Polycrystals", submitted to *Hyperfine Interact.*
11. Edmonds, A.R. (1996) *Angular Momentum in Quantum Mechanics*, Princeton, New Jersey.
12. Nakai, Y., Ooi, Y., and N. Kunitomi, N. (1988) "Mössbauer Diffraction Spectra in Partially Ordered Ni$_3$Fe Alloy", *J. Phys. Soc. Jpn.* 57, 3172.
13. Fultz, B., Stephens, T.A., Lin, J.Y.Y., and Kriplani, U. (2002) "Mossbauer diffractometry on polycrystalline (Fe$_3$Al)-Fe-57" *Phys. Rev. B* 65, 064419.

14. O'Connor D.A. and Black, P.J., (1964) "The theory of the nuclear resonant and electronic scattering of resonant radiation by crystals" *Proc. Phys. Soc. London* **83**, 941.

15. We used the method described in: LeCaër, G. and Dubois, J.M. (1979) "Evaluation of hyperfine parameter distributions from overlapped Mössbauer spectra of amorphous alloys", *J. Phys. E* **12**, 1083.

16. Sturhahn, W. and Gerdau, E. (1994) Evaluation of time-differential measurements of nuclear-resonance scattering of x-rays, *Phys. Rev. B* **49**, 9285-9294.

17. Lin, J.Y.Y. and Fultz, B., unpublished results.

18. Kriplani, U., Regehr, M.W., and Fultz, B. "A Mössbauer Effect Powder Diffractometer", *Hyperfine Interact.*, in press.

19. Bolotnikov, A.E., Cook, W.R., Boggs, S.E., Harrison, F.A., Schindler, S.M. (2001) "Development of high spectral resolution CdZnTe pixel detectors for atronomical hard X-ray telescopes", *Nucl. Instr. Methods in Phys. Res. A* **458**, 585.

20. Lin, J.Y.Y., Fultz, B., and Kriplani, U., "Mössbauer Diffractometry on Chemical Sites of ^{57}Fe in Fe$_3$Al," submitted to *Hyperfine Interact.*

21. Gillian, M.J., and Wolf, D. (1985) "Point-Defect Diffusion from Coherent Quasielastic Neutron Scattering", *Phys. Rev. Lett.* **55**, 1299.

22. Wilmer, D., Funke, K., Witschas, M., Banhatti, R.D., Jansen, J., Korus, G., Fitter, J., Lechner, R.E. (1999) "Anion reorientation in an ion conducting plastic crystal - coherent quasielastic neutron scattering from sodium ortho-phosphate", *Physica B* **266**, 60.

23. Neumann, D.A., Copley, J.R.D., Cappelletti, M.L., Kamitakahara, W.A., Lindstrom, R..M., Creegan, K.M., Cox, D.M., Romanow, W.J., Coustel, N., McCauley, Jr., J.P., Maliszewskyj, N.C., Fischer, J.E., and Smith, III, A.B. (1991) "Coherent Quasielastic Neutron Scattering Study of the Rotational Dynamics of C$_{60}$ in the Orientationally Disordered Phase", *Phys. Rev. Lett.* **67**, 3808.

24. Carlsson, P., Zorn, R., Anderson, D., Farago, B., Howells, W.S., and Borjesson, L. (2001) "The segmental dynamics of a polymer electrolyte investigated by coherent quasielastic neutron scattering", *J. Chem. Phys.* **114**, 9645.

25. Singwi, K.S., and Sjölander, A. "Resonance Absorption of Nuclear Gamma Rays and the Dynamics of Atomic Motions", (1960) *Phys. Rev.* **120**, 1093.

26. K.Ruebenbauer, J. G. Mullen, G. U. Nienhaus, and G. Schupp "Simple model of the diffusive scattering law in glass-forming liquids", *Phys. Rev. B* **49**, 15607 (1994).

27. Sepiol, B. and Vogl, G. (1993) "Atomistic Determination of Diffusion Mechanism on an Ordered Lattice", *Phys. Rev. Lett.* **71**, 731. Vogl G., and Sepiol, B. (1994) "Elementary Diffusion Jump of Iron Atoms in Intermetallic Phases Studied by Mössbauer Spectroscopy — I. Fe-Al Close to Equiatomic Stoichiometry", *Acta Metall. Mater.* **42**, 3175.

28. Bodker, F., Hansen, M.F., Koch, C.B., Lefmann, K., and Morup, S. (2001) "Magnetic properties of hematite nanoparticles", *Phys. Rev. B* **61**, 6826. Hansen, M.F., Koch, C.B., and Morup, S. (2001) "Magnetic dynamics of weakly and strongly interacting hematite nanoparticles", *Phys. Rev. B* **62**, 1124.

EMISSION MÖSSBAUER SPECTROSCOPY IN CoO

U.D. WDOWIK AND K. RUEBENBAUER
Mössbauer Spectroscopy Division, Institute of Physics,
Pedagogical University, PL-30-084 Cracow, ul. Podchorążych 2, Poland

1. Introduction

Cobalt oxide, $Co_{1-x}O$, is an oxygen excess p-type semiconductor which has the NaCl structure with a lattice parameter of 4.26 Å [1]. $Co_{1-x}O$ exhibits a non-stoichiometry, x, which is due to the presence of cobalt sublattice defects, mainly cationic vacancies as dominant point defects [1,2]. The deviation from stoichiometry, x, increases with the increasing oxygen partial pressure. The Co^{2+} vacancies form acceptor levels in the band gap of 2.8 eV [1] and act as an electron trap centers. The Co vacancies can be un-ionized, single or doubly ionized [2]. The defect structure of cobalt oxide involving x non-interacting and randomly distributed Co vacancies can be described by the following equation (according to Kröger-Vink notation):

$$\tfrac{1}{2}O_2 \leftrightarrow V_{Co}^{\alpha'} + \alpha h^{\bullet} + O_O,$$ (1)

where $V_{Co}^{\alpha'}$, $h^{\bullet}$, and O_O denote cation vacancy ionized α times, electron hole, and an anion on the oxygen lattice site, respectively. Charge balance is maintained through the formation of electron holes having high mobility [2]. The condition of electrical neutrality is expressed as follows:

$$[h^{\bullet}] = [V_{Co}'] + 2[V_{Co}''],$$ (2)

where brackets denote molar concentration of particular defects. The total vacancy concentration [x] giving the oxygen excess is written as

$$[x] = [V_{Co}^{x}] + [V_{Co}'] + [V_{Co}''].$$ (3)

Due to the presence of neutral and singly charged vacancies some trivalent Co ions have to be present as well.

Cobalt oxide doped with ^{57}Co has been the subject of several Mössbauer investigations [3-14]. The results of these experiments have shown that ^{57}Fe exists in both divalent and trivalent states in CoO. Some controversial explanation of the presence of nucleogenic Fe^{2+} and Fe^{3+} has been given [3-14].

The main aim of the present paper is to report emission Mössbauer spectroscopy results obtained vs. elevated temperature and oxygen pressure.

M. Mashlan et al. (eds.), Material Research in Atomic Scale by Mössbauer Spectroscopy, 297–306.
© 2003 *Kluwer Academic Publishers. Printed in the Netherlands.*

2. Experimental

CoO single crystal plates of 80 μm thickness and with <111> direction perpendicular to the surface of dimensions 5 x 4.3 mm were prepared. Electron back-scattering diffraction did not revealed significant mosaicity. The sample contained neither Co_3O_4 nor Co_2O_3 phases as was confirmed by standard X-ray diffraction. A Mössbauer (^{57}Co)CoO source was made by diffusing in 10 mCi of carrier-free ^{57}Co leading to about 40 p.p.m. of radioactive Co and nucleogenic Fe altogether. Due to the phase diagram of CoO-Co_3O_4 system [15] the sample was slowly heated up to 1123 K at low oxygen pressure and subsequently the diffusion anneal was performed at 1333 K in air for 3 hours.

The sample was held in a Pt boat and was kept in a furnace equipped with an uniaxial goniometer allowing for the sample rotation *in situ*. The crystal remained uncovered from the top. Mössbauer spectra were measured vs. temperature, oxygen partial pressure, and sample orientation. Chemical pumps maintained oxygen pressure at 10^{-4} atm. or 10^{-7} atm. depending upon oxygen absorbing agent used. Additionally, measurements were performed at high temperature under oxidizing atmosphere (air). Temperature range covered extended from RT up to 1450 K. The temperature was controlled by Pt-Pt(10%Rh) thermocouple and stabilized to within 1 K by computer based virtual PID controller. Measurements were performed for the beam outgoing in the [110] plane along various directions. Here we report results obtained for <111> direction of the beam. A single line $K_4{}^{57}Fe(CN)_6 \times 3H_2O$ vibrating absorber in conjunction with Kr-filled proportional detector was used. Mössbauer data were collected using the equipment described in Ref. [16]. Spectra were fitted to multiple sites described by the hyperfine Hamiltonians assigned to each site. A Lorentzian approximation was used as the source resonant thickness is negligible [17]. Typical values of χ^2 ranged from 0.9 to 1.1, while MISFIT [18] ranged from 0.01% at low temperatures to 2% at the highest temperatures. Due to the fact that the sample became polycrystalline during the measurements spectra obtained in the high diffusivity region represent averages over the directions.

3. Results and Discussion

RT Mössbauer spectra of ^{57}Fe in CoO, Fig. 1, consist of two singlets corresponding to Fe^{2+} and Fe^{3+}, both in a high spin configuration. Additionally, a magnetically split Fe^{2+} component is observed yielding the hyperfine filed of 6 T at RT. CoO is an antiferromagnetic material with a Néel temperature T_N of about 291 K [6,7]. Since the RT measurements were performed in the vicinity of T_N (no temperature stabilization was applied) the single lines represent paramagnetic state, while the hyperfine pattern arises from the antiferromagnetic state. Divalent iron couples magnetically to the cobalt ordered spins more easily than trivalent iron.

3.1. Co$_{1-x}$O UNDER VARIOUS OXYGEN PARTIAL PRESSURES

Emission Mössbauer spectra were measured vs. temperature and oxygen partial pressure, the latter controlling the degree of non-stoichiometry, i.e., concentration of cation vacancies. Resultant spectra are shown in Figs 2–4. Ferrous and ferric lines observed at RT spectra tend to converge to a single mixed-valence state due to the electronic relaxation. The charge averaging leads to the IS and FWHM temperature dependence as shown in Figs 5-6. The mixed-valence iron was observed in the whole temperature/pressure range studied.

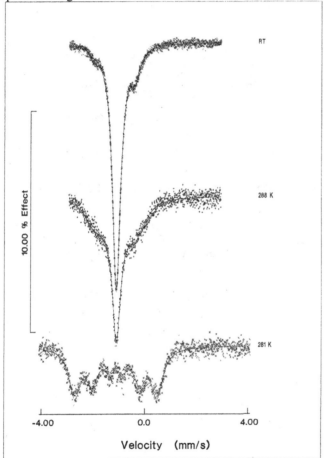

Figure 1. Mössbauer spectra of (^{57}Co)CoO as a function of temperature.

Cobalt vacancies and trivalent Co ions forming acceptor levels or quasi-continuous band in the energy gap can trap electrons coming either from Fe impurity or valence band. During the negative charge transfer between the iron and the vacancy level a trivalent iron can be created. Schematic diagram of Co$_{1-x}$O band structure is shown in

300

Fig. 7. The Fe impurity valence state is both temperature and pressure dependent. Increasing temperature leads to charge averaging of the daughter Fe. Due to the temperature dependence of the average isomer shift (Fig. 5), which tends to IS characteristic for 2+ state, the vacancy levels (band) are supposed to be located above the Fe level. The relaxation frequency is temperature dependent as well. Initially separated lines at RT corresponding to different valence states of Fe converge into a single mixed-valence line with increasing temperature. Relaxation is also responsible for the linewidth behavior vs. temperature especially up to 1280 K (Fig. 6). A diffusional line broadening is observed above this temperature.

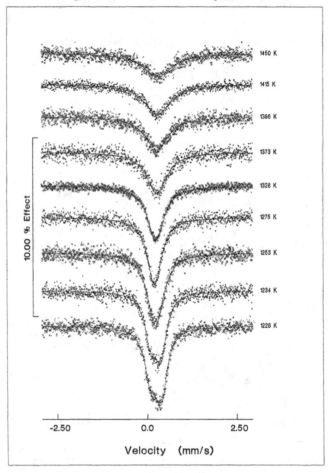

Figure 2. Mössbauer spectra of (^{57}Co)CoO vs. temperature under oxidizing atmosphere.

The average IS in the vicinity of 750 K increases rapidly (Fig. 5b) approaching the value characteristic for the iron in 3+ charge state. A similar behavior was previously encountered [5-7,10,13,14], but no satisfactory explanation was given. It is suggested that some kind of structural transformation occurring in the vicinity of 700 K accounts for the observed effect. Structural transformation may be due to decomposition of vacancy clusters (probably decorated with trivalent cobalts), the latter being created during the slow cooling of the CoO sample from high temperature. Hence, a vacancy band can broaden due to the vacancy clusters decomposition. The cluster's band might be located either just beneath the conduction band or can be completely submerged in the valence band. In both cases the cluster's band is not available for electrons. Single vacancy levels, which can also form a very narrow band are the sole electron trap centers participating in the relaxation. Such a behavior results in the nearly constant IS up to 750 K (Fig. 5b).

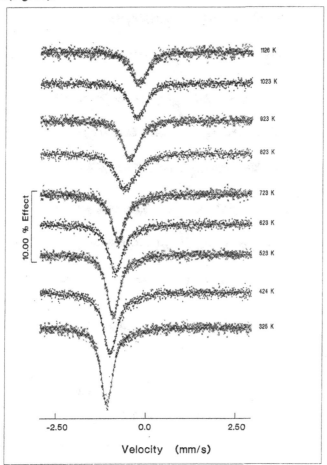

Figure 3. Mössbauer spectra of (^{57}Co)CoO vs. temperature under reduced oxygen partial pressure.

Broadening of the vacancy band generates some additional acceptor levels and probably increases charge mobility in this band leading to the effective lowering of the relaxation barrier. This results in the average IS to follow the 3+ iron valence state. In the temperature range showing the rapid increase of IS the linewidth broadening is also observed (see Fig. 6b) due to the relaxation. When the negative charge is transferred from the valence band to any empty level above, a chance to charge Fe to the 2+ state is increased. This process is particularly strong above ca. 800 K (see Fig. 5). This process is also observed in the diffusional temperature range. The existence of trivalent and divalent iron producing well separated lines in RT Mössbauer spectra can be explained by a slow relaxation occurring at low temperature. The increasing temperature increases the relaxation frequency leading to the charge averaging of the ^{57}Fe impurity.

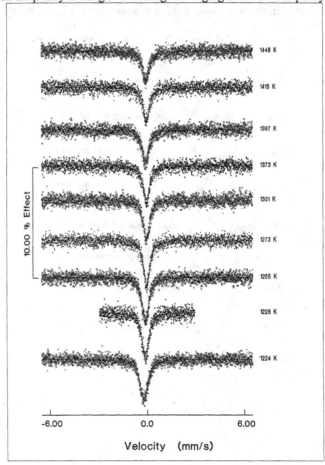

Figure 4. Mössbauer spectra of (^{57}Co)CoO vs. temperature under reduced oxygen partial pressure (continued).

Diffusionally broadened spectra measured in air were used to determine the temperature dependence of the diffusivity D. The diffusion coefficient follows the Arrhenius law as shown in Fig. 8. Activation energy of 1.42(3) eV stays in a good agreement with the literature data, the latter being 1.4 eV or 1.3 eV [14]. During the measurements at a very low oxygen partial pressure and at temperatures falling in the 700 K-800 K range the sample became polycrystalline. This excluded the possibility to measure the directional dependence of the line broadening, i.e., to study the microscopic diffusion mechanism. However, the data taken at the reduced oxygen partial pressure suggest that the impurity diffuses through the cobalt sublattice by the vacancy mechanism. Under reduced oxygen partial pressure $Co_{1-x}O$ goes to more stoichiometric CoO due to the lower vacancy concentration. Diffusion jumps are assumed to lead Fe impurity to the vacancies located at the nearest neighbors (NN).

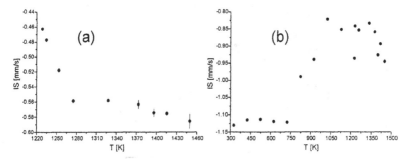

Figure 5. Temperature dependence of the average isomer shift vs. metallic iron for sample investigated in air (a) and sample kept under low oxygen pressure (b). IS values were corrected for the second order Doppler shift.

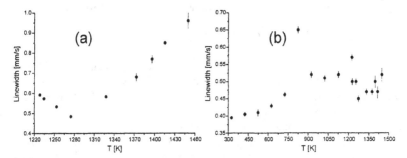

Figure 6. Temperature dependence of linewidth for sample investigated under oxidizing atmosphere (a) and for sample kept at low oxygen pressure (b).

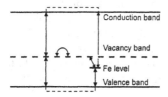

Figure 7. Band structure of $Co_{1-x}O$.

A small number of vacancies in the neighborhood results in the very small diffusional line broadening when the sample is examined under the reduced pressure, in contrast to measurements performed at oxidizing atmosphere (see Fig. 6). The vacancy mechanism of diffusion is characterized by the vastly different pre-exponential factors depending upon the oxygen pressure. Pre-exponential factors for CoO measured in air and under low oxygen pressure equal $7(1) \times 10^{-6}$ cm^2/s and $9(2) \times 10^{-7}$ cm^2/s, respectively. They could be used to estimate limiting frequency, ω_0, of the diffusion jumps to NN. Estimated values of ω_0 are 1.3 GHz and 55 MHz for the iron diffusing at oxidizing and reducing atmospheres, respectively.

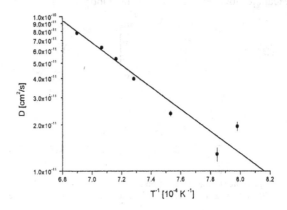

Figure 8. Arrhenius plot for the diffusion coefficient of the CoO sample at oxidizing atmosphere.

3.2. MÖSSBAUER SPECTROSCOPY IN Co$_3$O$_4$

Co$_3$O$_4$ source was obtained by oxidizing the (^{57}Co)CoO in air at 1123 K for about 6 hours. Subsequently the sample was cooled down to RT. Mössbauer spectra of Co$_3$O$_4$ were measured as a function of increasing temperature. Measurements were performed in air and in the temperature ranging from RT to 1123 K. Co$_3$O$_4$ has the normal spinel structure [19-21]. The spinel structure consists of two metal-ion sublattices held together by oxygens in the fcc arrangement. One sublattice has the cations tetrahedrally surrounded by oxygens, while the other sublattice has the metal ions surrounded octahedrally. There are twice as many octahedral (B) sites as tetrahedral (A) sites. Each oxygen atom has as nearest neighbors three metal ions at B sites and one metal ion at A site. A small trigonal distorsion of anion sublattice leads to an EFG at octahedral (B) sites, but not at tetrahedral (A) sites. The simplest spinel configuration involves cations in 2+ and 3+ charge states only. The normal spinel structure can be expressed as $(M_A)^{2+}[(M_B)_2^{3+}]O_4$. One can express the inverse structure as $(M_B)^{2+}[(M_A)^{3+}(M_B)^{3+}]O_4$. A mixed structure is partly normal and partly inverse, with the degree of inversion being dependent on preparation. Co^{2+} and Co^{3+} ions occupy A and B sites in Co$_3$O$_4$, respectively [19].

Since the effects of the decay are not sufficient to remove the nucleogenic Fe from the initial cobalt lattice position, the hyperfine parameters of Mössbauer spectra obtained between RT and 1073 K could be used to determine and verify the distribution of Co over the cobalt sublattices. The daughter Fe was found to occupy both tetrahedral and octahedral Co lattice sites. Tetrahedrally located Fe is in the 3+ valence state, even though the parent Co in A site is divalent. It is well known that the charge state of the daughter Fe produced by the Co decay is not always the same as that of the parent cobalt [16]. CoO is an example in which initially divalent Co decays to both Fe^{2+} and Fe^{3+} with the ratio of valences being dependent on the preparation and sample temperature [5-8,12] (see also previous subsection).

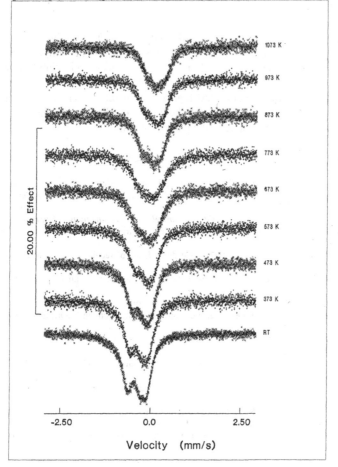

Figure 9. Mössbauer spectra of Co_3O_4 vs. temperature.

The Mössbauer spectra of ^{57}Fe in Co_3O_4 shown in Fig. 9 consist of a symmetrical doublet and a singlet. The doublet originates from the decay of Co^{3+} to Fe^{3+} at the B site having trigonal symmetry, while the singlet arises from Co^{2+} decaying to Fe^{3+} at A site being cubic. The isomer shifts of A-site line (-0.334(2) mm/s at RT) and B-site doublet (-0.413(1) mm/s at RT) are typical for the Fe^{3+} located at tetra- and octahedral spinel sites [19,21]. Relative contributions of the doublet and singlet to the Mössbauer spectrum (about 80% and 20% at RT, respectively) approach corresponding occupation by Co (1/3 of cations at the A sites and 2/3 of Co atoms at the B sites) above 573 K. Hence, some contribution to the doublet comes from tetrahedral iron at low temperature, probably due to some defects generated during the radioactive decay in the vicinity of the tetrahedral site. These defects anneal prior to formation of the excited Mössbauer level at elevated temperatures.

4. Conclusions

The main conclusion is that cation diffusivity in CoO strongly depends upon the stoichiometry controlled by the oxygen pressure at high temperatures. Cationic diffusion is quite fast far from stoichiometry and it occurs via simple mono-vacancy mechanism. Arrhenius law is well obeyed. Another important point is that iron impurity stays in the mixed-valence state, and that a charge relaxation occurs at intermediate temperatures. In the case of Co_3O_4 tetrahedrally coordinated daughter iron forms 3+ charge state despite divalent state of the parental cobalt.

References

1. Intersciences, New York.
2. Persels Constant, K., Mason, T.O., Rothman, S.J., and Routbort, J.L. (1992), Non-stoichiometry, electrical properties and cation diffusion in highly non-stoichiometric $Co_{1-x}O$-I. Experimental, *J. Phys. Chem. Solids* **53**, 405-411.
3. Kofstad, P. (1977), *Non-stoichiometry, diffusion and electrical conductivity in binary metal oxides*, Wiley Wertheim, G.K. (1961) Hyperfine structure of divalent and trivalent Fe^{57} in cobalt oxide, *Phys. Rev.* **124**, 764-767.
4. Triftshäuser, W. and Craig, P.P. (1967) Time dependence of recoil-free resonance following electron capture in Co^{57}, *Phys. Rev.* **162**, 274-285.
5. Mullen, J.G. and Ok, H.N. (1966) New results on the question of auger aftereffects, *Phys. Rev. Lett.* **17**, 287-290.
6. Ok, H.N. and Mullen, J.G. (1968) Evidence of two forms of cobaltous oxide, *Phys. Rev.* **168**, 550-562.
7. Ok, H.N. and Mullen, J.G. (1968) Magnetic properties of iron ions in CoO(I) and CoO(II), *Phys. Rev.* **168**, 563-574.
8. Schroeer, D. and Triftshäuser, W. (1968) Reinterpretation of data on multiple charge states of Fe in CoO, *Phys. Rev. Lett.* **20**, 1242-1245.
9. Trousdale, W. and Craig, P.P (1968) Limitations on relaxation rates of ^{57}Fe in CoO, *Phys. Lett.* **27**, 552-553.
10. Helms, W.R. and Mullen, J.G. (1971) Study of pure and doped cobaltous and nickelous oxide, *Phys. Rev. B* **4**, 750-757.
11. Bhide, V.G. and Shenoy, G.K. (1966) Temperature-dependent lifetimes of nonequilibrium Fe^{3+} ions in CoO from the Mössbauer effect, *Phys. Rev.* **147**, 306-310.
12. Song, C. and Mullen, J.G. (1976) Mössbauer evidence for relaxation valence averaging in CoO and NiO, *Phys. Rev. B* **14**, 2761-2768.
13. Harami, T., Loock, J., Huenges, E., Fontcuberta, J., Obradors, X., Tejada, J., and Parak, F. (1984) The influence of semiconductor properties on the Mössbauer emission spectra of ^{57}Co cobalt oxide, *J. Phys. Chem. Solids* **45**, 181-190.
14. Fontcuberta, J. (1987) Lattice dynamics study of polycrystalline ^{57}CoO by Mössbauer spectroscopy, *phys. stat. solidi b* **139**, 379-386.
15. Przybylski, K. and Smeltzer, W.W. (1981) High temperature oxidation mechanism of CoO to Co_3O_4, *J. Electrochem. Soc.* **128**, 897-902.
16. Wdowik, U.D. and Ruebenbauer, K. (2001), *Phys. Rev. B* **63**, 125101(8).
17. Mullen, J.G., Djedid, A., Schupp, G., Cowan, D., Cao, Y., Crow, M.L, and Yelon, W.B. (1988) Fourier-transform method for accurate analysis of Mössbauer spectra, *Phys. Rev. B* **37**, 3226-3245.
18. Ruebenbauer, K. (1981), Institute of Nuclear Physics, Cracow, Report No. 1133/PS.
19. Kündig, W., Kobelt, M., Appel, H., Constabaris, G., and Lindquist, R.H. (1969) Mössbauer studies of Co_3O_4; bulk material and ultrafine particles, *J. Phys. Chem. Solids* **30**, 819-826.
20. Cruset, A. and Friedt, J.M. (1971) Mössbauer study of the valence state of ^{57}Fe after ^{57}Co decay in $CoFe_2O_4$, *phys. stat. solidi b* **45**, 189-193.
21. Spencer, C.D. and Schroeer, D. (1974) Mössbauer study of several cobalt spinels using Co^{57} and Fe^{57}, *Phys. Rev. B* **9**, 3658-3665.

COMPARATIVE TRANSMISSION AND EMISSION MÖSSBAUER STUDIES ON VARIOUS PEROVSKITE-RELATED SYSTEMS

Z. Homonnay[1], Z. Klencsár[2], K. Nomura[3], G. Juhász[1], E. Kuzmann[2], G. Gritzner[4], A. Nath[5] and A. Vértes[1,2]

[1] *Department of Nuclear Chemistry, Eötvös Loránd University,*
Pázmány P. s. 1/A, 1117 Budapest, Hungary
[2] *MTA-ELTE Research Group on Nuclear Methods in Structural Chemistry,*
Pázmány P. s. 1/A, 1117 Budapest, Hungary
[3] *Graduate School of Engineering, the University of Tokyo,*
Hongo 7-3-1, Bunkyo-ku, Tokyo 113-8656, Japan
[4] *Institut für Chemische Technologie Anorganischer Stoffe, Johannes Kepler Universität,*
A-4040 Linz, Austria
[5] *Department of Chemistry, Drexel University,*
Philadelphia PA 19104, USA

1. Introduction

A whole group of compounds which have the ABO_3 stoichiometry and meet the appropriate conditions on the ratio of ionic sizes and metal-oxygen bond lengths is named after the well known mineral, perovskite ($CaTiO_3$). The basic structure is cubic as depicted in Fig. 1. The stability of this structure depends on the cations occupying the A and B sites. Variation of the cations at these sites produces an astonishingly wide variety of magnetic and electric properties [1]. For example, the originally dielectric $CaTiO_3$ becomes a ferroelectric material when Ca is substituted by Ba. Partial substitution of Ti by Zr, with Pb at the A site results in a piezoelectric substance. $(Ba,La)TiO_3$ is a semiconductor, while a wide range of cuprates like $YBa_2Cu_3O_{7-d}$ shows high temperature superconductivity. One of the hottest topic in perovskite chemistry and physics today is the colossal magnetoresistance effect which is found for example in $(La,Sr)MnO_3$, $(La,Sr)(Fe,Co)O_3$ and other compounds (although the perovskite structure is not a precondition for the CMR effect).

The great variability of the electric and magnetic properties of the perovskites is a consequence of the chemical properties of the constituent metals, mostly of those occupying the B site. The 12-fold oxygen-coordinated A site cations are typically alkaline earths (in the oxide perovskites) which have a strong ionic character and stable valence state. In contrast, the B site cations are often multivalent and have a tendency to form covalent bonds with the coordinated oxygen atoms. Thus one of the key factor in the determination of the physical and chemical properties of a given compound is the

M. Mashlan et al. (eds.), Material Research in Atomic Scale by Mössbauer Spectroscopy, 307–316.
© 2003 *Kluwer Academic Publishers. Printed in the Netherlands.*

balance between the covalency and ionicity of the B-O bonds. The valency of the B cations depends mostly on the oxygen stoichiometry which is determined by the conditions during the preparation step.

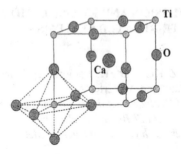

Figure 1. The cubic structure of perovskite (CaTiO$_3$) showing the octahedral oxygen coordination of the corner Ti^{4+} cations (site B).

If the B sites of the investigated compounds contain Fe, Mössbauer spectroscopy is the ultimate tool to study the electronic structure and the local environment at the B site. In this case, the analysis of the Mössbauer spectrum provides direct information on the electronic density at the ^{57}Fe nucleus, on the asymmetry of the charge distribution around it, on the magnetic field, etc. through the hyperfine interactions. The emission version of Mössbauer spectroscopy may be advantageously used if the B site is populated, partially or fully, by Co. The sample is then doped with ^{57}Co, mother nuclide of ^{57}Fe, and the spectrum is recorded with the help of a standard absorber. In this case, the interpretation of the Mössbauer spectra needs careful consideration because the spectrum is recorded on the nucleogenic ^{57}Fe produced by the ^{57}Co(EC)^{57}Fe decay. Faithful or false probing of the originally Co-populated site depends basically on the electronic relaxation which is always fast in systems with highly delocalized (metallic) states but is rather slow in insulators. Since the electric conductivity of the perovskites varies over a wide range, a general rule cannot be established and each case must be investigated carefully.

The most difficult situation is encountered when the target material does not contain a Mössbauer nuclide. Then, if there are suitable elements in the compound which can be substituted by Fe or Co (or other Mössbauer nuclides), impurity Mössbauer spectroscopy may be used. For such compounds, Fe- or Co-doping is also possible, and the latter has the advantage of much less necessary dopant quantity if the host material contains heavy elements.

In this paper, we show the application of emission Mössbauer spectroscopy on three main systems where this technique yielded unique information. Special attention was paid to the comparison of the results from emission and parallel transmission measurements. We will show how the deleterious effect of Pr on the superconductivity of YBa$_2$Cu$_3$O$_{7-d}$ could be successfully studied by impurity Mössbauer spectroscopy using ^{57}Co dopant. We studied Sr$_x$Ca$_{1-x}$Fe$_{0.5}$Co$_{0.5}$O$_{3-d}$ where the distribution of Fe and Co in the lattice could be investigated by comparison of the transmission and emission spectra. Finally, we report on a recent study of La$_{0.8}$Sr$_{0.2}$Co$_{0.95}$Fe$_{0.05}$O$_{3-d}$ CMR material using emission and transmission Mössbauer spectroscopy.

2. Emission Mössbauer Study of the $Y_xPr_{1-x}Ba_2Cu_3O_{7-d}$ System

The unit cell of $YBa_2Cu_3O_{7-d}$ can be derived from a tripled perovskite unit cell. The A sites are populated by Y and Ba while the B sites are taken by Cu. Due to the many empty O-sites, there are two kinds of Cu sites, namely the pyramidally coordinated Cu(2) sites (supposed to be responsible for superconductivity) and the square planarly coordinated Cu(1) sites, the latter forming Cu-O chains in the lattice (see Fig. 2). $YBa_2Cu_3O_{7-d}$ is a 90 K superconductor, although if Pr is substituted for Y, the superconducting critical temperature decreases gradually, and superconductivity completely disappears when 55 % of Y is substituted by Pr (in conventionally prepared material) [2].

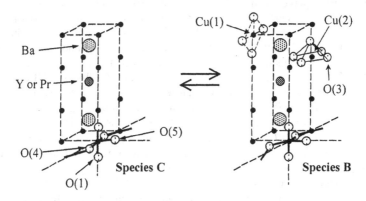

Figure 2. Transformation of the square pyramidally coordinated $[^{57}CoO_5]$ species (C) into a trigonal bipyramidal one (B) in the lattice of partially Pr-substituted $YBa_2Cu_3O_{7-d}$ For the sake of clarity, only one original Cu(1) and Cu(2) site for each are shown with their coordinated oxygens. Note that O(5) positions are populated only when Co is substituted for Cu.

The unit cell is orthorhombic at O_7 oxygen stoichiometry, and becomes tetragonal when it is reduced to O_6. The explanation why superconductivity is killed by Pr is of crucial interest because it may shed light on the mechanism of high temperature superconductivity.

We have found earlier [3] that in $YBa_2Cu_3(^{57}Co)O_{7-d}$ two species identified in the emission Mössbauer spectra convert into each other at about 280-290 K. A model was proposed to account for this phenomenon, knowing that the Cu(1)-O(4) chains in the lattice has a zigzag structure rather than a straight configuration. This zigzag structure assumes flexibility for the chains and it may have relevance to the mechanism of high-T_c superconductivity through enhancing electron-lattice interaction, a fundamental aspect of the classical BCS theory for low temperature superconductivity.

From among the four doublets identified in the Mössbauer spectrum of the superconducting phase, one was assigned to ^{57}Co (i.e.: nucleogenic ^{57}Fe) substituted for Cu(2) site, and three were assigned to the Cu(1) site. The multiple representation of the Cu(1) sites in the Mössbauer spectra was explained by the chance for population of the normally vacant O(5) site, and it could be easily explained by the preferred coordination number of Co, higher than that of Cu. The model was completed by the assumption of a four-coordinate [CoO$_4$]-species and two kinds of five-coordinate [CoO$_5$]-species at the Cu(1) sites. One of the five-coordinate species has a square pyramidal (C), the other one has a trigonal bipyramidal structure (B), and they are in thermal equilibrium (Fig. 2).

The essence of our finding [4] was that the intensity of the so called "B to C" interconversion proved to be dependent on the Pr substitution level at the Y sites. More interestingly, the B to C interconversion vanished when superconductivity vanished. Of course, direct correlation between the superconducting critical temperature and the B to C interconversion may not be expected since B to C interconversion takes place far above T_c. However, the mere chance for this interconversion and its dependence on the concentration of superconductivity killer Pr supports some correlation between the mechanism of high-T_c superconductivity and the vibrations of the Cu(1)-O(4) chains. Indeed, it was known that increasing Pr substitution at the Y site results in an increasing Cu(1)-Cu(1) distance along the chains. Actually in the pure Pr compound PrBa$_2$Cu$_3$O$_{7-d}$, this distance is larger than twice the Cu(1)-O(4) distance in the zigzag chain of YBa$_2$Cu$_3$O$_{7-d}$ [5]. This means that if the zigzag structure of the chains in YBa$_2$Cu$_3$O$_{7-d}$ is "forced" by the too short Cu(1)-Cu(1) distance, there is no more reason to be so in PrBa$_2$Cu$_3$O$_{7-d}$. Therefore, B to C interconversion cannot be observed.

In summary, our findings led us to the conclusion that one possible reason of killing superconductivity in YBa$_2$Cu$_3$O$_{7-d}$ by Pr is the quenching of a vibrational mode of the Cu(1)-O(4) chains. This is particularly interesting because the Cu(2)-O(3) sheets are believed to be hosting the superconducting charge carriers.

From the viewpoint of Mössbauer spectroscopy, it is puzzling why the B to C interconversion is *not* observed in transmission experiments [6] in spite of the facts that the same lattice positions are substituted by Fe, and that the same doublets are observed in the Mössbauer spectra. The most likely explanation is that in transmission experiments the applied concentration of the Fe dopant is at least by two order of magnitude higher than that of ^{57}Co in Co-doped samples, thus coalescence of the Fe species (clusterization) can be expected. If the Mössbauer parameters are sensitive to the first coordination sphere only, this process will not alter the parameters of the doublets only their relative intensity (as observed). The different character of the Fe-O and Co-O bonds cannot cause the different behavior because the 10^{-7} s that elapses from the EC decay to the emission of the Mössbauer gamma quantum is long enough to stabilize a chemical environment preferred by Fe. However, Fe prefers higher oxidation state and higher coordination number than Co as deduced from their chemistry. Thus a possibly higher O(5) concentration in the vicinity of Cu(1)-substituted Fe dopants may make the lattice too rigid to allow B to C interconversion. Thus we are witnessing a case when the chemical difference between Fe and Co widens the horizon of using Mössbauer spectroscopy to observe new phenomena not seen by the transmission technique only.

The significance of the chemical difference between Fe and Co in the interpretation of transmission and emission Mössbauer spectra could be well demonstrated in another perovskite system, (Ca,Sr)(Fe,Co)O$_{3-d}$.

3. Emission and Transmission Mössbauer Study of $Sr_xCa_{1-x}Fe_{0.5}Co_{0.5}O_{3-d}$

In contrast to the cuprates, $(Sr,Ca)(Fe,Co)O_{3-d}$ contains both Fe and Co, thus the localization of the ^{57}Fe and ^{57}Co dopants is straightforward and does not complicate unnecessarily the interpretation of the Mössbauer spectra. However, one has to take into account that in the case of doping with ^{57}Co, a nucleogenic ^{57}Fe will provide the Mössbauer signal from a Co-site.

We have studied two compounds with different Ca to Sr ratio but with a 1:1 Fe to Co ratio for both. The equal concentration of Fe and Co facilitated the interpretation of the results.

3.1. MÖSSBAUER STUDY OF $Sr_{0.5}Ca_{0.5}Fe_{0.5}Co_{0.5}O_{3-d}$

As was mentioned earlier, the ABO_3 type compounds may have the perovskite structure if some conditions on the relative ionic radii are met (tolerance factor). Thus while stoichiometric strontium ferrate, $SrFe^{IV}O_3$, can be relatively easily synthesized, replacement of Sr by Ca of smaller ionic radius but the same valence state gradually forces the perovskite structure to convert into that of an oxygen deficient perovskite, namely the brownmillerite (Ca_2AlFeO_5) structure which contains alternating octahedrally and tetrahedrally coordinated B cation layers. The valence state of the B cation must then decrease along with the oxygen loss and one gets $CaFe^{III}O_{2.5}$ with fully trivalent iron as a stable compound.

Mixed population of the B sites raises the question if there is any preference of the chemically different B cations for the crystallographically nonequivalent sites. In our case, one may expect that Co, which tends to acquire a lower valence state and a lower coordination number than Fe, will prefer the tetrahedral sites. Rigorously speaking, this statement is valid for high spin Co and Fe-ions. Some preference of Co for the tetrahedral sites was found earlier by Battle et al. in $SrCo_{0.5}Fe_{0.5}O_{2.5}$ using transmission Mössbauer technique [7].

The brownmillerite is an antiferromagnet, and the tetrahedral and octahedral cations have different hyperfine fields. The high value of the hyperfine fields is consistent with high spin states. The same two-sextet structure was observed by us in $Sr_{0.5}Ca_{0.5}Fe_{0.5}Co_{0.5}O_{3-d}$.

The transmission Mössbauer spectrum of $SrFeO_{2.5}$ contains the two sextets in almost equal intensity in perfect agreement with the structure [8]. (Deviation from the 1:1 ratio can be assigned to different f-factors as demonstrated by low temperature measurements.)

The transmission and emission Mössbauer spectra of $Ca_{0.5}Sr_{0.5}Fe_{0.5}Co_{0.5}O_{2.47}$ are shown in Fig. 3. The oxygen stoichiometry determination and preparation conditions are given elsewhere [9]. The different intensities of the sextets representing tetrahedral and octahedral B sites are clearly seen. The "tetrahedral sextet" is more intense in the emission spectrum, and the "octahedral sextet" is more intense in the transmission spectrum. Quantitatively: $I_{tet}/I_{oct}(EMS)=1.57$ and $I_{tet}/I_{oct}(TMS)=0.83$ where I stands for the sextet intensities and EMS and TMS refer to the emission and transmission experiments, respectively.

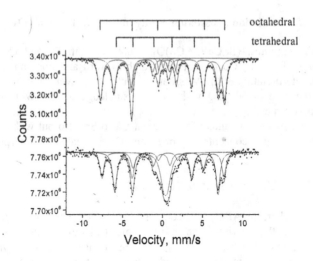

Figure 3. Transmission (top) and emission (bottom) Mössbauer spectra of $Sr_{0.5}Ca_{0.5}Fe_{0.5}Co_{0.5}O_{3-d}$ recorded at room temperature. The sextets assigned to the octahedral and tetrahedral Fe^{3+} at the B sites are indicated. The velocity scale refers to α-Fe.

In addition to the straightforward conclusion that the preferred occupancy of tetrahedral sites by Co is verified, the two values show an asymmetry: they are not exactly the reciprocals of each other. This may be attributed partially to the paramagnetic part of the spectra, which is rather large in the emission case and is believed to be due to more oxidized „normal" perovskite structure. On the other hand, as can be seen from the transmission spectra of $SrFeO_{2.5}$ recorded by Gallagher et al. [8], the f-factor is somewhat higher for the tetrahedral sites at room temperature, and this can itself cause the asymmetry in the relative intensities.

3.2. MÖSSBAUER STUDY OF $Sr_{0.95}Ca_{0.05}Fe_{0.5}Co_{0.5}O_{3-d}$

$Sr_{0.95}Ca_{0.05}Fe_{0.5}Co_{0.5}O_{3-d}$ shows a very different Mössbauer spectrum as compared to that of the previous Ca-rich compound. There is no magnetic structure just as it was not found either in the Ca-free compound by Gibb [10] in a wide Co/Fe range. The spectra contained only paramagnetic peaks as shown in Fig. 4.

The preparation conditions and oxygen stoichiometry determination are given elsewhere [11]. The decomposition of the spectra into individual doublets is not straightforward, and it was already experienced by Gibb for the Ca free $SrFeO_{3-d}$ with somewhat higher oxygen content [12].

The most important problem we addressed was whether this system is a random solid solution or there is an ordering of Co/Fe in the macroscopically cubic lattice. Although only about 1/3 of the oxygen vacancies allowed in the brownmillerite structure is populated by oxygen, the delocalized electron structure of the compound does not permit us to assign individual doublets to localized octahedral and tetrahedral sites. However, one may expect that because this mixed compound does exist, its structure must be fairly close to that of a random solid solution, and therefore the transmission and emission spectra should not be very different. Thus we searched for the simplest evaluation model that fits both the transmission and the emission spectra but still represents the obvious difference between them. The results of a three doublet evaluation are shown in Table 1. The only major difference between the EMS and TMS spectra is the quadrupole splitting of doublet C(C').

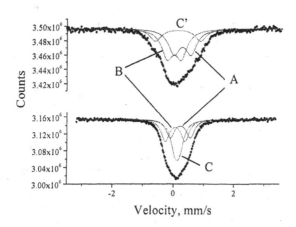

Figure 4. Emission (top) and transmission (bottom) Mössbauer spectra of $Sr_{0.95}Ca_{0.05}Fe_{0.5}Co_{0.5}O_{2.67}$ recorded at room temperature. The velocity scale refers to α-Fe.

Considering the case of a possible perfectly random solid solution of this compound of 1:1 Fe/Co ratio, the EMS and TMS spectra would have to match each other exactly because every Co has on the average three Fe plus three Co as nearest metal neighbors, and the same applies for Fe. In this case the population of O-sites should also be completely random.

TABLE 1. Mössbauer parameters of $Sr_{0.95}Ca_{0.05}Fe_{0.5}Co_{0.5}O_{3-d}$ measured at room temperature by transmission (TMS) and emission (EMS) technique. The isomer shifts refer to α-Fe.

Technique	Subspectra	Isomer Shift, δ, mm/s	Quadrupole Splitting, Δ, mm/s	Linewidth (FWHM, Γ), mm/s	Relative Area, %
	A	0.22	0.66		28.0
TMS	B	0.06	0.65	0.32	33.0
	C	0.12	0.16		39.0
	A	0.25	0.56		37.7
EMS	B	-0.02	0.46	0.45	42.7
	C'	0.11	1.49		19.7

314

The situation is depicted in Fig. 5, together with some other possible arrangements of Co and Fe in the lattice: three-dimensional alternating Co-Fe array, layered structure (like in brownmillerite) and complete phase separation.

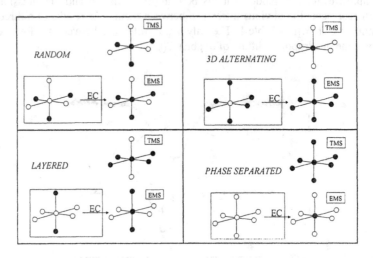

Figure 5. Metal nearest neighbor environment of regular and nucleogenic ^{57}Fe Mössbauer probe in $Sr_{0.95}Ca_{0.05}Fe_{0.5}Co_{0.5}O_{3-d}$ in various possible ordering scenarios. Full and open circles denote Fe and Co, respectively. Oxygens between the metal ions are not shown.

Since the evaluation of the spectra indicated a species with very different quadrupole splittings in the TMS and EMS case (C and C'), and only in the layered structure can one find symmetrically markedly different environments, it is concluded that this solid solution contains, most probably, microdomains in which the Co and Fe atoms are ordered in layers. Species A and B are assigned to delocalized states where this asymmetry is not seen due to either the delocalization itself or locally higher oxygen content. A more detailed discussion has been published in ref. [11].

In summary, by the comparison of emission and transmission Mössbauer spectra, it has been revealed that in addition to the generally expected preferential ordering of oxygen to Fe rather than Co in Fe-Co mixed perovskites, the Co and Fe atoms themselves have a short range order in $Sr_{0.95}Ca_{0.05}Fe_{0.5}Co_{0.5}O_{2.67}$.

4. Transmission and Emission Mössbauer Study of $La_{0.8}Sr_{0.2}Co_{0.95}Fe_{0.05}O_{3-d}$

$La_{0.8}Sr_{0.2}Fe_xCo_{1-x}O_3$ type perovskites show the colossal magnetoresistance (CMR) effect [13]. Since the sudden change of electrical resistivity occurs during magnetic ordering transitions, the study of the details of the magnetic transitions is of crucial importance.

In the $La_{0.8}Sr_{0.2}Fe_xCo_{1-x}O_3$ system, Fe seems to play a key role in the mechanism since in the iron free lanthanum-strontium-cobaltate the CMR effect is found only in a narrow temperature range around the Curie temperature, while the Fe-doped material shows intense CMR signal below 100 K and also above 150 K. It is even more interesting that for iron concentrations exceeding x = 0.025 the CMR effect at room temperature decreases monotonously with the iron content.

Transmission Mössbauer spectroscopy is a good tool to study the contribution of Fe in the magnetic ordering transitions in these compounds, while the emission technique may give information on the role of Co. The crucial question is that to what extent the nucleogenic ^{57}Fe acts as a faithful probe.

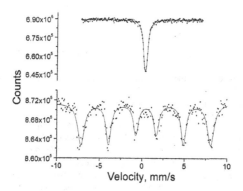

Figure 6. Emission Mössbauer spectrum of ^{57}Co-doped $La_{0.8}Sr_{0.2}Fe_{0.05}Co_{0.95}O_{3-d}$ recorded at room temperature (top) and at 4.2 K (bottom). The velocity scale refers to α-Fe at room temperature.

It was claimed by Bhide et al. [14] that in the perovskite $LaCoO_3$ the original spin state of cobalt can be observed in a ^{57}Co emission Mössbauer measurement at low temperature. In $La_{0.8}Sr_{0.2}Fe_xCo_{1-x}O_3$, Co is also believed to be in the low spin state (Co^{3+}_{LS}), and the low temperature CMR effect can be assigned qualitatively to the Co^{3+}_{LS} to Co^{3+}_{HS} (Low-Spin to High-Spin) transition.

We have studied $La_{0.8}Sr_{0.2}Fe_{0.05}Co_{0.95}O_{3-d}$ at 4.2 K as well as at 290 K by both emission and transmission technique to see if there is any difference observed between the electronic and magnetic states of the regular and nucleogenic Fe probes. Experimental details will be published elsewhere.

Figure 6 shows the emission spectra indicating one single species at both temperatures. Comparison of the Mössbauer parameters (δ = 0.32 mm/s at RT and H = 47.3 T at 4.2 K) to those obtained in our own transmission measurements and to other literature data for the same compounds but with higher Fe content [15] revealed the same high spin Fe^{3+} state.

There are two major conclusions that can be drawn from this observation. On the one hand, since the two spectra representing macroscopic (as in TMS) and extremely low (as in EMS) Fe-concentrations are the same, it proves that there is no preferred Fe-Fe interaction in the system (up to 5 % Fe), and Fe is randomly distributed in the perovskite solid solution. On the other hand, the high spin Fe^{3+} state observed even at low temperature shows that, in contrast to the case of the semiconductor $LaCoO_3$, the nucleogenic ^{57}Fe does not inherit the low spin state of Co^{3+}, and it represents the equilibrium chemical state of Fe in the lattice of $La_{0.8}Sr_{0.2}Fe_{0.05}Co_{0.95}O_{3-d}$, the same as that observed in TMS experiments.

316

Acknowledgement

Part of this work has been supported by the Intergovernmental Hungarian-Austrian Science and Technology Program, Project No. A-22/01, the Hungarian National Science Fund (OTKA T034839 KM1, T029537, F 034837) and the Hungarian Academy of Sciences (AKP 2000-143 2,4).

References

1. West, A.R. (1999) *Basic Solid State Chemistry, 2nd Edition*, John Wiley and Sons Ltd., Chichester, New York Weinheim, Brisbane, Singapore, Toronto, p.57.
2. Radousky, H.B. (1992) A review of the superconducting and normal state properties of $Y_{1-x}Pr_xBa_2Cu_3O_7$, *J. Mater. Res.* **7**, 1917-1945.
3. Nath, A. and Homonnay, Z. (1989) Emission Mössbauer Studies of Nonrigidity of Copper-Oxygen Chain in $YBa_2Cu_3O_{7-d}$, *Physica C* **161**, 205-208.
4. Homonnay, Z., Klencsár, Z., Chechersky, V., Vankó, Gy., Gál, M., Kuzmann, E., Tyagi, S., Peng, J.-L., Greene, R.L., Vértes, A., and Nath, A.(1999) The Effect of Praseodymium on the Lattice Dynamics and Electronic Structure of the Cu(1)-O(4) chain in $Y_{1-x}Pr_xBa_2Cu_3O_{7-\delta}$, *Phys. Rev. B* **59**, 11596-11604.
5. Guillaume, M., Allenspach, P., Henggeler, W., Mesot, J., Roessli, B., Staub, U., Fischer, P., Furrer, A., and Trounov, V. (1994) A systematic low-temperature neutron diffraction study of the $RBa_2Cu_3O_x$ (R = yttrium and rare earths; x = 6 and 7) compounds, *J. Phys.: Condens. Matter* **6**, 7963-7976.
6. Homonnay, Z. (1997) Mössbauer spectroscopy of high-T_c superconductors, in A. Vértes and Z. Homonnay (eds.), *Mössbauer spectroscopy of sophisticated oxides*, Akadémiai Kiadó, Budapest, pp.159-305.
7. Battle, P.D., Gibb, T.C., and Nixon, S. (1988) The formation of nonequilibrium microdomains during the oxidation of Sr_2CoFeO_5, *J. Solid State Chemistry* **73**, 330-337.
8. Gallagher, P.K., MacChesney, J.B., and Buchanan, D.N.E. (1964) Mössbauer effect in the system $SrFeO_{2.5-3.0}$, *J. Chem. Phys.* **41**, 2429-2434.
9. Homonnay, Z., Nomura, K., Juhász, G., Kuzmann, E., Hamakawa, S., Hayakawa, T., and Vértes A. (2002) Microstructure and CO_2-absorption in $Sr_{0.95}Ca_{0.05}Co_{0.5}Fe_{0.5}O_{3-\delta}$ and $Sr_{0.5}Ca_{0.5}Co_{0.5}Fe_{0.5}O_{3-\delta}$ as studied by Emission Mössbauer Spectroscopy, *J. Radioanal. Nucl. Chem.* accepted.
10. Gibb, T.C. (1987) Magnetic exchange interactions in perovskite solid solutions. Part 8. A study of the solid solution $SrFe_{1-x}Co_xO_{3-y}$ by iron-57 Mössbauer spectroscopy, *J. Chem. Soc.. Dalton Trans.* 1419-1423.
11. Homonnay, Z., Nomura, K., Juhász, G., Gál, M., Sólymos, K., Hamakawa, S., Hayakawa, T., and Vértes A. (2002) Simultaneous probing of the Fe- and Co-sites in the CO_2-absorber perovskite $Sr_{0.95}Ca_{0.05}Co_{0.5}Fe_{0.5}O_{3-\delta}$: a Mössbauer study, *Chemistry of Materials* **14**, 1127-1135.
12. Gibb, T.C. (1985) Magnetic Exchange Interactions in Perovskite Solid Solutions.Part 5. The Unusual Defect Structure of $SrFeO_{3-y}$, *J. Chem. Soc., Dalton Trans.* **7**, 1455-1470.
13. Fontcuberta, J. (1999) Colossal magnetoresistance, *Physics World*, February, 33-38.
14. Bhide, V.G., Rajoria, D.S., Rama Rao, G., and Rao, C.N.R. (1972) Mössbauer studies of the high-spin-low-spin equilibria and the localized-collective electron transition in $LaCoO_3$, *Phys. Rev. B* **6**, 1021-1032.
15. Cziráki, Á., Gerőcs, I., Köteles, M., Gábris, A., Pogány, L.; Bakonyi, I., Klencsár, Z., Vértes, A., De, S.K., Barman, A., Ghosh, M., Biswas, S., Chatterjee, S., Arnold, B., Bauer, H.D., Wetzig, K., Ulhaq-Bouillet, C., and Pierron-Bohnes, V. (2001) Structural features of the La-Sr-Fe-Co-O system, *European Physical Journal B* **21**, 521-526.

TRENDS IN MÖSSBAUER POLARIMETRY WITH CIRCULARLY POLARIZED RADIATION

Mössbauer polarimetry

K. SZYMAŃSKI[1], L. DOBRZYŃSKI[1,2], D. SATUŁA[1]
AND B. KALSKA-SZOSTKO[3]
[1]*Institute of Experimental Physics, University of Białystok, 15-424 Białystok, Poland*
[2]*The Soltan Institute for Nuclear Studies, 05-400 Otwock-Świerk, Poland*
[3]*The Freie Universität Berlin, Institut für Experimentalphysik, Arnimallee 14,*

Abstract

In Mössbauer Spectroscopy (MS) the use of circularly polarised radiation adds new capabilities not available in conventional MS. The resonant absorption becomes dependent on the relative orientation of the photon angular momentum and direction of the hyperfine magnetic field acting on the nucleus. In complex systems, either with mixed hyperfine interactions or chemical or magnetic disorder, this new option allows one to determine magnetic structure of such systems in complementary way to the neutron diffraction. Sometimes its sensitivity to the Mössbauer nucleus leads to the unique possibility of identifying individual contributions of various species to the net magnetisation. The formalism of so-called velocity moments for the $I=3/2 \rightarrow 1/2$ nuclear transitions permits one to extract some averages of hyperfine fields irrespective of the type of the interactions and eventual shape of their distribution. The applicability of the method is presented for a number of experimental situations. Construction of the source used in these experiments is also presented.

1. Introduction

Determination of the angular distribution of local magnetic moments in disordered materials is generally rather difficult. However, the fact that magnetic moment is a pseudovector quantity which changes sign upon the time reversal suggests that the use of circularly polarised radiation may be particularly attractive. Indeed, its use offers site selective method for investigation of direction of hyperfine magnetic field (h.m.f.).

Let us consider nuclear absorption of a photon with angular momentum +h in magnetic field parallel to the wave vector of photon. The allowed transitions and

M. Mashlan et al. (eds.), Material Research in Atomic Scale by Mössbauer Spectroscopy, 317–328.

polarisation of involved photons are shown schematically in Fig. 1a. A photon whose projection of an angular momentum on the direction of movement is +h (Fig. 1b), will be absorbed only in transitions with $\Delta m=+1$. These processes result in the spectrum shown in Fig. 1c. Rotation of the magnetic field by π (Fig. 1c) changes magnetic numbers of sublevels from m to -m, and the photon will be absorbed in transitions previously forbidden. This will result in the absorption spectrum shown in Fig. 1e. The two spectra (Fig. 1d, e) differ substantially, so it is easy to determine the sense of the h.m.f. with respect to the wave vector of photon.

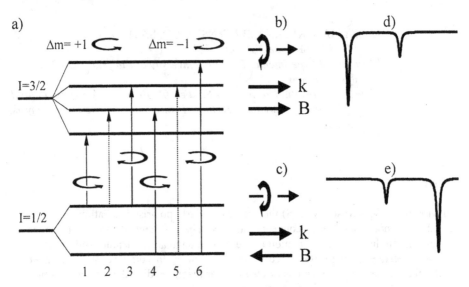

Figure 1. a) hyperfine splitting of ^{57}Fe nuclear levels by a magnetic field (polarisation of transitions shown schematically), b), c) orientations of the h.m.f. **B** and wave vector **k** of photon (polarisation state of photon shown schematically), d), e) shape of the spectra measured with polarised radiation.

If the iron magnetic moment is tilted by an angle θ with respect to the direction of photon, the relative line intensities for circularly polarised beam are [1]:
$$i_1:i_2:i_3:i_4:i_5:i_6 = 3(1-\cos\theta)^2 : 4\sin^2\theta : (1+\cos\theta)^2 : (1-\cos\theta)^2 : 4\sin^2\theta : 3(1+\cos\theta). \qquad (1)$$
The sum of the line intensities in (1) is constant, and does not depend on the orientation of h.m.f.

The h.m.f. (and magnetic moments) usually exhibit an angular distribution due to magnetic texture. The line intensities are thus given by an appropriate average of the right hand side of Eq. (1) over the ensemble of the iron atoms in the absorber. Measurements with circularly polarised radiation allow one [2] to find two important angular averages: the average cosine $\langle \gamma \cdot m \rangle$ and the average cosine square $\langle (\gamma \cdot m)^2 \rangle$. Here γ and m are unit vectors which are parallel to the photon wave vector and the vector of h.m.f., respectively.

The average cosine square $\langle (\gamma \cdot m)^2 \rangle$ is a measure of the transverse (to the direction of photon) component of hyperfine field, ranging from zero (*m* perpendicular to γ) to unity (*m* parallel or antiparallel to γ). For randomly oriented hyperfine fields, $\langle(\gamma \cdot m)^2\rangle = 1/3$. The average cosine square in thin absorber approximation is given by:

$$\langle(\gamma \cdot m)^2\rangle = \frac{4-z}{4+z}, \tag{2}$$

where z is defined as

$$z = \frac{3(i_2 + i_5)}{i_1 + i_6} = \frac{i_2 + i_5}{i_3 + i_4}. \tag{3}$$

In a standard experiment with unpolarised radiation line intensities, this ratio is usually abbreviated as:

$$i_1:i_2:i_3:i_4:i_5:i_6 = 3:z:1:1:z:3. \tag{4}$$

The macroscopic magnetisation μ, is a vector quantity. The component of μ in the direction of unit vector γ can be expressed as a weighted sum of individual atomic magnetic moments:

$$\gamma \cdot \mu = \mu \gamma \cdot m = \sum_i c_i \mu_i \langle\gamma \cdot m_i\rangle, \tag{5}$$

where c_i is concentration of the i-th element, and $\mu_i = \mu_i$ m_i is the atomic magnetic moment along unit vector m_i. It is clear from Eq. (5) that the average cosine $\langle\gamma \cdot m_i\rangle$ is related to the contribution of i-th element to total magnetisation. For an isotropic sample or for antiferromagnetic order of magnetic moments μ_i, the average cosine $\langle\gamma \cdot m_i\rangle$ is zero. The average cosine $\langle\gamma \cdot m_i\rangle$ itself carries information about the average direction of the magnetic moment μ_i with respect to the γ vector. A uniqueness of ^{57}Fe Mössbauer spectroscopy with circularly polarised radiation is its ability to provide $\langle\gamma \cdot m_{Fe}\rangle$ through an asymmetry between the left- and right-hand parts of a Zeeman sextet, namely (again in thin absorber approximation):

$$\langle\gamma \cdot m_{Fe}\rangle = \frac{4a}{(4+z)p}, \tag{6}$$

where the asymmetry a depends on intensities for lines 3 and 4 (or 1 and 6) :

$$a = \frac{i_4 - i_3}{i_4 + i_3} = \frac{i_1 - i_6}{i_1 + i_6}, \tag{7}$$

and p is the degree of circular polarisation of the beam.

2. Concept of the Velocity Moments

Polarimetric methods as described in section 1 can be applied to systems in which distributions of hyperfine interactions are not so wide, so absorption lines can be separated with reasonable precision. In some cases, for example close to the magnetic transition points or for systems in which the magnetic splitting is of the order of the quadrupole splitting or of the order of natural width, the spectra overlap so much that the method of section 1 cannot be used. However, the shape of such overlapped spectra is still sensitive to the applied magnetic field or to the orientation of the photon wave vector with respect to the sample magnetisation. In this section we present a formalism, which allows one to get some averages of hyperfine interactions over the whole sample.

Analytical expressions for intensities of $I=3/2\rightarrow I=1/2$ nuclear transitions were presented in [3,4]. Mössbauer spectrum can be considered as a weighted sum of the intensities $A_{\alpha\beta\zeta}$ with appropriate lineshapes, centered at Doppler velocities $v_{\alpha\beta}$. Indices α and β refer to the excited and ground states of ^{57}Fe nucleus, while $\zeta=\pm 1$ is polarisation state of the photon. Let us consider a quantity W_ζ^n as:

$$W_\zeta^n = \sum_{\alpha\beta} v_{\alpha\beta}^n A_{\alpha\beta\zeta} \Big/ \sum_{\alpha\beta} A_{\alpha\beta\zeta},\qquad(8)$$

This quantity, called the n-th velocity moment, can be extracted directly from experimental data: one has to sum absorption line intensities multiplied by a power of the velocity. Using the explicit expressions for $A_{\alpha\beta\zeta}$ [3] one arrives at:

$$W_\zeta^1 = \delta + \frac{eQc}{4E_\gamma}\gamma\cdot\hat{V}\cdot\gamma - \zeta\frac{c\mu_N}{4E_\gamma}\left(g_{1/2}-5g_{3/2}\right)B\cdot\gamma\qquad(9)$$

where V is the electric field gradient: $V_{ij}=-\partial^2\varphi/\partial x_i\partial x_j$ where φ denotes an electric potential at the nucleus. Q is electric quadrupole and $g_{1/2}$, $g_{3/2}$ magnetic dipolar nuclear moments. Quantities e, c and μ_N refer to elementary constants and E_γ is the energy of the transition.

The first moment (9) contains three types of hyperfine interactions, which can be separated experimentally in the following way:

- if one prepares a texture-free sample (randomly distributed grain orientations), the second and the third term sof (9) will average to zero, and the first moment will be a measure of the average isomer shift.
- using unpolarised radiation, or averaging measurements with polarised radiation over all photon polarisation states ζ, the third term in (9) is averaged to zero, and anisotropy of the first moment reflects symmetry of the electric field gradient (EFG) tensor.
- from two measurements with opposite photon polarisations $\zeta=\pm 1$ the component of the hyperfine magnetic field in the direction of the photon wave vector is obtained:

$$\frac{W_{-\zeta}^1 - W_\zeta^1}{2} = \frac{c\mu_N}{4E_\gamma}\left(g_{1/2}-5g_{3/2}\right)B\cdot\gamma\qquad(10)$$

Results (9, 10) are rigorous, and valid for any value of quadrupole interaction. We note that with (9, 10) one does not need to solve a problem of mixed electric quadrupole and magnetic dipole interaction for spin $I=3/2$, which usually involves secular equations of 4th order. The first moment (9) can be used for determination of the sign either dominating component of EFG (usually abbreviated by V_{zz}) or the h.m.f. Examples will be given in section 5.

3. Source of the Monochromatic, Circularly Polarised, Resonant Radiation

Developing ideas [5-8] we have found that a nonstoichiometric, ordered Fe$_3$Si alloy exhibits convenient parameters for resonant filter construction [9,10]. The alloy is chemically stable, magnetically soft, and may operate in connection with a commercially available ^{57}Co source in a Cr matrix.

The principle of operation is as follows. The resonant filter contains ^{57}Fe atoms in a soft ferromagnetic matrix. An external magnetic field produced by a permanent magnet aligns magnetic moments of the filter parallel (or antiparallel) to the photon wave vector. The filter must fulfil a basic condition, namely, the 3rd (or 4th) absorption line of the Zeeman sextet must be at zero Doppler shift with respect to the emission line of the source (see dashed vertical line in Fig. 2a). Under this condition, photons with well defined circular polarisation will be absorbed by the filter. The remaining photons of opposite polarisation leaving the filter are used for the Mössbauer measurement on the sample, see Fig. 2b. Details of the construction can be found in [11, 12].

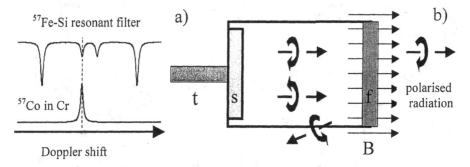

Figure 2. Conceptual design of the monochromatic source. a) the third line of the ^{57}Fe-Si filter is at resonance with emission line of ^{57}Co in Cr matrix (intensities of 2nd and 5th lines are zero because h.m.f . is parallel to the direction of photon). b) schematic layout of the construction, "s" denotes ^{57}Co in Cr matrix, element "t" is connected to the velocity transducer, "f" denotes Fe-Si filter in axial magnetic field "B".

Spectra of an α-Fe powder measured with circularly polarised radiation are shown in Fig. 3. The powder was aligned by an axial magnetic field. In comparison with theoretical lines in Fig. 1 d,e, one sees some residual intensity from lines 6 (Fig. 3a) and 1 (Fig. 3b), owing to imperfect circular polarisation of the photons. Traces of lines no 2 and 5 are also seen in Fig 3, which indicate an imperfect alignment of the iron magnetic moments.

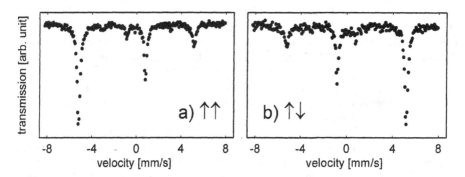

Figure 3. Mössbauer spectra of α-Fe powder oriented by external magnetic field, measured with circularly polarised radiation. The arrows ↑↑ and ↑↓ indicate the two opposite polarisation states of the photons.

4. Complex Magnetic Structure of Amorphous Er-Fe-B Alloy

Some of rare earth and 3d elements exhibit the property of antiferromagnetic ordering of the 3d and rare earth subsystems. At a so-called "compensation temperature" their net magnetisation may be equal to zero. Mössbauer polarimetry is an efficient tool for studying how this happens. As an example, we present results of investigations on an amorphous $Fe_{0.66}Er_{0.19}B_{0.15}$ alloy. The ribbons with thickness of about 40 μm were prepared by rapid quenching in an inert atmosphere. X - ray diffraction patterns were measured to ensure there was no crystalline phase. A randomly oriented absorber was made in the form of a pellet containing about 12 mg of nonenriched $Fe_{0.66}Er_{0.19}B_{0.15}$ powder per cm².

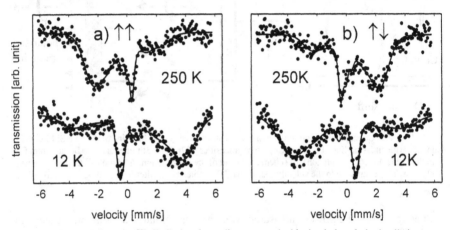

Figure 4. Spectra of Er-Fe-B aborphous alloy measured with circularly polarised radiation at temperatures above (top) and below (bottom) the compensation temperature T_{comp}. The arrows ↑↑ and ↑↓ indicate two opposite polarisation states of the photons.

Magnetisation measurements reveal zero magnetisation, and a maximum of coercivity at T_{comp}=120 K, indicating the possibility of ferrimagnetic order of Er and Fe subsystems. Mössbauer spectra measured at temperatures above and below the compensation point are shown in Fig. 4. There is clear evidence that above compensation point (T=250 K), the asymmetry of the spectrum is similar to that of α-Fe (cf. Fig. 4a, T=250 K with Fig. 3a and Fig. 4b with Fig. 3b), showing that iron moments in Fe-Er-B are oriented parallel to the net magnetisation. Asymmetry below the compensation temperature is reversed, which means that iron moments become oriented antiparallel to the net magnetisation.

To obtain details of the magnetic structure, the measured spectra were fitted with Zeeman components broadened by a Gaussian distribution of h.m.f. Spectra measured in a magnetic field with unpolarised radiation (not shown here) were simultaneously fitted with data obtained with polarised beam (see solid lines in Fig.4). The average cosine and the average cosine square determined by use of Eq. (6) and (2) are given in the Table 1. To shorten notation we dropped the subscript Fe in m_{Fe}.

To describe the actual spin structure let us assume that the magnetic moments are distributed homogeneously within a cone at apex angle θ (e.g $\theta = \pi/2$ correspond to a half-sphere), and that the symmetry axis of the cone is along the direction of γ. In such a case

$$\langle \gamma \cdot m \rangle = \frac{1 + \cos\theta}{2}, \qquad (11)$$

$$\langle (\gamma \cdot m)^2 \rangle = \frac{1 + \cos\theta + \cos^2\theta}{3}. \qquad (12)$$

TABLE 1. Average cosine square $\langle(\gamma \cdot m)^2\rangle$ and average cosine $\langle \gamma \cdot m \rangle$ obtained from Mössbauer measurements. Deduced angle θ of the cone for Fe magnetic moments (Eq. 11 and 12) and minority fraction δ (Eq. 13). A minus sign in line 3 indicates antiparallel orientation of the average moment with respect to net magnetisation.

T [K]	12	250	295
$\langle(\gamma \cdot m)^2\rangle$	0.82(3)	0.86(3)	0.92(6)
$\langle \gamma \cdot m \rangle$	$-(0.79 \div 0.59)$	$0.62 \div 0.83$	$0.75 \div 0.99$
θ_{Fe} [deg]	36(3)	31(4)	22(10)
δ	$0.08 \div 0.19$	$0.07 \div 0.17$	$0 \div 0.1$

Comparison of (11) and (12) with experimentally determined values in Table (1) indicate that at room temperature, Fe spin structure can roughly be considered as distributed homogeneously in a cone with an angle $\theta = 22 \pm 10$ deg. At lower temperatures, however, one cannot find a single value of θ which would satisfy observed values $\langle(\gamma \cdot m)^2\rangle$ and $\langle \gamma \cdot m \rangle$, neither at T=12 K nor at T=250K. To explain such a situations we assumed that some minority part δ of Fe magnetic moments is oriented antiparallel with respect to the majority part $(1-\delta)$, see Fig. 5. An antiparallel orientation of the minority moments does not change the value of $\langle(\gamma \cdot m)^2\rangle$ while it reduces the $\langle \gamma \cdot m \rangle$ value, and in our simple model of homogeneously distributed cone we have:

$$\langle \gamma \cdot m \rangle = (1 - 2\delta)\frac{1 + \cos\theta}{2}. \qquad (13)$$

An estimate of δ at different temperatures is given in Table 1.

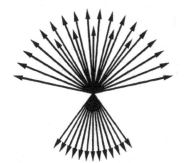

Figure 5. Schematic representation of a possible spatial arrangement of the Er and Fe magnetic moments which are homogeneously distributed within 3D-cones. Longer and shorter arrows correspond to Er and Fe atomic magnetic moments, respectively.

5. Cr-Mn-Fe Disordered Alloys - Low Temperature State

In a certain concentration range Cr-Mn-Fe alloys exhibit a coexistence of ferro- and antiferromagnetic orders [13]. The magnetisation curve of a sample of $Cr_{79}Fe_{20}Mn_5$ measured at T=12K shows a knee at magnetic field of about 0.1T typical for ferromagnetic materials. Saturation is not reached at the maximum of the measuring field of 1.2T [14]. Mössbauer spectra measured at T=12K consist of broad, overlapped absorption lines, see Fig. 6a, in agreement with earlier investigations [15-17].

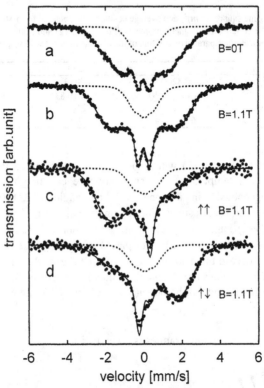

Figure 6. Mössbauer spectra of $Cr_{79}Fe_{20}Mn_5$ measured at T=12K with unpolarised radiation a) in zero external field, b) in an axial field of B=1.1 T. c), d) spectra measured with circularly polarised radiation. The arrows ↑↑ and ↑↓ indicate two opposite polarisation states of the photons. Results of fit with two Zeeman components (see text) are shown by solid lines. Dashed line above each spectrum correspond to the component not sensitive to the applied magnetic field.

An applied magnetic field changes the angular distribution of magnetic moments, resulting in a change of the Mössbauer spectrum (compare Fig.6 a and b). To obtain detailed information about the behaviour of magnetic moments under an applied field, measurements with circularly polarised radiation were performed. Spectra measured with two opposite polarisations are shown in Fig. 6 c,d. Spectra displayed in Fig. 6 b,c,d were simultaneously fitted with two Zeeman components, each one having a Gaussian distribution of h.m.f. Results of the fits show that components having h.m.f. smaller than about 4 T are insensitive to the polarisation state of photon, see dotted lines in Fig. 6. One concludes that the magnetic moments of iron characterised by the low field component are not ordering under the influence of the field of 1.1 T. The opposite is true for iron magnetic moments related to the high field component (about 12T). Only this component is responsible for the clear asymmetry between the lines 1 and 6 (and 4 and 3) in the spectra shown in Fig. 6 c and 6 d.

Further analysis and experiments on samples with other compositions revealed that the relative area under Zeeman components with smaller h.m.f. is proportional to the calculated probability of iron surrounded by 8 Cr atoms in the first coordination shell [14]. This strongly indicates that the antiferromagnetic phase in the sample originates from Cr-rich clusters.

6. Determination of Velocity Moments

To illustrate the sensitivity of the first moment to the orientation of the electric field gradient, we performed measurements with unpolarised radiation of FeB textured sample. The sample was prepared from fine powder mixed in epoxy glue and solidified under an applied magnetic field. The magnetic field oriented the powder grains, inducing axial texture. Mössbauer measurements were performed under different angles θ between wave vector of photon and texture axis. Results are shown in Fig. 7a.
S

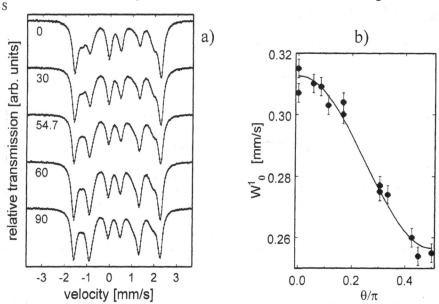

Figure 7. a) Spectra of FeB absorbers prepared from grains oriented by a magnetic field. Measurements were performed in zero applied field. The angles θ between the directions of photon and the field orienting the powder are labelled. b) The first velocity moment determined from unpolarised radiation data.

The spectra consist of two Zeeman components. The dominating one, with larger hyperfine field, originates from iron atoms in an unperturbed orthorombic B27 type of structure [18, 19]. The second weaker component results from a perturbation at iron from nearest neighbour boron atom [20]. The variation of line intensities with the angle θ results from the preferred orientation of crystallites. This is also seen in the angular variation of the first moment W_0^1, see Fig. 7b. Since W_0^1 is a tensor component of the second order, the angular change of the average W_0^1 should be quadratic in $\cos\theta$. The experimental results were fitted by $W_0^1=0.201(5)+0.112(8)\cos^2\theta$, see the solid line in Fig. 7b. It was demonstrated that in FeB the h.m.f vector is inclined from the direction of dominating EFG component by 38(2) degrees [21]. We expect that during in-field orientation the directions of V_{zz} axes are, on the average, oriented parallel to the applied field direction. This in turn should result in a maximum of W_0^1 for $\theta=0$ when V_{zz} is positive, and a maximum of W_0^1 for $\theta=\pi/2$ for V_{zz} negative. Results shown in Fig. 7b clearly indicate positive V_{zz} in FeB, in agreement with earlier reports [21].

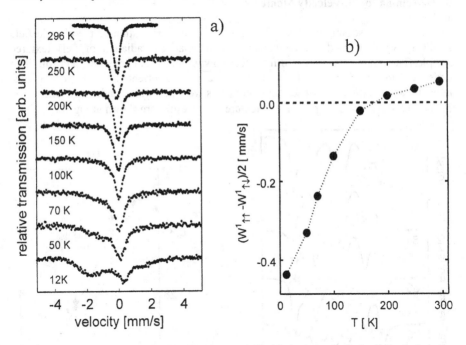

Figure 8. a) Mössbauer spectra of $Cr_{79}Fe_{20}Mn_5$ measured with circularly polarised radiation (polarisation state of photons abbreviated by ↑↑, see paragraph 2) b) Temperature dependence of the difference of the first velocity moments (see Eq. 10).

As the last example we present results of the temperature measurements on the Cr-Mn-Fe alloy. The Curie temperature of the sample was determined by magnetisation measurements to be about 90K [14]. Mössbauer measurements reveal magnetic splitting below 50K, while between room temperature and T=50 K only constant line broadening was observed (FWHM about 0.4 mm/s). It was found that the shape of the spectra were very sensitive to the external magnetic field, showing splitting below T=150K. Because the observed spectra are strongly collapsed, in order to determine the first velocity moments we performed measurements with circularly polarised radiation (see Fig. 8). Experiments with polarised photons show that the projection of the average h.m.f. is antiparallel to the direction of magnetisation up to about T=150 K. At temperatures above 150K, the difference of the first moments change sign, and at high temperatures the contribution from only applied external field should be observed. Since the applied field is about 1.1T, it follows from (10) that the difference should attain the value of 0.13 mm/s, in agreement with experiment.

7. Conclusions

The use of circularly polarised photons in Mössbauer spectroscopy allows for a more complete understanding of structurally and magnetically complex systems. The examples shown by far do not exhaust all phenomena which can be observed with this relatively new technique. We believe that its use may open up very new fields of investigations of magnetic materials that exhibit even weak ferromagnetic properties. This technique may bring unique results for studies of magnetisation distributions in the vicinity of the phase transition temperatures, spin glasses of various types, and the influence of weak ferromagnetic particles on superconductivity. The concepts are also applicable to work with synchrotron radiation, which is highly polarised

References

1. Frauenfelder H., Nagle D. E., Taylor R. D., Cochran D. R. F. and Visscher W. M. (1962) Elliptical Polarisation of Fe[57] Gamma Rays, *Phys. Rev.* **126** 1065-75.
2. Szymański K. (1998) Magnetic texture investigation by means of the monochromatic circularly polarized Mössbauer spectroscopy, *NIM* **B 134** 405-12.
3. Szymański K. (2000) Theoretical Treatment of Mixed Hyperfine Interactions Resulting from a Monochromatic, Circularly Polarized Mössbauer Source, *Hyp. Int.* **126** 431-434.
4. Szymański K. (2000) Explicit expression for the Intensity tensor for 3/2-1/2 transitions and solution of the ambiguity problem in the Mössbauer spectroscopy, *IOP Condens. Matter* **12** 7495-507.
5. Shtrikman S. (1967) Mössbauer spectroscopy with polarized monochromatic radiation, *Solid State Comm.* **5** 701-3.
6. Shtrikman S. and Somekh S. (1969) Mössbauer spectroscopy with Monochromatic Circularly polarized radiation, *Rev. Sci. Instr.* **40** 1151-3.
7. Stampfel J. P. and Flinn P. A. (1971) A simple monochromatic, polarized Fe[57] Mössbauer source, *Mössb. Effect Meth.* **6**, 95-107.
8. Varret F., Imbert P., Jehanno G., and Saint-James R., (1975) Compact linearly polarized source for Mössbauer [57]Fe studies, *phys. stat. sol.* **(a) 27** K99-101.
9. Szymański K. (1995) [57]Co in Cr matrix as a single line polarised source, *Proc. Conf. of ICAME 95* Rimini, ed. I. Ortálli, pp 891-4.
10. Szymański K. (1996) Doped Fe₃Si as a filter in Mössbauer spectroscopy, *Hyp. Int.* (C) **97/98**, 573-6
11. Szymański K., Dobrzyński L., Prus B. Cooper M. J. (1996) A Single line circularly polarised source for Mössbauer spectroscopy, *Nucl. Instr. Meth.* **B 119** 438-41.
12. Szymański K., Dobrzyński L. Prus B. Mal'tsev Yu., Rogozev B. and Silin M. (1997) Double sided, Doppler tunable, monochromatic circularly polarized Mössbauer source, Proc. of ICAME'97, Rio de Janeiro, ed. E. Baggio-Saitovich., pp 256-268.
13. Das A. Paranjpe S. K., Honda S., Murayama S. and Tsuchiya Y. (1999) Neutron depolarization measurements on the reentrant spin glass bcc Cr-Fe-Mn, *J. Phys. Condens. Matter* **11** 5209-17.

328

14. Szymański K., Satuła D., Dobrzyński L. Perzyńska K., Biernacka M. and Zaleski P. (2001) Spin alignment and related properties of bcc Cr-Fe-Mn system, *JMMM*, **263** 56-70.
15. Xu, W. M., Steiner W., Reissner, M. Pösinger A., Acet M. and Pepperhoff W. (1992) Magnetic and Mössbauer investigations on $Cr_{75}(Fe_xMn_{1-x})_{25}$ alloys, *JMMM* **104-107** 2023-4.
16. Tsuchiya Y., Nakamura, H., Murayama, S., Hoshi K., Shimojyo Y., Morii Y. and Hamaguchi Y. , (1998) Neutron diffraction and Mössbauer measurements for magnetism of BCC Cr-Fe-Mn alloys, *JMMM* **177-181** 1447-8.
17. Shiga M., Nakamura Y. (1976) Mössbauer Study of B.C.C. Fe-Cr Alloys near the Critical Concentration, *phys.stat.sol.*(a) **37** K89-92.
18. Pearson W. B. (1967) in: *Handbook of lattice spacing and structure of metals* vol 2, Oxford, Pergamon.
19. Shinjo T., Itoh F., H. Takaki, H., Nakamura Y. and Shikazono N. (1964) ^{57}Fe Mössbauer effect in Fe_2B, FeB and Fe_3C, *J. Phys. Soc. Jap.* **19** 1252-1252.
20. Takacs L., Cadeville M. C. and Vincze I. (1975) Mössbauer study of the intermetallic compounds
21. $(Fe_{1-x}Co_x)_2B$ and $(Fe_{1-x}Co_x)B$, *J. Phys. F: Met. Phys.* **5** 800-11.
22. Jeffries J. B. and Hershkowitz N.(1969) Temperature dpendence of the hyperfine Interactions in FeB, *Phys. Lett.* **30A** 187-8.

NEW TRENDS IN MÖSSBAUER SPECTROSCOPY FOCUSED ON NANOSTRUCTURED MAGNETIC MATERIALS

Evaluation, Theory and Methodology

M.A. CHUEV[1], A.M. AFANAS'EV[1], J. HESSE[2] AND O. HUPE[2]

[1] *Institute of Physics and Technology, Russian Academy of Sciences, Nakhimovskii pr. 34, 117218 Moscow, Russia*
[2] *Institut für Metallphysik und Nukleare Festkörperphysik, Technische Universität, Mendelssohnstr. 3, 38106 Braunschweig, Germany*

1. Introduction

Mössbauer spectroscopy proved to be a powerful technique for characterisation of iron-based nanocrystalline magnetic alloys (NCMA) due to, first of all, its local character which allows one to elucidate the nature of hyperfine interactions of the iron nuclei in different crystallographic sites and to probe the nature of their immediate surroundings. Mössbauer spectra of NCMA and amorphous materials as a whole consist usually of a great number of overlapping lines which are due to a variation of hyperfine parameters from site to site, so that extracting the parameters of hyperfine structure requires a corresponding mathematical processing.

In this case the conventional χ^2 method well-known from the mathematical statistics leads to the ill-posed problem [1] which is usually solved in Mössbauer spectroscopy by means of taking into consideration a continuous distribution of the hyperfine field, H_{hf} [2-6]. Recent applications of this method in analysing the Mössbauer spectra of NCMA has allowed researchers to restore the temperature-dependent H_{hf} distribution for nanograins (NG) and amorphous phase in these materials as well as to justify the presence of so-called interface zones between NG and amorphous matrix [7-13]. However, it is hardly possible to extract a really quantitative information about partial contributions from different magnetic phases to the resulting H_{hf} distributions.

An alternative approach for the solution of ill-posed problems has been recently proposed by Afanas'ev and Chuev [14], which is based on the relatively simple principle that the line number density over a spectrum is limited by its statistical quality. The method allows one to find models with the maximum possible number of lines for a given spectrum that is why it was called as the DISCVER ('*Discrete Versions of Mössbauer spectra*') method. We have applied this method for treatment of complex ^{57}Fe Mössbauer spectra of the nanostructured $Fe_{86-x}Cu_1Nb_xB_{13}$ alloys [15-17]. As a result, the temperature evolution of the mean H_{hf} values and the fractions of iron atoms in different sites of the samples with indication of their mean-square errors has been derived within the analysis providing a detailed quantitative information about the nanograins, amorphous residual phase and interface zones.

On the other side, NCMA can be regarded as systems with superparamagnetic

329

M. Mashlan et al. (eds.), Material Research in Atomic Scale by Mössbauer Spectroscopy, 329–338.
© 2003 *Kluwer Academic Publishers. Printed in the Netherlands.*

particles and, due to rather small (5-15 nm) sizes of NG in these materials, the relaxation time of their magnetic moments can be within the Mössbauer time window (10^{-11}-10^{-6} s for ^{57}Fe) at finite temperatures. In this case the superparamagnetic relaxation may be dominant factor in forming the Mössbauer lineshape of NCMA. Beginning with theoretical work of Wickman [18], the two-level relaxation (TLR) model is used to analyse the relaxation Mössbauer spectra of materials with superparamagnetic particles (see, for instance, [19-24]) including also NCMA [11]. However, the conventional TLR model often fails to describe experimental spectra even qualitatively, even if a H_{hf} distribution is simultaneously taken into account. The most intriguing puzzle in the field is a specifically asymmetric Mössbauer lineshape observed in almost each second work dealt with ultra small magnetic particles (see, e.g., [7-13,15-17,19-24] and references cited therein). The lineshape could be evaluated by no way, but introducing a rather wide H_{hf} distribution. The first qualitative explanation of the asymmetric lineshape with no H_{hf} distribution has been recently proposed by Afanas'ev and Chuev in [25] where a generalisation of the TLR model within the presence of interparticle interaction has been performed.

In spite of enhanced facilities of the new *generalised two-level relaxation* (GTLR) *model*, especially as combined with introducing a H_{hf} distribution, it may be difficult in some cases to distinguish between contributions from both the factors into Mössbauer spectra (even if the spectra are taken at different temperature). One of possible ways to solve this problem for NCMA is methodological: Mössbauer spectroscopy under radiofrequency (rf) field excitation proved to be rather informative [26]. Using this technique, the transformation of the spectra could be traced as a function of the rf field amplitude and frequency as well as the relaxation parameters. Facilities of the method are strongly enhanced with prediction of new type of resonant phenomena: *relaxation-stimulated resonances* revealed in the Mössbauer spectra of NCMA in a rf magnetic field [27].

In the present paper we will discuss the above-mentioned theoretical approaches to the analysis of Mössbauer spectra of NCMA in order to combine the latest achievements in the evaluation methods, theory and methodology for better understanding of magnetic and structural properties of these materials.

2. DISCVER Method for Evaluation of Mössbauer Spectra

The DISCVER method for analysis of Mössbauer spectra has been proposed several years ago and provides finding spectral models with maximum possible number of lines and quantitative representation of the spectra with well-defined average values and dispersions of the parameters derived [14]. The method is based on a physical idea that at a given level of measurement precision it is not meaningful to introduce a line with an intensity (area) A_i when its error due to the statistics of the measured spectrum is larger than this intensity (*amplitude criterion*):

$$\Delta A_i < A_i. \tag{1}$$

This criterion has a mathematical consequence for the smallest possible value of distance, q_{ij}, between two single lines in the spectrum, which can be distinguished in the sense of fulfilment of the criterion (1) for both the lines (*splitting criterion*):

$$q_{ij} > q_0(Q_{ij}) \qquad (2)$$

where q_0 depends on the signal-to-noise ratio, Q_{ij}, at the positions of these lines in the spectrum as well as on the line shape which, for the absorption experiment, is naturally determined by the doubled natural linewidth and additional broadening in the source. The effective estimates of a lower bound for q_0 between two Lorentzian lines have been derived in [14]. This mathematical formulation of the physical principle (1) enables one to establish an upper limit on the line number density over a spectrum even before the fitting procedure is carried out, and also to combine two lines into one according to certain criteria during the fitting process. Thus, the spectral model is not prescribed ahead of time, but is derived directly in the course of fitting the spectrum, so that the number of lines appears to be an adjustable parameter.

The first step of analysis within the DISCVER method is the arrangement of lines over the spectrum in accordance with the condition (2) in the manner shown in Figure 1. The first line is placed at the spectral point v_i with maximum absorption. The range of velocities around this point, $(v_i - q_0(Q_i), v_i + q_0(Q_i))$, within which no lines are expected, is excluded from further consideration. The next line is placed at the spectral point v_j with maximum absorption within the rest spectral range and again the range of $q_0(Q_j)$ around the new point is excluded. This procedure is repeated until the whole range of velocities is exhausted. The resulting set of line positions is fixed and the linear χ^2 method is used to find the intensities of lines in these locations. The next stage of analysis consists of fitting the spectrum with variations of both the intensities and the positions of the lines. The previous model is used as an initial approximation. In the course of iterations lines can appear with large errors in the intensities so that the criterion (1) does not hold for them. Such lines are eliminated from further consideration and the fitting procedure is repeated. In the course of this procedure the resultant number of lines decreases essentially.

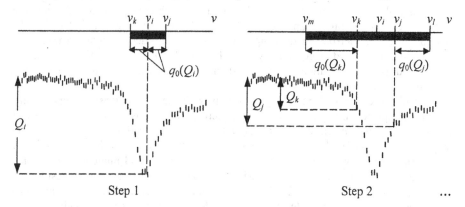

Figure 1. Arrangement of initial positions of lines using the splitting criterion (2).

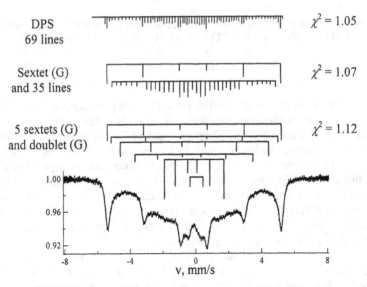

Figure 2. Intermediate stages of the analysis of the room-temperature Mössbauer spectrum of the $Fe_{79}Cu_1Nb_7B_{13}$ NCMA (one hour annealing at 748K) within the DISCVER method.

Thus, in contrast to the conventional χ^2 procedure, in our approach the number of lines in the model varies in the course of fitting. The resulting spectral model automatically delivered by the DISCVER program consists of a rather large number of only single lines. Since none additional line can be added to this model without violating the criterion (1), this model can be called as '*the densest possible solution*' (DPS) for the corresponding spectrum (Figure 2, top). This first spectral model not yet containing any physical interpretation results in a perfect fit with the χ^2 parameter very close to unity. This χ^2 parameter now is our measure in the model finding procedure.

In spite of very good description of the spectrum, it is hardly possible to interpret all the single lines in this discrete representation and DPS can be regarded only as a mathematical approximation for the real spectra in which continuous line distributions may exist. It is natural to suppose that the distributions are due to, first of all, H_{hf} distributions in the ferromagnetic sample. That is why, in order to get any real physical information, at the next stage of analysis we have performed an extraction of magnetic sextets and assumed variations of the additional Gaussian widths for their lines so that the resulting line shape is the Voigt profile.

The identification of sextets begins usually with the most intensive pairs of lines which are expected to be outer lines of a magnetic sextet, that is followed by fitting with restrictions for parameters of the corresponding lines. A situation may arise when the amplitude criterion (1) doesn't hold for some unbounded lines and they are eliminated from consideration with further fitting, which reduces the total number of lines. The resulting model consisting of a sextet with Gaussian broaden lines and the residual 35 single lines is shown in Figure 2. The identification of sextets among the residual lines can be continued and such a step-by-step procedure results in the model with a number of sextets (with different line widths involved into the Voigt profile) and some single lines which can not be bounded into a sextet. A similar procedure can be used in order to extract a quadrupole doublet among the residual single lines. The final model for the spectrum analysed consists of 5 sextets and a quadrupole doublet with Gaussian broaden lines as shown in Figure 2. In each step of searching for the most proper spectral model the χ^2 parameter must be observed and the current model should be accepted only if its χ^2 value is very close to unity.

The credibility of the final model for the room-temperature spectrum is usually to be checked by collecting and fitting spectra taken at different temperature. There may arise situations in fitting such a series of spectra when the amplitude criterion (1) again doesn't hold for some lines. For instance, such a situation occurs inevitably for the Mössbauer spectra taken at temperature close to the Curie temperature of one of magnetic phases so that the hyperfine structure corresponding to this phase collapses into a central line or a quadrupole doublet (see Figure 3, left). In these cases initial model is modified just within the DISCVER program so that instead of the sextet a single line or a doublet are introduced into the current model. Correctness of this procedure is again verified by the resulting χ^2 value which were rather good in our case for all the spectra.

The advantage of the DISCVER procedure is that it provides spectral models with the maximum meaningful number of lines and a quantitative representation of the spectrum with well-defined average values of the parameters together with their errors. That is also true for the H_{hf} distribution, since the additional Gaussian broadening of lines of a magnetic sextet can be regarded as a good estimate for the H_{hf} distribution over different sites in the sample [17]. Then, the total hyperfine field distribution can be expressed as a sum over all sextets with Gaussian broadened lines (see Figure 3, right).

Besides, the temperature dependence of the mean H_{hf} values and the fractions of Fe atoms in different sites of the samples are derived in the entire temperature range of measurements providing a detailed information about the iron nanograins, the amorphous residual phase and the interface zones between them [15,16]. Thus, the new strategy of analysis within the DISCVER method allows one to successfully analyse the extremely complicated Mössbauer spectra of NCMA.

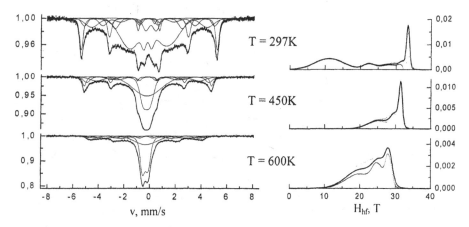

Figure 3. Experimental and calculated Mössbauer spectra of $Fe_{79}Cu_1Nb_7B_{13}$ (left) and corresponding hyperfine field distributions (right).

3. Mössbauer Spectra within Generalised Two-level Relaxation Model

However, the results obtained within the approach of Section 2 are ambiguous to some extent. First, the widths of H_{hf} distribution requires for their explanation a rather wide distribution over particle's size, which is not confirmed by complementary methods. In the other side, the sizes of iron NG in NCMA are regarded to be as small as 5-15 nm, so that the particles may demonstrate superparamagnetic relaxation at finite temperatures and an alternative approach for analysis of the Mössbauer spectra of NCMA is to treat them within relaxation effects.

Beginning in the 1960s, the TLR model is used to analyse the relaxation Mössbauer spectra of materials with superparamagnetic particles [18]. In this simple model only two energy states corresponding to opposite directions of the particle's magnetic moment along the easy magnetisation axes are considered, so that jumps from one state to the other are determined by the energy barrier U_0 between them and the temperature-dependent transition rate, p, is usually described by the Néel formula [28]:

$$p = p_0 \exp(-U_0/k_B T) \qquad (3)$$

where p_0 is a constant. Typical absorption spectra calculated within this model are shown in Figure 4, left.

In the slow relaxation regime when $p < \omega_{Mm} = M\omega_e - m\omega_g$ (M and m are the nuclear spin projection onto the H_{hf} direction; ω_e and ω_g are the Larmor frequencies in the hyperfine field for the nucleus in the excited and ground states, respectively), a well-resolved magnetic hyperfine structure with lines identically broadened ($\Delta\Gamma = 2p$) and slightly shifted (proportionally to $(p/\omega_{Mm})^2$) is observed. As the relaxation rate (temperature) increases and p reaches the corresponding value ω_{Mm}, the line pairs with opposite signs of spin projections M and m collapse successively. In the fast relaxation

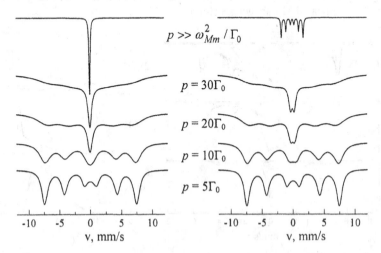

$$p \gg \omega_{Mm}^2 / \Gamma_0$$

$$p = 30\Gamma_0$$

$$p = 20\Gamma_0$$

$$p = 10\Gamma_0$$

$$p = 5\Gamma_0$$

-10   -5   0   5   10   -10   -5   0   5   10
v, mm/s   v, mm/s

Figure 4. ^{57}Fe Mössbauer spectra calculated in the conventional TLR model (left) and in the presence of interaction $\Delta E = k_B T$ (right). The scale for the top curves is decreased tenfold. Here and below $\omega_{3/2,1/2} = 75\Gamma_0$.

regime ($p >> \omega_{Mm}^2 / \Gamma_0$ where Γ_0 is the natural linewidth), the spectrum collapses into a single central line (or a quadrupole doublet).

This simple model appeared to be very popular and many investigators have used it to describe spectra of superparamagnetic particles in different materials [11,19-24]. But only few examples could be found where experimental data are satisfactorily described within this model. Moreover, in a great number of studies lines of asymmetrical shape with sharp outer and strongly extended inward wings are observed in the spectra like those shown in Figure 3. One usually uses rather artificial particle's distribution over sizes in order to describe spectra of the kind. And no alternative explanation had been found until we have suggested a generalisation of the TLR model [25].

The principal idea of the GTLR model is that the relaxation between the particle's states with opposite directions of its magnetic moment never occurs as a transition between the states of the same energy because even weak interaction with the environment should inevitably smear out the energy levels. The energy levels of each particle at a certain time prove to be separated by a certain gap ΔE and the average value of ΔE may be comparable to temperature. Such a separation results in different energy barriers for jumps between two states (in the case of weak interaction $U_{2,1} = U_0 \pm \Delta E$) and the relaxation rates becomes different:

$$p_{21,12}(\Delta E) = p \exp(\pm\Delta E /k_B T). \qquad (4)$$

As seen in Figure 4, right, the asymmetric fluctuations lead to remarkable changes in the temperature evolution of the shape of Mössbauer spectrum [18,25,29]. First of all, the inclusion of interaction results in a substantial slowing down of the relaxation process. However, the most representative is the fact that in the fast relaxation limit the spectrum do not collapse into a single line and exhibit a well-resolved hyperfine structure (top curve in Figure 4). In this case the nucleus should 'feel' the stochastically averaged hyperfine field defined by the difference of the equilibrium populations, $w_1(\Delta E)$ and $w_2(\Delta E)$, which is determined from the detailed balance principle:

$$w_1(\Delta E) - w_2(\Delta E) = \tanh(\Delta E /k_B T). \qquad (5)$$

Since the interaction with environment in a system like NCMA with a great number of degrees of freedom should be random in character, the energy shifts ΔE is to be scattered over a certain interval σ with a probability function $P(\Delta E,\sigma)$. Then, the absorption cross-section is determined by the sum of subspectra with different ΔE weighed by $P(\Delta E,\sigma)$ [25]. As a result of the generalisation of the TLR model, the shape of Mössbauer absorption spectra and its temperature evolution changes drastically (see Figure 5). And the most salient feature of the GTLR spectra is just the appearance of asymmetrically shaped lines with extended inward wings, which are so often observed in experiments.

Thus, the GTLR model actually represents an alternative approach to the analysis of the NCMA spectra taking into account the interparticle interaction in not so complicated form which can be easily adopted by experimentalists. The first successful application of this model could be found in the contribution of O.Hupe *et al.* in this Proceedings.

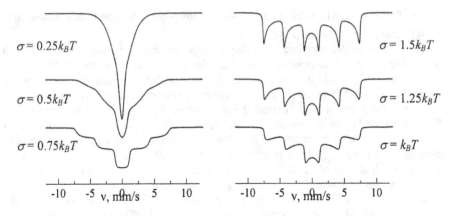

Figure 5. ^{57}Fe Mössbauer spectra calculated within the GTLR model in the fast relaxation limit for the Gaussian shape of $P(\Delta E, \sigma)$.

4. Relaxation-stimulated Resonances in Mössbauer Spectra under RF Field Excitation

Because of a formal similarity of the lineshape calculated within the GTLR model and using the H_{hf} distribution method, it may be difficult in some cases to distinguish between contributions of both the factors into Mössbauer spectra (even if the spectra are taken at different temperature). On the other side, Mössbauer spectroscopy under rf field excitation proved to be rather informative [26], providing a rich transformation of the spectral shape as a function of the rf field amplitude and frequency. Influence of an external rf field on NCMA results in changes in magnetic moments of both single particles and the sample as a whole so that the intrinsic magnetic dynamics should reveal in Mössbauer spectra. Generally speaking, a theory developed in [30-32] allows one to calculate the relaxation Mössbauer spectra as a function of the rf field frequency and amplitude and by that to extract information about the magnetic properties of the sample studied. However, it takes a huge computer time and a substantial optimisation of the fitting procedure.

Effectiveness of the method are strongly enhanced with the prediction of the relaxation-stimulated resonances (RSR) in Mössbauer spectra under rf field excitation [27]. Resonant effects of the kind are still unknown in physics, they are realised not at the frequencies of transitions between the energy sublevels of the ground (ω_g) or excited (ω_e) nuclear states, which takes place in the conventional NMR, but at frequencies being a combination of the Larmor frequencies ω_e and ω_g. The most pronounced effect should manifest as a sharp, resonance narrowing of particular hyperfine components to the frequencies of which the radio frequency is tuned:

$$\omega_{rf} = \left| M\omega_e - m\omega_g \right| / n,$$

$$(6)$$

where n is integer (see Figure 6, left).

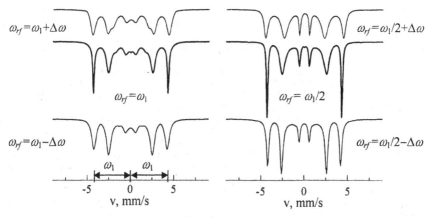

Figure 6. Relaxation-stimulated resonances in ^{57}Fe Mössbauer spectra in a rf field with amplitude $H_0 \leq H_c$ (left) and an additional static field with strength $H_s = H_0 \leq H_c/2$ (right). H_c is the critical field strength, $\omega_1 = \omega_{3/2,1/2} = 2\pi \times 50$ MHz, $\Delta\omega/2\pi = 5$ MHz. A non-polarised γ-beam transverse to the rf field direction has been assumed.

Non-trivial effects of a static magnetic field applied along with a rf field on RSR have been also predicted [33]. As shown in Figure 6, a shift of the resonance radio frequency toward lower range should be observed. This effect is non-linear and threshold in character and can be very important for applications. Tuning to the resonance can be also realized by applying a static magnetic field, which may essentially simplify arrangement of corresponding experiments.

Acknowledgements

We are grateful to the "Internationales Büro des BMBF" and the Russian Foundation Sponsoring the Domestic Science for supporting our collaboration.

References

1. Tikhonov, A.N. and Arsenin, V.Ya. (1977) *Solution of Ill-Posed Problems*, Winston, Washington, DC.
2. Hesse, J. and Rübartsch, H. (1974) Model independent evaluation of overlapped Mössbauer spectra, *J. Phys. E: Sci. Instrum.* **7**, 526-532.
3. Mangin, P., Marshal, G., Piecuch, M. and Janot, C. (1976) Mössbauer spectra analysis in amorphous system studies, *J. Phys. E: Sci. Instrum.* **9**, 1101-1105.
4. Le Caer, G. and Dubois, J.M. (1979) Evaluation of hyperfine parameter distributions from overlapped Mössbauer spectra of amorphous alloys, *J. Phys. E: Sci. Instrum.* **12**, 1083-1090.
5. Wivel, C. and Mørup, S. (1981) Improved computational procedure for evaluation of overlapping hyperfine parameter distributions in Mössbauer spectra, *J. Phys. E: Sci. Instrum.* **14**, 605-610.
6. Brand, R. A. and Le Caër, G. (1988) Improving the validity of Mössbauer hyperfine parameter distributions: the maximum entropy formalism and its applications, *Nucl. Instrum. and Meth.* **B34**, 272-284.
7. Miglierini, M. and Grenèche, J.-M. (1997) Mössbauer spectrometry of Fe(Cu)MB-type nanocrystalline alloys: I. The fitting model for the Mössbauer spectra, *J. Phys.: Condens. Matter* **9**, 2303-2319.
8. Miglierini, M., Skorvanek, I. and Grenèche, J.-M. (1998) Microstructure and hyperfine interactions of the Fe$_{73.5}$Nb$_{4.5}$Cr$_5$CuB$_{16}$ nanocrystalline alloys: Mössbauer effect temperature measurements, *J. Phys.:*

338

Condens. Matter **10**, 3159-3176.

9. Suzuki, K. and Cadogan, J.M. (1998) Random magnetocrystalline anisotropy in two-phase nanocrystalline systems, *Phys.Rev. B* **58**, 2730-2739.

10. Hernando, A. (1999) Magnetic properties and spin disorder in nanocrystalline materials, *J. Phys.: Condens. Matter* **11**, 9455-9482.

11. Kemény, T., Kaptás, D., Balogh, J., Kiss, L.F., Pusztai, T. and Vincze, I. (1999) Microscopic study of the magnetic coupling in a nanocrystalline soft magnet, *J. Phys. Condens. Matter* **11**, 2841-2847.

12. Balogh, J., Bujdoso, L., Kaptás, D., Kemény, T., Vincze, I., Szabo, S. and Beke, D. (2000) Mössbauer study of the interface of iron nanocrystallites, *Phys. Rev. B* **61**, 4109-4116.

13. Miglierini, M., Schaaf, P., Skorvanek, I., Janickovic, D., Carpene, E. and Wagner, S. (2001) Laser-induced structural modifications of FeMoCuB metallic glasses before and after transformation into a nanocrystalline state, *J. Phys.: Condens. Matter* **13**, 10359-10369.

14. Afanas'ev, A.M. and Chuev, M.A. (1995) Discrete forms of Mössbauer spectra, *JETP* **80**, 560-567.

15. Hupe, O., Bremers, H., Hesse, J., Afanas'ev, A.M. and Chuev, M.A. (1999) Structural and magnetic information about a nanostructured ferromagnetic FeCuNbB alloy by novel model independent evaluation of Mössbauer spectra, *Nanostructured Mater.* **12**, 581-584.

16. Hupe, O., Chuev, M.A., Bremers, H., Hesse, J., and Afanas'ev, A.M. (1999) Magnetic properties of nanostructured ferromagnetic FeCuNbB alloys revealed by a novel model independent evaluation of Mössbauer spectra, *J. Phys.: Cond. Matter* **11**, 10545-10556.

17. Chuev, M.A., Hupe, O., Bremers, H., Hesse, J., and Afanas'ev, A.M. (2000) A novel method for evaluation of complex Mössbauer spectra demonstrated on nanostructured ferromagnetic FeCuNbB alloys, *Hyperfine Interact.* **126**, 407-410.

18. Wickman, H.H. (1966) Mössbauer Paramagnetic Hyperfine Structure, in I.J. Gruverman (ed.), '*Mössbauer effect methodology*', v.2, Plenum Press, New York, pp. 39-66.

19. Reid, N.M.K., Dickson, D.P.E. and Jones, D.H. (1990) A study of the parametrisation of the uniaxial model of superparamagnetic relaxation, *Hyperfine Interact.* **56**, 1487-1490.

20. Mørup, S. (1994) Superferromagnetic nanostructures, *Hyperfine Interact.* **90**, 171-185.

21. Mørup, S. and Tronc, E. (1994) Superparamagnetic relaxation of weakly interacting particles, *Phys. Rev. Lett.* **72**, 3278-3281.

22. Tronc, E., Prené, P., Jolivet, J.P., d'Orazio, F., Lucari, F., Fiorani, D., Godinho, M., Cherkaoui, R., Noguès, M. and Dormann, J.L. (1995) Magnetic behaviour of γ-Fe₂O₃ nanoparticles by Mössbauer spectroscopy and magnetic measurements, *Hyperfine Interact.* **95**, 129-148.

23. Dormann, J.L., D'Orazio, F., Lucari, F., Tronc, E., Prené, P., Jolivet, J.P., Fiorani, D., Cherkaoui, R. and Noguès, M. (1996) Thermal variation of the relaxation time of the magnetic moment of γ-Fe₂O₃ nanoparticles with interparticle interactions of various strengths, *Phys. Rev. B* **53**, 14291-14297.

24. Tronc, E., Ezzir, A., Cherkaoui, R., Chanéac, C., Noguès, M., Kachkachi, H., Fiorani, D., Testa, A.M., Grenèche, J.-M. and Jolivet, J.P. (2000) Surface-related properties of γ-Fe₂O₃ nanoparticles, *J. Magn. Magn. Mater.* **221**, 63-79.

25. Afanas'ev, A.M. and Chuev, M.A. (2001) New relaxation model for superparamagnetic particles in Mössbauer spectroscopy, *JETP Lett.* **74**, 107-110.

26. Hesse, J., Graf, T., Kopcewicz, M., Afanas'ev, A.M. and Chuev, M.A. (1988) Mössbauer experiments in radio frequency magnetic fields: A method for investigations of nanostructured soft magnetic materials, *Hyperfine Interact.* **113**, 499-506.

27. Afanas'ev, A.M., Chuev, M.A. and Hesse, J. (2000) Relaxation-stimulated resonances in Mössbauer spectra under rf magnetic field excitation, *J. Phys.: Cond. Matter* **12**, 623-635.

28. Néel, L. (1949) Theorie du trainage magnetique des ferromagnetiques en grains fins avec applications aux terres cuites, *Ann. Geophys.* **5**, 99-136.

29. Rancourt, D.G. (1996) Analytical methods for Mössbauer spectral analysis of complex materials, in G.J. Long and F.Grandjean (eds.), '*Mössbauer spectroscopy applied to magnetism and materials science*', v.2, Plenum Press, New York, pp. 105-124.

30. Afanas'ev, A.M., Chuev, M.A. and Hesse, J. (1997) Relaxation Mössbauer spectra under rf magnetic field excitation, *Phys. Rev. B* **56**, 5489-5499.

31. Afanas'ev, A.M., Chuev, M.A. and Hesse, J. (1998) Collapse in the model of the non-interacting Stoner-Wohlfarth particles, *JETP* **86**, 983-992.

32. Afanas'ev, A.M., Chuev, M.A. and Hesse, J. (1999) Mössbauer spectra of Stoner-Wohlfarth particles in rf fields in a modified relaxation model, *JETP* **89**, 533-546.

33. Afanas'ev, A.M., Chuev, M.A. and Hesse, J. (2001) Shift of relaxation-stimulated resonances in Mössbauer absorption spectra with applying of static magnetic field, *JETP Lett.* **73**, 519-523.

THE USE OF THE WAVELET TRANSFORM FOR MÖSSBAUER SPECTRA FITTING

R. RYVOLA AND M. MASHLAN
*Department of Experimental Physics, 17. Listopadu 50a, 772 07
Olomouc, Czech Republic*

1. Introduction

Several programs have been written for the analysis of the Mössbauer data [1]. They usually suppose that the spectrum consists of n peaks arising from absorption and scattering events suffered by the Mössbauer photons, each peak having a Lorentzian form. For n absorption lines the computed content of channel i, represented by N_i will be given by

$$N_i = b_i - \sum_{k=1}^{n} \frac{N_k^o}{1 + \left[2\left(i - i_k^o\right)/\Gamma_k\right]^2},$$

where i is the channel number, i^o is the channel for maximum absorption and N^o the number of photons recorded in this channel, b_i is the number of photons recorded in the absence of Mössbauer absorption and Γ is the line width.

The program then changes i^o, N^o and Γ to reach a minimum of the function

$$\chi^2 = \sum_{i=1}^{m} W_i \left(N_i^{obs} - N_i\right)^2.$$

W_i is a weighting factor of $1/N_i^{obs}$, allowing for the error in N^{obs}, which is determined by the Poisson distribution of the radioactive decay events, m is the number of channels in the data collection system. After each iteration, yielding a new value of χ^2, the values of $d\chi^2/dq$, where q represents one of the variables i^o, N^o or Γ, are calculated and the iteration proceeds until $d\chi^2/dq \Rightarrow 0$. The first estimation of the values of i^o, N^o and Γ is necessary for starting of the iteration process. Moreover, the accuracy of their estimation determines the final number of iteration, i.e. the calculation time and the convergence of the iteration process to the correct values of i^o, N^o and Γ.

The application of the wavelet transform procedure to estimate the i^o, N^o and Γ parameters is described in this paper.

M. Mashlan et al. (eds.), Material Research in Atomic Scale by Mössbauer Spectroscopy, 339–346.
© 2003 *Kluwer Academic Publishers. Printed in the Netherlands.*

2. Wavelet Transform

The wavelet transform (WT) in the real range is defined as follows

$$W(a,b)(f) = \int_{-\infty}^{\infty} f(t)\Psi_{a,b}(t)dt, \tag{1}$$

where $f(t)$ is a transformed signal, $W(a,b)$ is the WT of that signal, $\Psi_{a,b}(t)$ are the daughter wavelets derived from a mother wavelet according to this equation

$$\Psi_{a,b}(t) = \frac{1}{\sqrt{|a|}}\,\Psi\left(\frac{t-b}{a}\right), \qquad a,b \in R, \ \ a \neq 0, \tag{2}$$

where $\Psi(t)$ is the mother wavelet. The mother wavelet is a basis function of WT. Owing to the conditions imposed on the mother wavelet it has significantly non-zero values only in a small range close to the origin and it creates a small wave in this range. There is an example of the wavelet in the figure 1. It is the second derivative of a gauss function also called "The Mexican hat".

The daughter wavelet $\Psi_{a,b}(t)$ is created from the mother wavelet $\Psi(t)$ by a dilatation (given by the parameter a) and a translation (given by the parameter b). The WT can be understood as a correlation of signal $f(t)$ with a dilated and translated mother wavelet $\Psi(t)$. The WT of a function of 1 variable is a function of 2 variables – the frequency represented by the parameter a and the position represented by b. The WT gives us information about which frequencies are situated in which position of the processed signal. It means that the WT has a frequency (or time) and a space resolution. This can be used for processing of non-periodic signals because of different frequencies in different places. The WT can be also viewed as an extension to the Fourier transform about the space resolution.

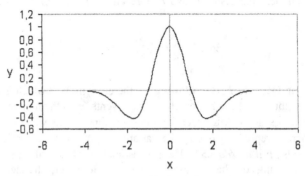

Figure 1. „Mexican hat" wavelet.

3. Applications of Wavelet Transform

The WT can be used for studying nonstationarity in time series. For example it is useful for processing meteorological data [2] where a physical quantity (e.g. temperature) changes the period over the time. The WT gives us also information about changing the intensity over the time.

Noise reduction is another wavelet application [3]. Noise reduction algorithms are based on the fact that the WT of a noiseless signal consists of small nonzero wavelet coefficients - that is $W(a,b)$ - but the WT of a noisy signal contains many coefficients only slightly above zero. Thresholding these coefficients and computing an inverse WT can reduce the noise level in the signal.

A data compression can also be done using the WT. Thresholding wavelet coefficients and saving only those coefficients that are significantly above zero takes much less memory than the original data. This can be used also for an image compression [4]. The new JPEG2000 standard uses the wavelet compression [5]. This is loss compression but there is also a loss less version of the wavelet compression.

An application of the wavelet transform technique in chemical analysis was reviewed in [6].

4. Algorithm

The WT of a signal can give us information about what wavelets the signal consists of. That means what their halfwidths and positions are. Let us consider the Mössbauer spectrum as a sum of Lorentz curves. The WT of this spectrum using the Lorentz curve as mother wavelet could give us the parameters (position, halfwidth and amplitude) of the curves the spectrum consists of. Unfortunately, not every function can be used as a mother wavelet. It must satisfy some conditions and the Lorentz function does not satisfy them. Thus we must find a function satisfying the conditions for use as a mother wavelet and similar enough to the Lorentz function to use it for our purposes. We use the second derivative of the Gauss function (fig. 1). Figure 2 shows the WT of a noisy signal that consists of 2 Lorentz peaks. The white areas mean high correlation of the signal with the mother wavelet (the second derivative of the Gauss function) with appropriate parameters (position and halfwidth). Knowing the position and intensity of maximum in the WT, we can calculate the position, amplitude and halfwidth of the appropriate Lorentz peaks that create the original signal.

There is one disadvantage. If there are two or more peaks in the signal very close to each other, then the maximums of those peaks in the WT create only one maximum and we cannot recognize the individual peaks in this manner. Let us consider that the maximum is created from two peaks. If we recreate the signal from this one maximum we obtain a slightly different progress of function than the original signal since it is created from one Lorentz peak instead of the original two peaks. Discounting these two signals from each other (the original signal and the recreated one) we can obtain the difference between them. The WT of this difference will produce two maximums corresponding approximately to the sought two peaks (let us call them p_1 and p_2). Now we discount the signal created from one of these maximums (e.g. p_1) from the original signal. We should obtain a signal containing only peak p_2. The WT of this signal will contain only one maximum corresponding to peak p_2 and this maximum is closer to the real maximum than the one counted from the difference. More exact parameters of peak p_1 are calculated in the same way as finding the maximum in the WT of the difference between the original signal and the reconstructed signal containing only peak p_2. This process is repeated until parameters of those peaks, which are exact enough, are obtained. Absolutely exact values cannot be reached since the mother wavelet is different from what we are seeking in the signal but we can obtain those parameters exactly enough to use them as the initial parameters in the least-squares method to obtain the exact values.

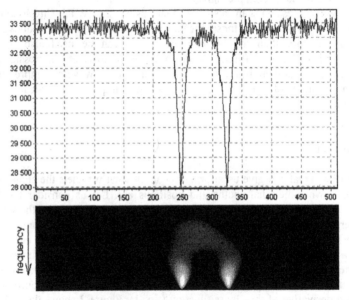

Figure 2. WT of two noisy Lorentz peaks.

The block diagram on figure 3 shows this process.

5. Example of Extracting Two Close Peaks from a Signal

Let us suppose we have a signal created from two Lorentz peaks close enough to each other to form one deformed peak (fig. 4). The WT of this signal contains only one maximum corresponding to the Lorentz peak that is similar to the processed signal (fig. 5). The WT of the difference (fig. 6) between the real signal (fig. 4) and the reconstructed one (fig. 5) will produce two peaks corresponding approximately to the original two small peaks creating the entrance signal (fig. 7,8). The thicker curve is the peak reconstructed from the WT compared to the real peak (thinner curve). Now let us take the original signal and discount from it the right reconstructed peak (fig. 8). We should obtain approximately the left peak. Its WT and reconstruction returns more exact

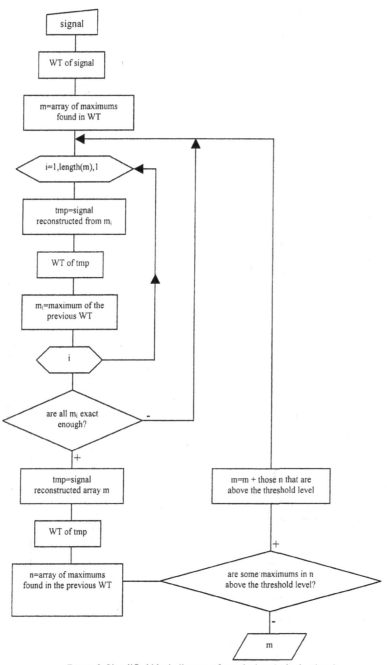

Figure 3. Simplified block diagram of searched peaks in the signal.

344

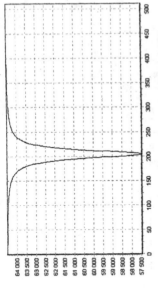

Figure 5. Reconstructed peak from the first WT.

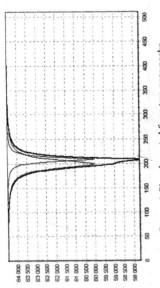

Figure 4. Signal created from two near peaks.

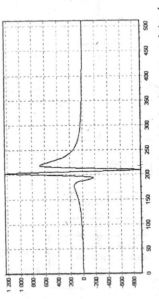

Figure 6. Difference between the original and reconstructed signal.

345

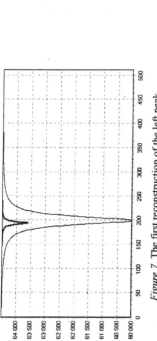

Figure 7. The first reconstruction of the left peak.

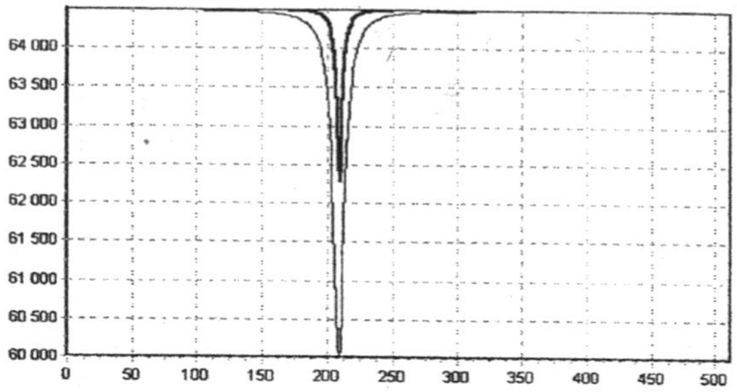

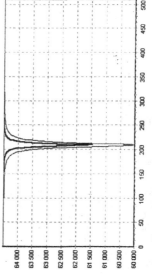

Figure 8. The first reconstruction of the right peak (thick)
and original peak (thin).

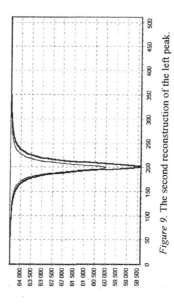

Figure 9. The second reconstruction of the left peak.

Figure 10. The second reconstruction of the right peak.

parameters of the left peak (fig. 9). The same for the other peak can be seen in figure 10. These reconstructed peaks are closer to the real ones. If no noise is added to the signal then repeating this process several times will result in exact values of the forming peaks.

Acknowledgements

Financial support from the Ministry of Education under Projects MSM 153100007 is acknowledged.

References

1. Vandenberghe, R.E., De Grave, E. and De Bakker, P.M.A. (1994) On the methodology of the analysis of Mössbauer spectra, *Hyperfine Interactions* **83**, 29-49.
2. Torrence, C. and Compo, B.P. (1998) *A Practical Guide to Wavelet Analysis*, American Meteorological Society, (http://paos.colorado.edu/research/wavelets/bams_79_01_0061.pdf)
3. Fligge, M. and Solanki, S.K. *Noise reduction in astronomical spectra using wavelet packets* (http://www.edpsciences.com/articles/astro/pdf/1997/12/ds1226.pdf)
4. http://www.c3.lanl.gov/~brislawn/FBI/FBI.html
5. http://www.jpeg.org/JPEG2000.htm
6. Leung, A.K., Chau F. and Gao J. (1998) A review on application of wavelet transform techniques in chemical analysis: 1989-1997, *Chemometrics and Intelligent Laboratory Systems* **43**, 165-184.

DOSE MEASUREMENTS "IN" MÖSSBAUER SPECTROSCOPY AND DOSE MEASUREMENTS "BY" MÖSSBAUER SPECTROSCOPY

G. PEDRAZZI[1,2], S. VACCARI[2], M. GHILLANI[1] AND E. PAPOTTI[2]
[1] *University of Parma, Department of Public Health - Section of Physics and INFM Udr Parma, Plesso Biotecnologico Integrato, Via Volturno 39, 43100 Parma-Italy*
[2] *University of Parma, Health Physics Service, Plesso Biotecnologico Integrato, Via Volturno 39, 43100 Parma-Italy*

Abstract

The present work deals with ^{57}Co-Mössbauer spectroscopy and radiation doses. We have taken into consideration two opposite sides of the technique, simplifying we could call them the "producer" and "consumer" faces of the method.

In the context of radiation safety, Mössbauer spectroscopy is a "producer" of radiation doses since it makes use of radioactive sources. On the other side, Mössbauer spectroscopy can also be viewed as a "consumer" since it can be used to measure the effective radiation dose that has been absorbed by a medium after irradiation by x-rays or gamma-rays (radiations to measure radiation doses). This will be shown in the last part of the present paper.

1. Introduction

This study was undertaken mainly because of a personal curiosity, in fact, speaking with colleagues and friends we have matured the conviction that apart a general idea of radiation risks and doses few people have a correct perception of the real amount of radiation dose (and so of risks) involved in the laboratory procedures. We have therefore considered the absorbed dose that is involved in the normal laboratory practice and, more in details, when uncommon or special situations occur. This is the case, for instance, of the arrival of a new strong source that we needs to mount on the spectrometer bench. We have calculated the absorbed dose for many different geometries, operations and accidents, however, because of the limited space in the present document only some relevant cases will be discussed.

Regarding the measurements of absorbed dose by Mössbauer spectroscopy we have investigated the oxidation of ferrous ions (Fe^{2+}) to ferric ions (Fe^{3+}) as the result of the irradiation of a solution of ferrous ammonium sulfate (modified Fricke solution). Fe^{2+} ions act like scavengers for the active products of the radiation chemistry of water. The G value is in principle known for this reaction, therefore, the relative amount of Fe^{2+} and Fe^{3+} can be used to estimate the effective dose absorbed by the medium [1].

347

M. Mashlan et al. (eds.), Material Research in Atomic Scale by Mössbauer Spectroscopy, 347–356.
© 2003 *Kluwer Academic Publishers. Printed in the Netherlands.*

2. Quantities and Units

The physical quantities used in radioprotection to express the absorption of energy from ionizing radiations are the "absorbed dose" (D), the "Equivalent Dose" (H) and the "Effective Dose" (E). These latter are related to the biological effects [1,2,3,4].

The absorbed dose, defined as $D = d\bar{\varepsilon}/dm$, represents the mean energy imparted to an irradiated mass. Its unit, the gray, is $1\ Gy = 1\ J\ kg^{-1}$. An absorbed dose is always a dose in a material. Therefore we have to specify the material, for example: dose in air; dose in water; dose in tissue.

The Equivalent Dose, H, is the absorbed dose in tissue or organ T weighted for the type and quality of radiation R: $H_{T,R} = w_R D_{T,R}$ where w_R is the weighting factor for radiation R and $D_{T,R}$ is the absorbed dose averaged over tissue or organ T. The physical unit for the Equivalent Dose is the sievert (Sv).

The Effective Dose is the sum of the weighted equivalent doses in all the tissues and organs of the body from internal and external radiations :

$$E = \sum_T w_T H_T = \sum_T w_T \sum_R w_R D_{T,R}$$ where w_T is the weighting factor for tissue or organ T. The radiation weighting factor, w_R, for gamma and x-rays is assumed to be 1, therefore, it follows imediately that $1\ Gy = 1\ Sv$ for X and gamma photons.

The absorbed dose for gamma radiation and x-rays at a point x in space, if we consider a point source, can be calculated by

$$D = \sum_i \varphi_i (x) y_i E_i \left(\frac{\mu_E}{\rho} \right)_i \tag{1}$$

where $\varphi_i (x)$ is the fluence of the i-th radiation, E_i is the energy of the i-th radiation, y_i is the mean number of events per nuclear transformation, $(\mu_E/\rho)_i$ is the mass energy absorption coefficient for the i-th radiation. Since an extended source may be thought as a combination of point sources, and since all the above quantities, except $\varphi_i (x)$, are tabulated, the problem to estimate the dose is reduced to evaluate the flux at the point of interest.

2.1. FLUENCE, FLUX DENSITY AND DOSE CALCULATION

Due to the normal way a ^{57}Co–Mössbauer source is manufactered and shipped, and given the common laboratory practice, we can consider three main flux configurations : 1) point source ; 2) disk - on the axis ; 3) disk - from the border.

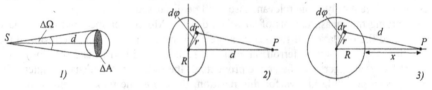

Figure 1. Different flux configurations.

These configurations may correspond to real situations where, for instance, 1) the operator is far from the source ; 2) and 3) the operator is holding the source with his fingers or small tools.

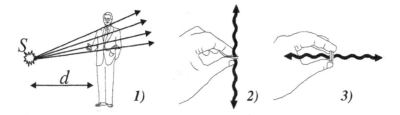

Figure 2. Some possible geometries of irradiation.

In the following we will calculate the equivalent dose and effective dose for the above situations. Let us assume for our ^{57}Co source :

S   (source strenght) $\equiv$ activity (Bq) $\rightarrow$ 3.7 GBq (100 mCi)
S_A activity per unit area ; R (radius of the disk source) = 3 mm

Case 1) Point source

The definition of flux density is $\Phi = (S/\Delta A)(\Delta\Omega/4\pi)$ where $\Delta\Omega$ describes the solid angle from which particles or photons from the point source can pass a sphere with the area of great circle ΔA. With $\Delta\Omega = \Delta A/d^2$ it follows that $\Phi = S/(4\pi d^2)$.

Case 2) Disk – P along the axis

Decomposition of the area source into point sources of source strength $S_A dA = S_A r dr d\varphi$ yields to a contribution to the flux density at P (distant d along the axis of the disk) of $d^2\Phi = S_A r dr d\varphi/\left[4\pi\left(r^2 + d^2\right)\right]$. The flux can therefore be derived from

$$\Phi = \int_0^R \int_0^{2\pi} \frac{S_A r dr d\varphi}{4\pi\left(r^2 + d^2\right)} \rightarrow \Phi = \frac{S_A}{2}\ln\left(\frac{\sqrt{R^2 + d^2}}{d}\right) = \frac{S}{4\pi R^2}\ln\left(\frac{R^2 + d^2}{d^2}\right) \qquad (2)$$

Case 3) Disk – out of axis

The general case of the flux at a point P (out of the axis, distant x from the center and at height h, figure 4) can be calculated as follows : $d^2 = h^2 + r^2 + x^2 - 2rx\cos\varphi$;

$$d\Phi^2 = S_A \frac{r dr d\varphi}{4\pi d^2} = \frac{S_A r dr d\varphi}{4\pi\left(h^2 + r^2 + x^2 - 2rx\cos\varphi\right)} \qquad \text{the flux can therefore be obtained by}$$

double integration [5]

$$\Phi = 2\int_{r=0}^R \int_{\varphi=0}^{\pi} \frac{S_A r dr d\varphi}{4\pi\left(h^2 + r^2 + x^2 - 2rx\cos\varphi\right)} = \frac{S}{2\pi}\int_{r=0}^R r dr \int_{\varphi=0}^{\pi} \frac{d\varphi}{\left(h^2 + r^2 + x^2 - 2rx\cos\varphi\right)} \qquad (3)$$

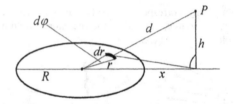

Figure 3. Disk configuration , general case.

to obtain

$$\Phi = \frac{S_A}{4} \ln \left[\frac{R^2 + h^2 - x^2 + \sqrt{\left(R^4 + 2R^2\left(h^2 - x^2\right) + \left(h^2 + x^2\right)^2\right)}}{2h^2} \right] \qquad (4)$$

The same procedure can be applied in case $h = 0$ (just on the disk plane). This leads to $\Phi = \dfrac{S_A}{2\pi} \displaystyle\int_{r=0}^{R} r\,dr \int_{\varphi=0}^{\pi} \dfrac{d\varphi}{\left((R+d)^2 + r^2 - 2r(R+d)\cos\varphi\right)}$ and to the final result

$$\Phi = -\frac{S_A}{4} \ln\left(1 - \frac{R^2}{(R+d)^2}\right) \qquad (5)$$

2.2. DOSE, DOSE RATE

For a point source the dose rate can be determined [4] by

$$\dot{D} = \sum_i \frac{S_i}{4\pi d^2} B_i\left(\mu_i, d\right) e^{-\mu_i d} \left(\frac{\mu_E}{\rho}\right)_i E_i \qquad (6)$$

where the various terms have the same meaning as above and the index i spans over all the radiations emitted during the decay process; μ_i and $B_i\left(\mu_i, d\right)$ represent the linear attenuation coefficient and the build-up coefficient for i-th radiation to take into account the attenuation and scatter of radiation, respectively.

The decay scheme of ^{57}Co, following the International Commission on Radiological Protection (ICRP) [6] is reported in figure 4.

We can observe that the relevant radiations to be considered for safety purposes are γ_1, γ_2, γ_3, γ_9, and the various $K_{\alpha,\beta}$–X-rays. All the other radiations provide negligible contributions and can be omitted.

In a first approximation we could also consider $B_i\left(\mu_i, d\right) = 1$. The results of the calculation are reported in table II, where a comparison with other sources of radiation is shown.

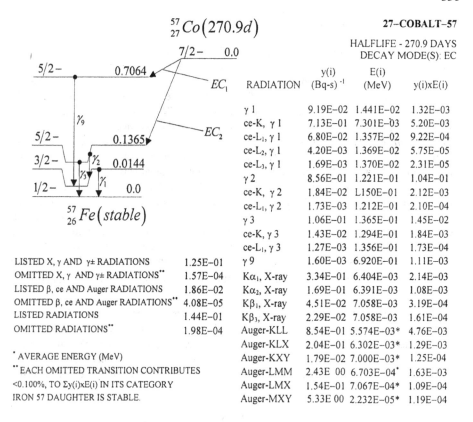

$$^{57}_{27}Co\left(270.9d\right)$$

HALFLIFE - 270.9 DAYS
DECAY MODE(S): EC

RADIATION	y(i) (Bq-s)$^{-1}$	E(i) (MeV)	y(i)xE(i)
γ 1	9.19E–02	1.441E–02	1.32E–03
ce-K, γ 1	7.13E–01	7.301E–03	5.20E–03
ce-L₁, γ 1	6.80E–02	1.357E–02	9.22E–04
ce-L₂, γ 1	4.20E–03	1.369E–02	5.75E–05
ce-L₃, γ 1	1.69E–03	1.370E–02	2.31E–05
γ 2	8.56E–01	1.221E–01	1.04E–01
ce-K, γ 2	1.84E–02	L150E–01	2.12E–03
ce-L₁, γ 2	1.73E–03	1.212E–01	2.10E–04
γ 3	1.06E–01	1.365E–01	1.45E–02
ce-K, γ 3	1.43E–02	1.294E–01	1.84E–03
ce-L₁, γ 3	1.27E–03	1.356E–01	1.73E–04
γ 9	1.60E–03	6.920E–01	1.11E–03
Kα₁, X-ray	3.34E–01	6.404E–03	2.14E–03
Kα₂, X-ray	1.69E–01	6.391E–03	1.08E–03
Kβ₁, X-ray	4.51E–02	7.058E–03	3.19E–04
Kβ₃, X-ray	2.29E–02	7.058E–03	1.61E–04
Auger-KLL	8.54E–01	5.574E–03*	4.76E–03
Auger-KLX	2.04E–01	6.302E–03*	1.29E–03
Auger-KXY	1.79E–02	7.000E–03*	1.25E–04
Auger-LMM	2.43E 00	6.703E–04*	1.63E–03
Auger-LMX	1.54E–01	7.067E–04*	1.09E–04
Auger-MXY	5.33E 00	2.232E–05*	1.19E–04

LISTED X, γ AND γ± RADIATIONS	1.25E–01
OMITTED X, γ AND γ± RADIATIONS**	1.57E–04
LISTED β, ce AND Auger RADIATIONS	1.86E–02
OMITTED β, ce AND Auger RADIATIONS**	4.08E–05
LISTED RADIATIONS	1.44E–01
OMITTED RADIATIONS**	1.98E–04

* AVERAGE ENERGY (MeV)
** EACH OMITTED TRANSITION CONTRIBUTES
<0.100%, TO Σy(i)xE(i) IN ITS CATEGORY
IRON 57 DAUGHTER IS STABLE.

Figure 4. ^{57}Co decay scheme (from ICRP [6]).

TABLE 1. Conversion factors D/Φ , D(0.07)/D , and D(10)/D as function of photon energy E.
* Dose values at depths of 0.07 mm and 10 mm [2, 7]; photon incidence as broad directed beam.

E (MeV)	D/Φ (10^{-12} Gy · cm^2)	D(0.07) / D *	D(10) / D *
0.010	7.45	0.919	0.010
0.015	3.13	0.982	0.264
0.10	0.371	1.56	1.64
0.15	0.599	1.43	1.48
0.20	0.856	1.38	1.41
0.30	1.38	1.31	1.33
0.40	1.89	1.25	1.26
0.50	2.38	1.23	1.23
0.60	2.84	1.21	1.21
0.80	3.69	1.17	1.17
1.0	4.49	1.15	1.15

3. Absorbed Doses in the Lab: Results

From the dose estimations we may derive the following results :

a) point source. If an operator stands 1 m away from a 3.7 GBq unshielded ^{57}Co source we are allowed to use the point source approximation. The effective dose rate he/she can receive amounts to about <u>0.1 mSv /h</u> . One hour of exposition to this radiation field produces nearly the same dose of a chest radiography. The dose rate, obviously, varies with the source activity and the inverse of the square of the distance ;

b) disk source. Let suppose a researcher holds the source with his/her fingers as shown in figure 2 [case 2) and 3)]. Are you thinking this is not a realistic case ? Well, it is. We know more than one colleague that have done this operation at least one time.

In this extreme situation we need to separate further the case where the disk axis is directed as reported in case 2) or case 3). In fact, there could be a large dose difference in the two situations due to the material used to seal the source. If the external material is Al, or Ti, it produces a nearly total attenuation of the low energy photons; on the contrary the berillium window on the top will let them to pass. In case 2) we have more low energy photons passing through the fingers than in case 3). We have assumed $B_i(\mu_i, d) = 1$ for simplicity, and the attenuation coefficients, μ_i, have been taken from ref. [2, 3, 4, 7, 8].

Essentially, we have estimated that case 2) provides the operator an equivalent dose that might reach **1 mSv /min** (> 50 mSv/h) to the basal cells of epidermis (0.07 mm).

For case 3) the dose equivalent is about 20 times less, about **0.05 mSv /min** to the basal cells of epidermis, lower than the previous case but still to be considered.

The dose rate to the body, of course much lower, can be estimated to be of the order of **1 mSv/h** .

4. BERT and DARI Concepts

Even having defined the equivalent dose and the effective dose it is unlikely a non specialist would be completely satisfied to hear "that operation will give you an effective dose of about 1 mSv." It remains a quantity difficult to link with common life.

One method that has been proposed to make easier to understand if the dose is large or small is to convert the effective dose into the amount of time it would take us to accumulate the same effective dose from background radiation.

Since the average background rate in many countries is about 3 mSv per year, the answer in the case above would be "about four months". This method of explaining radiation is called Background Equivalent Radiation Time or BERT [9]. The idea is to convert the effective dose from the exposure to the time in days, weeks, months or years to obtain the same effective dose from background. This method has also been recommended by the U.S. National Council for Radiation Protection and Measurement (NCRP).

With nearly the same idea in mind Nobel prize G. Charpak and his colleague R.L. Garwin introduced the concept of the DARI [10] from the French for "Dose Annuelle due aux Radiations Internes"– annual dose from internal radioactivity. It is equal to the irradiation dose provided to a human being by the naturally occurring radioactivity of human tissues during the period of one year. The use of this unit for expressing the individual's radiation dose from an incident or an accident involving radioactive materials would facilitate a proper judgment of its impact, and would avoid unwarranted concerns. The DARI is to be defined as <u>0.2 mSv</u>, precisely, although the annual dose itself is about 10% less.

TABLE 2. Typical Effective Doses, BERT and DARI for some cases of interest. BERT is defined as the time required to get a certain dose by natural background; 1 DARI is the annual dose produced by the radionuclides inside human body (mainly ^{40}K and ^{14}C).

	EFFECTIVE DOSE (mSv)	BERT (time to get same dose from nature)	DARI (annual dose due to internal radioactivity)
Natural Background (US) / year	3	1 year	15
Body Internal Dose / year	0.2	24 day	1
European Safety Standards against the dangers arising from ionizing radiations			
population	< 1 /year	4 month	5
Workers (B cat.)	< 6 / year	2 year	30
Workers (A cat.)	< 20 / year	7 year	100
X-ray Study			
Dental, intra-oral	0.06	1 week	0.3
Chest x-ray	0.08	10 days	0.4
Thoracic spine	1.5	6 month	7.5
Lumbar spine	3	1 year	15
Upper GI series	4.5	1.5 year	22.5
Lower GI series	6	2 years	30
Single CT scan	8	2.5 year	40
Aircraft and Spacecraft Flights			
Short range ($h = 8$ km)	0.006	17.5 hour	0.03
Long range ($h = 12$ km)	0.06	1 week	0.3
Passengers (200 h/y)	1.0	1 year	5
Air crue / year	4.5	1.5 year	22.5
Astronauts is space	750	250 year	3750
Astronauts - solar flears case	2000	670 year	10 000
Lethal dose for a human	~ 5000	1670 year	25 000
Dose delivered as local irradiation to treat cancer	80 000	26 000 year	400 000
Mössbauer Spectroscopy (3,7 GBq – ^{57}Co source)			
source (no shielding) 1 m away / hour	0.1	12 day	0.5
Mount source on optical bench Max dose – basal cells (0.07 mm) **equivalent dose** / minute	1	4 month	5
Max dose (1 cm) **equivalent dose** / minute	0.2	24 day	1

354

According to the International Commission on Radiation Protection (ICRP), exposure to a DARI conveys a probability of incurring lethal cancer of ten parts in one million. If a lethal cancer corresponds to 20 years of life shortening, each DARI then costs the individual one hour of life expectancy. This calculation is of course rejected by who believe in the existence of a threshold below which there is no harm from radiation.

The natural variability from place to place amounts to five DARI or more.

Table II reports the values, in the different units, of some cases if interest including some Mössbauer operations, and table III compares different human activities from the point of view of risks.

TABLE 3. Fatality risk factor coefficients for human activities

Type of activity	FR "nominal fatality probability coefficient"
smoking (20 sigarette/day)	5×10^{-3}
Driving	1.5×10^{-4}
building worker	2×10^{-4}
housework	10^{-4}
air pollution	10^{-4}
generic employee	5×10^{-5}
1 mSv – whole body, single dose	5×10^{-5}
Leukemia, natural incidence	10^{-5}
thyroid tumor, natural incidence	10^{-6}

5. Measuring Absorbed Dose by Mössbauer Spectroscopy

To our knowledge, the idea of using Mössbauer spectroscopy to measure absorbed dose has not been tried before the present work. There are many practical reasons that may explain why, among them, the first is with no doubts the time required to give an answer. The same procedure used here, in fact, can be applied in combination with optical measurement and the desired result may be obtained in a minute. Anyway, the fields of science are large so we are confident that there will be interesting situations where Mössbauer spectroscopy may provide original contributions.

The idea of measuring radiation dose with Mössbauer spectroscopy goes back to the Fricke dosimeter [11, 12, 13]. It is made of an acidic solution (H_2SO_4 - 0.4 M) of ferrous ammonium sulfate, $Fe(NH_4)_2(SO_4)_2(6H_2O)$.

After the irradiation of an appropriate solution, ferrous ions, Fe^{2+}, act as scavengers for the active products of the radiation chemistry of water (table IV) leading to Fe^{3+} formations. The chemical equations are the following :

$$H\cdot + O_2 \rightarrow HO_2\cdot \tag{7}$$

$$HO_2\cdot + Fe^{2+} \rightarrow HO_2^- + Fe^{3+} \tag{8}$$

$$HO_2^- + H^+ \rightarrow H_2O_2 \tag{9}$$

$$OH\cdot + Fe^{2+} \rightarrow OH^- + Fe^{3+} \tag{10}$$

$$H_2O_2^{\cdot} + Fe^{2+} \rightarrow HO^- + Fe^{3+} + OH\cdot \qquad (11)$$

$$H\cdot + H_2O \rightarrow OH\cdot + H_2 \qquad (12)$$

$G\left(Fe^{3+}\right) = 15.6$ in the presence of oxygen and 8.3 in the absence of oxygen.

The G value is the number of molecules formed or destroyed as the result of the deposition of 100 eV from secondary electrons.

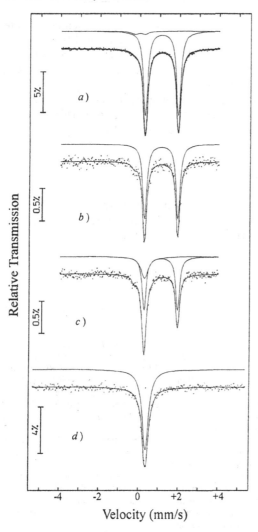

Figure 5. Mössbauer spectra relative to the Fricke dosimetric system. a) powder of Fe(NH₄)₂(SO₄)₂(6H₂O) before the preparation of the dosimeter; b) powder obtained from the solution not irradiated; c) powder obtained from the solution irradiated with x-rays, 100 Gy -100 kVp); d) powder from the solution completely oxidized. Mössbauer parameters for the 2 components are : (δ = 1.24 mm/s, Δ = 1.72 mm/s) and (δ=0.48 mm/s), respectively.

TABLE 4. Yields of Primary Radiolytic Products of water at neutral pH.

Product	G Value
e_{aq}	2.6
H	0.6
OH·	2.6
H_2	0.45
H_2O_2	0.75

Some preliminary results on a modified Fricke solution are reported in figure 5. The details of the preparation will be published elsewhere [14].

As it can be seen it is possible to observe the formation of Fe^{3+} following irradiation by x-rays. From the G values and the relative area it is also possible to calculate the dose imparted to the medium.

We have observed the tranformation of ferrous to ferric ions also in solution with composition and pH different from the original one. This opens too the possibility to find out new dosimetric systems based on Fe^{2+}/Fe^{3+}.

Acknowledgments

A kind aknowledgment is addressed to Prof. Ida Ortalli, Head of the Department of Public Health and Director of the Health Physics Service of the University of Parma for support and encouragement to the present work.

References

1. Alpen, L.P. (1990) *Radiation Biophysics*, Prentice-Hall, Inc., London, and references therein.
2. Dörschel, B., Schuricht, V., Steuer, J. (1996) *The Physics of Radiation Protection*, Nuclear Technology Publishing, Ashford.
3. Cember, H. (1996) *Introduction to Health Physics. III Ed.*, McGraw-Hill, New York.
4. Pelliccioni M. (1989) *Fondamenti Fisici della Radioprotezione*, Pitagora Editrice, Bologna.
5. Beyer, W.H. (1984) *Standard Mathematical Tables, 27-th Edition*, CRC Press Inc., Bota Raton, Florida
6. The International Commission on Radiological Protection (1983) ICRP Publication n. 38. *Radionuclide Transformations. Energy and Intensities of Emissions*, Pergamon Press, Oxford-New York- Frankfurt.
7. The International Commission on Radiological Protection (1996) ICRP Publication n. 74. *Conversion Coefficients for use in Radiological Protection Against External Radiations*, Pergamon Press, Oxford-New York- Frankfurt.
8. The International Commission on Radiological Protection (1987) ICRP Publication n. 51. *Data for Use in Protecting Against External Radiation*, Pergamon Press, Oxford-New York- Frankfurt.
9. Cameron, J.R (1999) *Medical Physics World* 15, 20-21.
10. Charpak, G., and Garwin, R.L. (2002) The Dari. A unit of measure suitable to the practical appreciation of the effect of low doses of ionizing radiation. *Europhysics News* 33,14-17.
11. Fricke, H., and Hart, E.J. (1966) *Radiation Dosimetry*, Academic Press.
12. McLaughlin, W.L., Boyd, A.W., Chadwick, K.H., McDonald, J.C., and Miller, A. (1989) *Dosimetry for Radiation Processing*, Taylor and Francis, London.
13. American Society for Testing and Materials (ASTM) (1995). *Standard Practice for Using the Fricke Reference Standard Dosimetry System. Designation E1026-95.*
14. G. Pedrazzi, S. Vaccari, M. Ghillani, E. Papotti , manuscript in preparation.

Author Index

357